Innovation Policy and Governance in High-Tech Industries

Johannes M. Bauer • Achim Lang •
Volker Schneider
Editors

Innovation Policy and Governance in High-Tech Industries

The Complexity of Coordination

Springer

Editors
Prof. Dr. Johannes M. Bauer
Michigan State University
Quello Center for Telecommunication
Management and Law
409 Communication Arts and Sciences
East Lansing, MI 48824
USA
bauerj@msu.edu

Dr. Achim Lang
University of Konstanz
Department of Politics and Public
Administration
Universitätsstraße 10
78464 Konstanz
Germany
achim.lang@uni-konstanz.de

Prof. Dr. Volker Schneider
University of Konstanz
Department of Politics and Public Administration
Universitätsstraße 10
78464 Konstanz
Germany
volker.schneider@uni-konstanz.de

ISBN 978-3-642-12562-1 e-ISBN 978-3-642-12563-8
DOI 10.1007/978-3-642-12563-8
Springer Heidelberg Dordrecht London New York

Library of Congress Control Number: 2011942897

Printed on acid-free paper

Springer is part of Springer Science+Business Media (www.springer.com)

Preface

Since this project started a few years ago, high-tech policy has been elevated to a very prominent position in the economic policy agenda of many countries. However, the conceptual foundations for designing institutional and legal frameworks conducive to high-tech industries and for shaping specific measures that support advanced technologies have changed considerably during the past decades. Most countries now embrace a differentiated, non-linear approach to high-tech policy. Decision-makers have gradually recognised the complexity of coordinating the many policy components required to succeed in the dynamic and often risky endeavor of developing and adopting advanced technologies. We hope that the articles in this volume will further uplift the academic and practical discussions on the prospects, conditions, and possible limitations of high-tech policy.

Early versions of most of the chapters in this volume were presented at a conference organised by the Chair for Empirical Theory of the State at the University of Konstanz, Germany. Meersburg's baroque castle, overlooking Lake Constance, offered an inspiring backdrop for spirited debate. The conference was arranged under the auspices of the German Political Science Association's working group on "Politics and Technology." It took place prior to the economic crisis that engulfed the global economy in 2008, but its main concerns have only gained in importance and practical relevance. All authors generously contributed additional time and effort to revise and update their initial papers. Consequently, the chapters assembled in this volume are all up-to-date, in some cases reflecting developments into early 2011.

The conference and subsequent work was facilitated by funding from several sources. A TransCoop grant by the Alexander von Humboldt Foundation, Bonn, enabled research collaboration between Volker Schneider at the University of Konstanz and Johannes Bauer at Michigan State University. Dedicated to an exploration of complex adaptive systems theory and its contribution to a better understanding and modeling of technologically dynamic industries, the grant helped build a conceptual foundation that has informed and unified the chapters in this book. Johannes Bauer gratefully acknowledges the opportunity to spend the summer semester 2010 at the Department of Politics and Public Administration at the University of Konstanz. We would also like to recognise in-kind and financial support of the Quello Center for Telecommunication Management and Law at Michigan State University and Synthesys, Inc., a research firm located in East Lansing, Michigan.

The project benefitted from the help of many individuals and we are grateful for their support. We would like to particularly thank Micha Bächle at the University of Konstanz for assistance in formatting the chapters and Dakota Morgan at Michigan State University for help in editing the contributions. At Springer, Barbara Fess supported the project from its inception.

East Lansing, Michigan, USA
Konstanz, Germany

Johannes M. Bauer
Achim Lang
Volker Schneider

Contents

List of Figures

List of Tables

List of Abbreviations

AAL	Ambient assisted living
ABPI	Association of the British Pharmaceutical Industry
ADAS	Arbeitsgemeinschaft Deutscher Apotheken-Softwarehäuser (Association of German pharmacy software houses)
AEV	Arbeiter-Ersatzkassen-Verband (Federation of workers' alternative health insurance funds)
AEV/VdAK	Arbeiter-Ersatzkassen-Verband (Workers' compensation funds Association and association of employee health insurance)
AOK	Allgemeine Ortskrankenkasse (Federal association of local health insurance funds)
ARPA	Advanced Research Projects Agency
ARPA-E	Advanced Research Projects Agency-Energy
ARPANET	Advanced Research Projects Agency Network
ARPE-ED	Advanced Research Projects Agency-Education
ARRA	American Recovery and Reinvestment Act of 2009
AT&T	American Telephone and Telegraph company
ATP	Advanced Technologies Program
AZUR	German research satellite
BAG	Bundesarbeitsgemeinschaft Selbsthilfe von Menschen mit Behinderung und chronischer Erkrankung und ihren Angehörigen (Federal association of self help for people with disability and chronic illness and their families)
BÄK	Bundesärztekammer (German medical association)
BDI	Bundesverband der Deutschen Industrie (Federation of German industry)
BfdBdP	Beauftrager der Bundesregierung für die Belange der Patientinnen und Patienten (Government commissioner for the concerns of patients)
BfdDi	Bundesbeauftragter für den Datenschutz und die Informationssicherheit (Federal commissioner of data protection and freedom of information)

BITKOM	Bundesverband Informationswirtschaft, Telekommunikation und neue Medien (Federal association for information technology, telecommunications and new media)
BKK	Betriebskrankenkasse (Company health insurance fund)
BLK	Bundesverband der landwirtschaftlichen Krankenkassen (Federal Union of Agricultural Insurance)
BMBF	Bundesministerium für Bildung und Forschung (Federal Ministry of Education and Research)
BMF	Bundesministerium der Finanzen (Federal Ministry of Finance)
BMFT	Bundesministerium für Forschung und Technologie (Federal Ministry of Research and Technology)
BMG	Bundesministerium für Gesundheit (Federal Ministry of Health)
BMI	Bundesministerium des Innern (Federal Ministry of the Interior)
BMJ	Bundesministerium der Justiz (Federal Ministry of Justice)
BMVBS	Bundesministerium für Verkehr, Bau und Stadtentwicklung (Federal Ministry of Transport, Building and Urban Development)
BMVg	Bundesministerium der Verteidigung (Federal Ministry of Defence)
BMWi	Bundesministerium für Wirtschaft und Technologie (Federal Ministry of Economics and Technology)
BPL	Broadband over powerline
BPtK	Bundespsychotherapeutenkammer (Federal association of psychotherapists)
BRL	Ballistics Research Laboratory
BSE	Bovine spongiform encephalopathy
BSI	Bundesamt für Sicherheit in der Informationstechnik (Federal Office for Security for Information Security)
BTX	Bildschirmtext (Interactive videotex)
BZÄK	Bundeszahnärztekammer (German dental association)
CAN	Climate Action Network
CASA	Construcciones Aeronáuticas S.A.
CdTe	Cadmium telluride
CDU	Christlich-Demokratische Union Deutschlands (Christian democratic union)
CHF	Swiss Francs
CIG	Council for Innovation and Growth
CIS	Copper indium dieseline
CME	Coordinated market economy
COP	Conferences of the parties
CSNET	Computer science network
CSU	Christlich-Soziale Union (Christian social union)
CTI/KTI	Kommission für Technologie und Innovation (Commission for technology and innovation)
D21	Initiative D21

DARPA	Defense Advanced Research Projects Agency
DASA	Daimler Aerospace
DAV	Federation of German Associations of Pharmacists
DGT	Directorate General for Telecommunications
DIHT	Deutscher Industrie-und Handelstag (Federation of German chambers of industry and commerce)
DKG	Deutsche Krankenhausgesellschaft (German Hospital Federation)
DM	Deutsche Mark
DNA	Deoxyribonucleic acid
DOCSIS	Data over cable service interface specification
DRAM	Dynamic random access memory
DSL	Digital subscriber line
EADS	European Aeronautic Defence and Space Company
EEG	Erneuerbare-Energien-Gesetz (Renewable energy sources act)
EFTA	European Free Trade Association
eHC	Electronic health card
ELDO	European Launcher Development Organization
ENIAC	Electronic numerical integrator and calculator
EOG	Ecology of games
ERI	Education, research and innovation
ERINDEX	Embryo research index
ESA	European Space Agency
ESRO	European Space Research Organization
ETH	Eidgenössische Technische Hochschule (Swiss Federal Institute of Technology)
EU KOM	European Commission
EU	European Union
EUREKA	European Research Coordination Agency
Eurosolar	European Association for Renewable Energies
FAZ	Frankfurter Allgemeine Zeitung
FCC	Federal Communications Commission
FDP	Freie Demokratische Partei (Liberal democratic party)
FHD	Fachhochschule Dortmund (University of Applied Sciences Dortmund)
FITs	Federal Institutes of Technology (Switzerland)
GDP	Gross domestic product
GEM	Global entrepreneurship monitor
GKV	Gesetzliche Krankenversicherung-Spitzenverband (Umbrella association of German health insurance funds)
GMES	Global monitoring for environment and security
GPS	Global Positioning System
GSM	Global Standard for Mobile Communications
GVG	Gesellschaft für Versicherungswissenschaft und –gestaltung (Society for insurance science)

GWEC	Global Wind Energy Council
HART	Human assisted reproductive technology bill
HPC	Health professional card
HTS	High-tech strategy for Germany
IAP	International Action Program
ICANN	Internet Corporation for Assigned Names and Numbers
ICT	Information and communications technology
ICUN	International Union for Conservation of Nature and Natural Resources
IEA	International Energy Agency
IETF	Internet Engineering Task Force
IG	Metall Industriegewerkschaft Metall (Industrial union of metalworkers)
IKK	Bundesverband der Innungskrankenkassen in Deutschland (Federation of guild health insurers)
ILEC	Incumbent local exchange carriers
IMS	Intelligent manufacturing system
IPCC	Intergovernmental panel on climate change
IRENA	International renewable energy agency
ISA	Industry-science research alliance
ISE	Fraunhofer Institute for Solar Energy Systems
ISP	Internet service provider
IT	Information technology
ITIF	Information Technology and Innovation Foundation
ITU	International Telecommunication Union
IVF	In-vitro-fertilization
KBV	Kassenärztliche Bundesvereinigung (Federal association of panel doctors)
KfW	Kreditanstalt für Wiederaufbau (Reconstruction Credit Institute)
KIBS	Knowledge intensive business services
KNAPP	Bundesknappschaft (Social miners' and mine-employees' insurance)
KOF	Konjunkturforschungsstelle (Institute for Business Cycle Research at the ETH Zurich)
KTT	Knowledge and technology transfer
kWh	Kilowatthour
KZBV	Kassenzahnärztliche Vereinigung (Federal Association of Panel Dentists)
LME	Liberal market economy
LTS	Large technical system
MBB	Messerschmitt-Bölkow-Blohm (German aerospace company)
MDS	Multidimensional scaling
MITI	Ministry of International Trade and Industry
MW	Megawatt

MWp	Megawatt peak
NASA	National Aeronautics and Space Administration
NCCR	National Centers of Competence in Research
NGO	Non-governmental organization
NIS	National innovation system
NNI	National nanotechnology initiative
NSF	National Science Foundation
NSFNET	National Science Foundation Network
NTIA	National telecommunications and information administration
OECD	Organisation for Economic Co-operation and Development
OSTP	Office of Science and Technology Policy
P&T	Posts and telecommunications
PC	Personal computer
PD	Prisoner's dilemma
PHP	Hypertext preprocessor
PISA	Programme for international student assessment
PKV	Verband der privaten Krankenversicherung (German Association of private health insurance funds)
PPP	Public-private partnership
PTT	Post, telegraph and telephone (company)
PV	Photovoltaics
R&D	Research and development
R&E	Research and experimentation
RBOC	Regional Bell Operating Company
REN21	Renewable energy policy network for the 21st century
SAGE	Semi-automatic ground environment
SBA	Small Business Administration
SBIR	Small Business Innovation Research Program
SBTT	Small Business Technology Transfer Program
SECO	Staatssekretariat für Wirtschaft (State Secretariat for Economic Affairs)
See-KK	Social insurance for seafarers
SEMATECH	Semiconductor Manufacturing Technology
SGB	V Sozialgesetzbuch Fünftes Buch (Social security code book V)
SME	Small and medium-size enterprises
SNSF	Swiss National Science Foundation
SPD	Sozialdemokratische Partei Deutschlands (Social Democratic Party of Germany)
TCP/IP	Transmission control protocol/internet protocol
TDF	Télédiffusion de France
TI	Texas Instruments
TMF	Technologie- und Methodenplattform für die vernetzte medizinische Forschung (Umbrella organization for networked medical research)

TV	Television
UAS	Universities of Applied Sciences
UN	United Nations
UNEP	United Nations Environment Programme
UNESCO	United Nations Educational, Scientific and Cultural Organization
UNFCCC	United Nations Framework Convention on Climate Change
UniF	University of Freiburg
U.S.	United States
VAT	Value-added tax
VdAK	Verband der Angestellten-Krankenkassen (Union of Health Insurance Companies for Employees)
VdK	Sozialverband VdK Deutschland (Social association VdK Germany)
VDMA	Verband Deutscher Maschinen- und Anlagenbau (Association of the machinery and equipment producers)
VET	Vocational education and training
VHitG	Verband der Hersteller von IT-Lösungen für das Gesundheitswesen (Association of IT solution suppliers for the healthcare industry)
VoC	Varieties of capitalism
VZBV	Verbraucherzentrale Bundesverband (Federation of German Consumer Organisations)
W3C	World Wide Web consortium
WBCSD	World Business Council of Sustainable Development
WCRE	World Council for Renewable Energy
WEC	World Energy Council
WEF	World Economic Forum
WMO	World Meteorological Organization
WSSD	World Summit on Sustainable Development
WTO	World Trade Organization
WWEA	World Wind Energy Association
WWF	World Wide Fund for Nature
WWW	World Wide Web
ZVEI	Zentralverband Elektrotechnik- und Elektronikindustrie (German electrical and electronic manufacturers' association)

Chapter 1
Innovation Policy and High-Tech Development: An Introduction

Johannes M. Bauer, Achim Lang, and Volker Schneider

1.1 Introduction

The continued prosperity of high and medium income countries hinges on their ability to sustain comparative advantage through continuous innovation. Advanced technologies are an important component of such strategies, but the conditions for successful policies supporting them are undergoing rapid transformation. A widely shared view holds that, in a globally interconnected economy, routine and low-skill production activities will be relocated to low-cost emerging economies. Sustaining high living standards in high-income countries will depend on a strong presence in knowledge-intensive, cutting-edge industries. National and regional governments are therefore actively facilitating the development, production and application of advanced technologies. In the aftermath of the economic recession that started in 2008, national governments have further intensified their efforts to promote industries with high knowledge content.

However, advanced technology industries are organised in more complicated ways than this policy orthodoxy recognises. As Saxenian (e.g., 2007, p. 660) documents aptly, high-tech activities have become internationally mobile and are migrating to emerging economies. A growing number of entrepreneurs from Silicon Valley and other centres of leading edge technological innovation are repatriating to their countries of origin, including Taiwan, India, China, and Israel. More and more multinational firms, including industry leaders such as Nokia, locate parts of their research and development operations in emerging economies. These changing global conditions, in addition to the intrinsic challenges of continuous innovation in

J.M. Bauer (✉)
Michigan State University, East Lansing, MI, USA
e-mail: bauerj@msu.edu

A. Lang • V. Schneider
University of Konstanz, Konstanz, Germany
e-mail: achim.lang@uni-konstanz.de; volker.schneider@uni-konstanz.de

J.M. Bauer et al. (eds.), *Innovation Policy and Governance in High-Tech Industries*,
DOI 10.1007/978-3-642-12563-8_1, © Springer-Verlag Berlin Heidelberg 2012

high-tech industries, create daunting challenges for high-tech policy. Governments of advanced industrial countries and emerging economies are applying a broad spectrum of policies to support high-tech innovation, many of which are ambitious and risky. Whereas some may succeed, many may not bring the expected outcomes. The complexity of policy coordination and the conditions of successful high-tech policies are the main concerns of this book.

In this introductory chapter we will outline high technology development and related policies, the object of our investigation, and the broad analytical frame and the various methodological approaches that the contributions to this book are using. In a first step, we will discuss the topic of high-tech industries and the various efforts to conceptualise and measure this intricate complex. In a second step, we will present a systemic and actor-centred analytical framework in which governance and the challenges to policy coordination are key concepts in the study of advanced innovation processes and technology development. In a final step, we will give a preview of the different sections and chapters of this book.

1.2 Conceptualizing and Measuring the High-Tech Complex

After a relative lack of interest during the 1990s, innovation policy is experiencing a rejuvenation, as indicated by national and international initiatives to design forward-looking and comprehensive policy programmes (OECD 2010b). Surveys by the Organisation for Economic Co-operation and Development (OECD) reveal that practical policy is slowly becoming more aware of the complex coordination tasks of innovation policy, although not all countries have succeeded in developing workable approaches (OECD 2005). Early attempts at supporting innovation relied on a "linear" model, assuming that support for basic research was sufficient and would translate into applied research and development (R&D), and eventually into innovations that could survive market tests (Braun 2008; OECD 2005).

When research and practical experience suggested that this model might not reflect the dynamics of innovation sufficiently well, a second generation of approaches evolved during the 1980s: the national systems of innovation framework (Lundvall 1992; Nelson 1993; see also Werle, Chap. 2). This framework paid more attention to the systemic interactions between stakeholders and areas of policy. Nonetheless, many of the policy recommendations were generic and did not particularly well take the unique nature of national systems into account (Dodgson et al. 2010). With the continued development of technology, the weaknesses of the national systems approach have become more visible, contributing to the beginning of yet another theoretical extension. Many of the new efforts are inspired by systems theory and the theory of complex adaptive systems (Ahrweiler 2010; Frenken 2006). Most importantly, the challenges of coordination, typical for many advanced technology projects, were recognised. Systems theory is open to the possibility that multiple constellations of private and public sector roles are feasible that are capable of generating good innovation performance. The emerging "third generation" innovation research and policy therefore shifted the emphasis even further to the challenges and mechanisms of coordination, including the potential limits of governance.

The distinction between high-tech and other industrial activities and the relative standing of countries in high-tech industries is not clear-cut and can be operationalised in several ways. A generic definition of high-tech products, services and industries relies on the relative importance of scientific knowledge in the production process and/or the specificity of a product. Until the 1970s, most national statistics and international statistics generated by the OECD relied on input measures, such as the amount spent on basic research and development (Godin 2002). Up until the 1970s, only the U.S. made reasonable efforts to classify such industries. An early classification of industries as high, medium, and low-tech was based on U.S. conventions and was later modified to reflect the conditions of other countries. Because of the conceptual weaknesses of input-based data, the OECD started to also collect output-based data in the 1980s, including data on patents, the technological balance of payments (TBP) between countries, and trade in high-tech products and services. Each of these indicators has conceptual strengths and weaknesses, but they continue to form the backbone of the OECD's biannual publication of Main Science and Technology Indicators. Based on the share of resources spent on research and development, the OECD classifies sectors into "high-tech" (with a R&D intensity of 8.5% or higher) and "medium-high-tech" (with a R&D intensity of more than 3.5% but less than 8.5%). In international trade statistics, the OECD also uses a product-based definition (see Table 1.1). Industry and product-based definitions, however, do not yield the same results, so that aggregation from products to industries is not possible. Sector and product-based

Table 1.1 Alternative high-tech and innovation indicators

	First generation		Second generation	Third generation
	Sector	Products		
Main focus	Inputs	Inputs	Inputs, outputs	Multiple indicators
Examples	Aerospace technology, artificial intelligence, biotechnology, energy, nanotechnology, robotics, optoelectronics, telecoms, nuclear physics, instrumentation	Aerospace, computers and office machinery, electronics, pharmacy, chemistry, scientific instruments, electrical machinery, non-electrical machinery, armaments	Patents, technological balance of payments, international trade in high-tech products	Patents and trademarks, new to market innovators, modes of innovation, collaboration in innovation, clusters, innovation hotspots

Source: Godin (2002), OECD (2010a)

approaches have become very influential in international comparisons. Often the global competitiveness of a country is judged solely on the share of these sectors in its exports.

However, the concept of treating whole societal sectors as more or less advanced has come under increasing criticism. Research points to the fact that most advanced technologies develop as complex and interrelated systems, in which some parts may be highly advanced and cutting edge, but other parts may use low-tech components with important complementary functions. For example, advanced computing systems may be composed of massive parallel arrays of processors that in and of themselves do not qualify as high-tech. There are also sectors that are overall classified as low-tech in which highly advanced technologies play an important role. For example, agriculture may use sophisticated communication systems to coordinate business processes. Moreover, important synergies and interactions exist between high-, medium-, and low-tech industries. Hauknes and Knell (2009), in a study of the direct and indirect flows of knowledge between different knowledge-intensive industries, show that such industries are complementary to each other. Knowledge exchange between these segments is essential for the production, diffusion and use of technology. Countries with a presence in all segments may therefore do better than countries with only a presence in high-tech.

However, neither input nor output-based measures fully capture the processes of innovation and technology development in various sectors that constitute modern economies. Many advanced technologies, products and services require the complex combination of multiple technological artefacts and components, and hence, the collaboration of multiple specialised firms from various sectors at different levels.

Useful conceptual schemes of various technology configurations and their combination possibilities have been proposed by Shenhar (1993) and Hobday (1998). Their multidimensional perspectives extend the conventional system perspective, in which systems consist of components and relations that are separated from environments (Bunge 1996), into a "system-of-systems" perspective, with multiple layers and various degrees of nestedness (Maier 1998). Figure 1.1 outlines this multi-level perspective in a two-dimensional space. Technological products are distinguished with respect to their sophistication (horizontal axis) and with respect to their systemic levels (vertical axis).

Shenhar (1993) differentiates between low and high technology not just on the basis of R&D input but also through qualitative criteria relating to specific features of advanced technologies such as novelty and technological uncertainty. Low-technology products are based on established technologies, whereas medium-technology products incorporate some new features. High-technology products consist mostly of recently developed technology, such as advanced computing hardware, and software or new biotechnologies. In addition, there are super high-technologies that are based on completely new artefacts, skills and materials. Super-high-tech also involves high levels of uncertainty, risk and new investment (e.g., spacecraft and satellite systems).

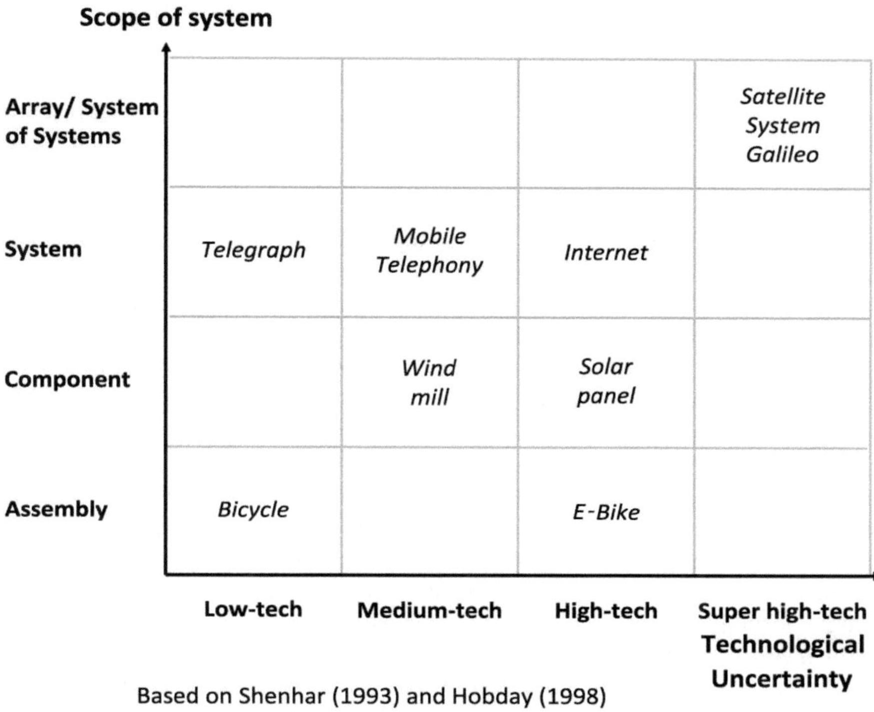

Fig. 1.1 Types of technologies

The vertical dimension differentiates between levels of integration. An "assembly", the category at the lowest level, is a mass produced stand-alone product that performs a single function and is not a part of a wider system (e.g., a bicycle) unless it is connected by a network. By contrast, components are always embedded and functionally integrated in a larger system. Systems integrate components, relations, control mechanisms, and are built to perform common goals (e.g., communication, defence). Finally, an array is a system of interrelated or nested systems, each performing independent functions that are integrated into a common super system (Hobday 1998).

This nested systems perspective acknowledges that the degree of "advancement" of a technology is not only related to the sophistication of its individual components but also to the complexity of its relational and functional integration. A technological masterpiece thus is not only demonstrated by the construction of its various parts but also through the coordination and integration of a heterogeneous complex of multiple components and subsystems into an overall super system. Super-high-tech systems of this kind may be airplanes (e.g., the Airbus 380), supersonic transport systems, or satellite navigation (e.g., GPS, Galileo), but can also be smart cards used in the financial, security, and health sectors which integrate hybrid systems. In this view, technological advancement does not only relate to technological artefacts in the narrow sense, but also includes arrays and networks of social technologies, such as organizational patterns, logistical systems, and complex forms of social coordination.

Since the 1980s, such integrated technological systems have also been conceptualised as "large technical systems" based on multiple technical and social components. These structural features – social and technical heterogeneity and interdependence – have specific implications for their developmental logic (see Hughes 1983; Mayntz and Hughes 1988; Coutard 1999). In order to gain "momentum" and "viability," these systems have to solve a number of critical problems ("reverse salients") which generally requires governmental support (Hughes 1983). Early examples of state sponsored large technical projects can be found centuries ago, but focused support on most advanced technologies are symbols of the technology race since World War II, of which the most famous projects have been the Manhattan Project and the Apollo Program (Steinberg 1985).

Since the 1970s and 1980s, such strategies of "High-tech Colbertism" have proliferated among some advanced industrial countries (Cohen 1992). Governments have provided support for advanced technological projects, and new programmes to promote high-tech have emerged during the past decade. Examples include initiatives to promote the deployment and use of (then) state-of-the-art communications systems, such as the French Minitel initiative in the 1970s or the German *Bildschirmtext* (BTX) project (see Dutton, Schneider and Vedel in this book). Similarly ambitious projects have been the Concorde or high-speed rail. Some of the projects have not succeeded (e.g., BTX), some only in a limited way (e.g., the Concorde), and others in a fully unanticipated fashion (e.g., the emergence of the Internet from the early research funded by the U.S. Department of Defense).

A third aspect of high-tech activities is the vital contribution of advanced cross-cutting infrastructures (e.g., information networks and systems) that support and enable other economic and social activities. Infrastructural systems are not like other conventional societal subsystems, such as economy or health, but are general purpose technologies on which the operation of many other systems hinges. For instance, advanced computing and communication networks are a prerequisite of innovation processes in many other sectors, ranging from advanced logistics to sophisticated financial services. A high-tech economy is critically dependent on these infrastructures and their near-universal availability. Röller and Waverman (2001) have found that information and communication technologies have the highest impact on productivity growth if they are widely diffused throughout the economy. These findings are also compatible with a new complexity perspective on growth and development economics in which diversity and product ubiquity, rather than sectoral specialization, are the key to competitiveness and successful economic development (Hidalgo et al. 2007).

Complex linkages between economic sectors and communication, transport and energy infrastructures are important prerequisites for innovation in general, and high-tech innovation in particular. Advanced data centres, for example, need a reliable high-quality energy supply. Not only material assets are needed, as education and continued professional training also could be considered immaterial

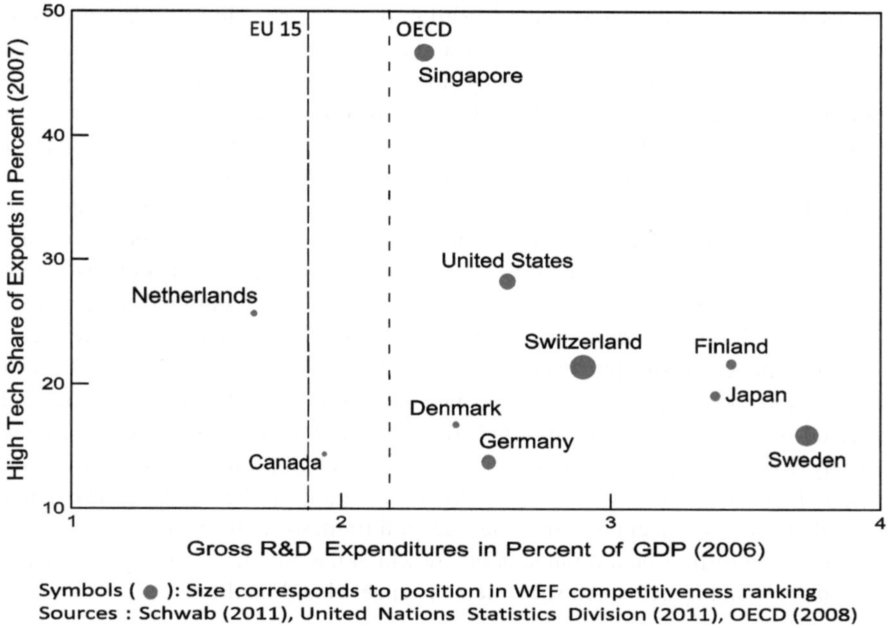

Fig. 1.2 R&D intensity and high-tech export shares (Schwab (2011), United Nations Statistics Division (2011), OECD (2008))

infrastructures. Whereas the systemic view of high-tech emphasises the pervasive complementarities between different economic activities, the infrastructure perspective highlights general preconditions for high-tech policy. Projects such as high-speed railways, the introduction of a new health identification card, or the wide deployment of e-government could consequently be seen as high-tech activities, even if they do not meet the R&D intensity thresholds used in the narrow product-based definition.

The relationships among these variables are more complex and multifaceted than linear statistical models can show. Contributors to this book thus use the case study method extensively. Several of the countries depicted in Fig. 1.2 are covered in-depth by examining high-tech sectors and policy programmes (particularly Germany, the United States, and Switzerland).

1.3 High-Tech Policy and Governance

The central theme of this book is how governments support high-tech industries and to what extent they succeed. Which strategies and programmes do they initiate in different economic sectors and at different socio-political levels? Which particular policy measures and instruments are applied, and which governance mechanisms are used to coordinate the growing spectrum of activities? However, a concentration on governmental activities alone would exclude many actors and contexts that are important in the innovation policy domain. During the last few decades, the policy perspective was thus extended to non-governmental actors and civic institutional mechanisms. When, for instance, policy analysis discovered that many governmental programmes had failed due to policy implementation problems, the focus shifted to specific governmental and non-governmental actor constellations in policy programmes (Mayntz 1983). When it was discovered that coordination and implementation problems could be avoided if private stakeholders were incorporated into the process of policy-formulation, the traditional state-centric actor constellation in policy-making perspective was broadened to the more inclusive governance perspective (Mayntz 2003; Schneider and Bauer 2009). In view of politics and society, policy networks play a major role in public policy-making and societal self-regulation (Scharpf 1997; Kenis and Schneider 1991). Governance is largely conceived as the process of governing, including all relevant actors and policy instruments that are involved in private and public policy-making. The major advantage of this concept is to provide a general framework to cover the broad array of actors and institutional arrangements by which the coordination, regulation and control of social systems and subsystems is enabled and facilitated.

1.3.1 Metatheory

Policy analysis and governance studies are research fields or research agendas rather than approaches. Both are primarily defined by a given research object and do not necessarily imply specific theoretical orientations. Recent overviews of the literature show that the landscape of policy theory is quite diverse and inclusive (Sabatier and Weible 2007; Schmidt 1993; Schneider and Janning 2006). Theories range from "grand theories" to "middle-range theories" and even "single-item theories" (Bunge 1996). The first group includes theories such as systems theory, institutionalism, and rational choice, to name but the most important ones. All of these grand architectures assume general laws in society and political life that account for all social phenomena, regardless of sectors, levels, and subsystems. The second group restricts its explanations to specific societal domains, such as specific governmental institutions, the interaction between interest groups and the state, or specific policy areas. Single-item theories – the third group – are focused on particular socio-political phenomena. For instance, the party difference hypothesis assumes that policy outputs are largely determined by party orientations of governments (Schmidt 1993).

In search of better explanations, many studies use approaches or frameworks that combine multiple theories and analytical perspectives. Widely accepted are the "Institutional Analysis and Development" approach (Ostrom 1999), the "Actor-centred Institutionalism" (Mayntz and Scharpf 1995), and the advocacy coalition framework (Sabatier 1988). Whereas the first two approaches stress actors, choice and institutional constraints in policy-making, the advocacy coalition framework emphasises policy discourses and belief systems.

Another framework that is broadly used in innovation policy studies is the "innovation systems" approach, which combines some middle-range theories in the area of science, technology and innovation studies in a fruitful way (see Werle, Chap. 2). Similar to the above mentioned institutionalist approaches, it tries to take into account major actors and institutional frameworks through which innovation is driven and policy-making is shaped. Main components of innovation systems are governmental institutions, public and private research organisations, firms and business associations, as well as networks between actors and institutions (OECD 1999). More refined perspectives also integrate consultancy firms, professional societies and industrial research organisations into the picture. These operate as intermediary institutions between industry and academic research (Metcalfe 1995). An innovation system's "organizational ecology" is populated by all actors and institutions that are involved in the production, accumulation and diffusion of knowledge, in education and training, technology development and its regulation (Kuhlmann et al. 2010). The regulatory policy space includes all actors shaping regulatory norms and standards, not only government, political parties and associations, but also relevant media. Since innovative technologies and their regulation are also affected by public discourse, innovation studies must take media actors into account, as they can play supportive but also obstructive roles in technology development (see Waldherr, Chap. 4). The innovation systems approach thus provides a rather inclusive and heterogeneous perspective on technology development. Several chapters in this book will apply this complex systems perspective.

1.3.2 Methods

With respect to methods, the state of the art in public policy analysis and governance studies is also manifold. It varies in two dimensions: qualitative versus quantitative studies, and Large-N versus Small-N studies. The strength of case studies is to study political processes in great detail in order to uncover causal mechanisms and complex context conditions. In policy analysis, case studies usually open the black box and trace policy developments from problem perception to agenda-setting, decision making and implementation of political programmes. The chapters on high-tech policy in Switzerland and in the U.S. adopt this approach.

Large-N studies as a rule use quantitative methods to test for the effects of one or more independent variables on a given policy outcome, controlling for the variation in context conditions. Fink's contribution to this book, for instance, uses this method to test for the relationship between the strictness of embryo research laws and innovation in the biotechnology sector. Most chapters in this book are small-N case studies or even concentrate on a single case. The recent methodological debate highlights the epistemic potential of case studies: They are not necessarily restricted to mere descriptions and hypothesis development but can also be used in an explanatory manner if they identify a supposed mechanism or trace a specific process pattern (Gschwend and Schimmelfennig 2007). From a mixed method perspective, case studies may also be combined with quantitative methods, as emphasised in the "nested analysis" approach (Lieberman 2005).

A particular mixed method type is the application of "social network analysis" to public policy-making, innovation and technology development. This approach is increasingly used also in the area of innovation, science and technology studies (Pittaway et al. 2004). Inquiry in this perspective concentrates on the relational dimension of the actor system and traces the exchange of resources, flows of communication, information diffusion, membership to policy committees, and other types of influence vectors. Policy network studies are quantitative case studies because their analysis concentrates on single cases (a policy domain or a policy process) which are studied in great detail at the level of actors and relations. Lang and Mertes (Chap. 11) and Dutton, Schneider and Vedel (Chap. 3) use this method.

1.3.3 Analytical Framework and Key Concepts

In the preceding section we presented a multi-dimensional view on high-technology development and high-tech sectors. A major point was that high-tech products and sectors cannot be reduced to isolated metrics and market segments, as their development and evolution is always embedded in complex "product spaces" and cross-cutting infrastructures.

The complexity of such product spaces is in most cases accompanied by complex organizational settings in which a myriad of organizations from different policy domains – such as education policy, research policy and economic policy, to name just a few – are involved. The technological and social complexity of innovation processes potentially constrains and undermines the feasibility of innovation governance. If the degree of complexity is too high, governance, seen as the coordination and regulation of interdependent organizations (Rhodes 1996) aimed "to craft order, thereby to mitigate conflict and realise mutual gains" (Williamson 2000), may be difficult or impossible to achieve. Moreover, the outcomes of governance decisions, even if they can be effectively coordinated, may be difficult to anticipate. Governance in highly complex systems, therefore, is, to a certain degree, experimentation.

Complexity theory provides an analytical framework to conceptually integrate these complex social settings. It focuses on the interdependence and adaptation of systems, and on the creation of order that makes it particularly suitable for analysing innovation policy and governance (Schneider 2012). Complexity theory first emerged in the natural sciences and then quickly diffused into other research fields such as economics (Arthur 1994; Beinhocker 2006) and sociology (Sawyer 2005). It emphasises self-organization processes in diverse research applications such as the evolution of species (Kauffman 1993), the emergence and coordination of collective action in bio-populations (Strogatz 2003), and the growth of network infrastructures such as the World Wide Web (WWW) and air traffic (Barabási 2002, 2003; Newman et al. 2006).

Most approaches within complexity science consider four basic elements (Bunge 1996; Butts 2000, 2001): (1) the number of parts of a system (the composition); (2) the relationships between elements of the system (the structure); (3) the relationship to external systems (the ecological dimension); and (4) the roles and positions of elements within and between systems (the function). The totality of the parts of a system indicates the number of subsystems that make up the larger system. The structure formed by the parts of the larger system includes the number and types of links between the different parts of the system. In political terms, these links can be interpreted as influence functions. These are modelled by political economists who measure the lobbying success of one interest group, given its lobbying efforts, in relation to other interest groups' lobbying efforts (Becker 1983). Another example is the advocacy coalition framework that focuses on changes in belief systems by inter-actor learning mechanisms (Sabatier and Weible 2007). From a governance perspective, links between actors denote mechanisms through which they mutually coordinate their behaviour. Basic coordination mechanisms include observation, influence and bargaining.

The relationship between different systems designates the interdependence between systems (the ecology of systems). Changes in one system trigger or inhibit changes in other systems. In the ecological dimension, the concept of interdependence plays a major role. Thompson (1967) distinguishes three forms of interdependence: pooled, sequential, and reciprocal. In a *pooled* interdependence setting, each subsystem contributes to the final outcome without the need for direct interaction and coordination between them. Total output is simply the sum of individual outputs (Saavedra et al. 1993). *Sequential* interdependence refers to settings in which each system performs a different role and performs different tasks. Total output requires the sequential task accomplishment by each system. Here, coordination of tasks is essential, although outputs flow only in one direction. Pooled and sequential interdependence are basic tenets of the varieties of capitalism approach that assumes complementary institutional settings that mutually account for each other's weaknesses (see Werle, Chap. 2). Sequential interdependence can also be found in many governance settings, when administrative and policy coordination is carried out by different actors at different points in time (see Wassermann and Fuchs, Chap. 10, and Lang and Mertes, Chap. 11).

In a social setting with *reciprocal* interdependence, each subsystem produces outputs that become inputs to other subsystems. Reciprocal interdependence requires high role specification and coordination of task. Role and functional differentiation is another dimension put forward by complexity theory and network science (Butts 2001). Most chapters in this volume deal with some kind of interdependence and role and functional differentiation. Orlowski (Chap. 7) explores the role differentiation in the German high-tech advisory network, while Lang and Mertes (Chap. 11) explain the delayed implementation of the German e-health card by invoking missing functional differentiation. Dutton Schneider and Vedel (Chap. 3) model different types of interdependence as an ecology of games.

Complexity issues are ubiquitous in innovation policy and governance. Supportive policy programmes affect a broad array of actors in different policy domains with diverse institutional backgrounds. Governmental agencies may be affected because they occupy important functional roles and positions in the political division of labour. Business firms and interest groups are mostly concerned about intended and non-intended effects of a policy initiative. Policy measures generate "stakes" in which social, economic, and political actors get interested and mobilised. Stakeholders strive for policy engagement, but their specific involvement is often constrained by institutional factors. Diverse political systems provide different opportunities for participation and policy involvement. In countries where policy areas are more differentiated (research policy, industrial policy, etc. versus an integrated innovation policy domain), political battlefields get populated by more numerous and heterogeneous actors. Diverse policy domains in general imply "competing rationales" based on different histories and institutional practices. Different ministries, for instance, have different views of innovation policy, its nature and its role (OECD 2005). As shown in the contribution of Dutton, Schneider and Vedel (Chap. 3), these competing logics of policy-making can be conceptualised as an "ecology of games." In this ecology, multiple games are simultaneously played in decision-making and resource mobilization. Policy outputs emerge in a largely uncontrolled and spontaneous way. Another factor that increases policy complexity is the recent transformation of modern political systems towards a "regulatory state" (Majone 1994). The creation and proliferation of more and more agencies increases fragmentation and segmentation of policy areas (Schneider 2004).

A logical consequence is that policy actors become more heterogeneous and interest fissures get multiplied. Policy-making takes place in multi-actor arenas (Kuhlmann 2001; Edler and Kuhlmann 2008). In reaction to this diversification and heterogeneity, governments have to put more and more emphasis on policy coordination to avoid redundancy in resource allocation, inconsistency in policy actions and multiplication of interest conflicts (Braun 2008).

Based on particular institutional traditions and policy-making styles, national governments use different social technologies for policy coordination. Since neither spontaneous adaptation nor hierarchical top-down control alone works, governments promote alternative and new forms of governance such as "networks" or "bargaining systems" in which common goals and strategies are bargained and deliberated (Scharpf 1997; Kenis and Schneider 1991). Coordination systems are often constrained by their size and scope: In small systems the actors may discuss and explore all policy options (positive coordination). However, because of time and resource constraints, such inclusive strategies are impossible in large coordination systems. As it was observed by Scharpf (1997) in German policy making, in large actor systems bargaining and debate often is reduced to a subset of policy options which imply no negative externalities on actors with veto powers (negative coordination).

In the recent literature, such arrangements have also been called forms of "collaborative governance," which is an "arrangement where one or more public agencies directly engage non-state stakeholders in a collective decision-making process that is formal, consensus-oriented, and deliberative and that aims to make or implement public policy or manage public programmes or assets" (Ansell and Gash 2008). In these networked forms of governance, governmental authorities at regional, national and supranational level play an important and sometimes central role, but not a "commanding role". Often their role is reduced to mediation and facilitation (Kuhlmann 2001).

1.4 Contents of the Book

This book aims at sharpening the understanding of researchers and policy makers for the complexity inherent in high-tech policy and the challenges it implies for policy design and implementation. To this end, the book is organised in four parts. In part one, the state of research on innovation systems and technology development is critically reviewed and new analytical perspectives are introduced. The contribution by Werle overviews and discusses the strengths and weaknesses of three conceptual approaches, and sets the state for a more comprehensive approach drawing on recent developments in comparative political economy, innovation studies and the sociology of socio-technical systems. The national innovation systems literature focuses primarily on R&D efforts by business firms, public sector and government actors, and the research and education communities. The "varieties of capitalism" (VoC) literature in comparative political economy acknowledges the systemic nature of innovation processes too, but underscores differences in institutional systems that promote distinctive types of innovations. Werle challenges the view often articulated in this tradition that liberal market economies tend to produce radical changes and novelties, whereas coordinated market economies provide favourable institutional conditions for incremental innovations. Techno-sociological institutionalism, the third perspective, understands technology development and innovation as a result of coordinated interaction in which various modes of governance – market, network and hierarchy – come into play. Based on the strengths and weaknesses of these approaches, Werle proposes an extension towards more differentiated and multi-level perspectives emphasizing the co-evolution of technology and society.

New theoretical perspectives in the analysis of innovation and technology developments are introduced by Dutton, Schneider and Vedel, as well as by Waldherr. This expanded view more explicitly recognises complex and inclusive networks of public and private institutions shaping technology development in high-tech sectors. Emphasis is put on the multi-layer nature of coordination processes and the nestedness of systems. Innovation policy is modelled as an intervention by public authorities and private collective actors attempting to influence technological change. So defined, innovation policy also includes elements of other policy fields, such as infrastructure policy, R&D policy, industrial policy, and technology policy. The unique lens of this approach allows conceptual advancements regarding the nature of innovation systems, the underlying knowledge creation process, and the role of the media in framing the debates on high-technologies, thereby shaping expectations in innovation processes and appropriate policy action.

Ecological thinking in the social sciences is a promising theoretical perspective on the construction of order and change in technology development, inspired by various branches of biology and environmental sciences. Ecosystem ideas and related concepts to change and adaptation are applied to social and technical systems. Ecologically inspired models put emphasis on: (1) the dynamic interdependencies and interactions between diverse actors; (2) the multiplicity of relations between the components and outcomes of these systems; and (3) the existence of multiple and relatively autonomous layers and levels in such systems, along with the emergent relations between these levels.

In the social sciences, most of these approaches have been developed in the sociology of organizations (population ecology of organizations; organizational ecology). In the political sciences, such approaches were applied to understand the development of local communities, policy sectors and interest group systems. A subtype of this perspective is the concept of an "ecology of games," which emphasises the complexity of nested (public and private) decision-making processes in the context of social and technical interdependencies and related conflicts. The contribution of Dutton, Schneider and Vedel demonstrates the fruitfulness of the "ecology of games" perspective as a framework for the study of technology development in the communications and information technology sector.

The chapter by Waldherr explores the influence of the mass media on high-tech policy decisions in innovation systems, a perspective that is missing in most innovation studies. Her article shows from a process, a functional, and a structural perspective that mass media are highly relevant actors in innovation systems and therefore also need to be considered as critical variables in political processes leading to high-tech strategies. From a process perspective, the mass media are influential in technology development by creating awareness of innovations, and even more in the attribution process by labelling new technologies as innovations. From a functional perspective, the mass media's impact on important functions in innovation systems includes coordinating functions like knowledge diffusion, selecting functions like guidance of the search, and legitimacy creation for new technologies. In a structural perspective, the mass media are seen as an important communication forum in the public sphere, mediating between the political, the economic and the research subsystems of society.

Part two of the volume explores the variety of governance mechanisms and public policies utilised in national, as well as sectoral, innovation systems. It looks at institutional settings in high-tech industries and highlights commonalities and differences among them. The German system of innovation is juxtaposed with that of two of its most successful competitors, namely the United States and Switzerland. These in-depth case studies yield a detailed analysis of the overall approaches pursued in these nations.

The U.S. has long been seen as a liberal market economy whose national innovation system is particularly geared toward radical innovation. Bauer presents a more multifaceted analysis. Whereas the U.S. did not have an overarching innovation policy for most of the second half of the twentieth century, many government initiatives were instrumental in promoting R&D and innovation. In a way, the U.S. system was organised in a non-linear fashion before this was recognised by other countries as supportive for innovation in advanced technologies. However, during the past decades, the U.S. lead has narrowed and, in some areas, been lost to other countries. In response, the Obama Administration has launched a national innovation strategy, based around principles of network governance and coordination. The contours of the historical approach and its recent changes are discussed in general and for the information and communications technology (ICT) sector, a central component of past, present, and future high-tech initiatives.

Hotz-Hart presents a detailed and in-depth analysis of innovation policy in Switzerland. The country regularly ranks among the top industrialised nations with respect to innovation performance. This is the outcome of a unique approach, a fourth way between other national models of innovation policy. Switzerland's model can best be characterised as decentralised network governance. Strengths of this system include a highly educated work force, high quality research and development, and global sourcing of knowledge inputs by Swiss firms. Risk aversion, unfavourable demographic trends, and stagnating public funding for R&D are some of the challenges. Given increased competition from abroad and these internal challenges, the chapter concludes that concerted efforts will be necessary to maintain past performance.

The chapter by Orlowski analyses the High-Tech Strategy for Germany, initiated in 2006 as an overarching national strategy that integrates efforts by all government departments. To assist in the coordinating tasks, the German Federal Government established two advisory bodies, the Council for Innovation and Growth (CIG) and the Industry-Science Research Alliance (ISA). Using interlocking directorate analysis, the chapter focuses on the role, composition and function of these two advisory bodies. Orlowski finds that some economic interests are better represented than other societal players. Moreover, as many of these players already had existing links, he concludes that the two new bodies might have weaker integrating functions that anticipated.

These national perspectives are complemented in part three by studies of the implementation of high-tech strategies in specific industrial sectors. A recurring theme in these case studies is the participation of a variety of actors that normally do not take part in (German) innovation policy. However, as high-tech industries raise new issues, new actors become involved, increasing the challenges of finding feasible policies. Examples include the biotech industry, in which economic reasoning and religious beliefs collide, the wind energy and photovoltaic industries, characterised by divergent preferences of economic actors and environmental groups, and the implementation of the electronic health card that brings ICT industry representatives together with the traditional players in health politics. The industry studies also explore the impact of different technical bases of sectors – large vs. small technical systems – on high-tech strategies and actor constellations.

Schneider and Weyer analyse the power struggles underlying the German space policy domain. They observe a rollback in policy orientations since 2007, when the EU and the national governments returned to the traditional procedure of state-driven construction of big technology, exemplified by the failure to assure industry participation in the Galileo satellite project. Innovation policy in the space domain consists of overlapping policy and strategic games played at the national, European and global level.

The interaction between political factors, embryo research laws and the innovative ability of national economies is assessed in Fink's chapter on biotech policy. He demonstrates that embryo research turned into a commercially beneficial enterprise at the end of the 1990s. At that point, it attracted the attention of the pharmaceutical industry and government regulators. In particular, left governments and strong players in the pharmaceutical industry mobilised to liberalise existing regulatory systems. However, in some countries the Catholic Church and Christian democratic parties thwarted the liberalization of the stem cell research, successfully opposing scientific and economic interests.

The contribution by Fuchs and Wassermann examines high-tech policy in the photovoltaics industry. The authors argue that the emergence and development of the photovoltaic industry in Germany was based on the establishment of a protected niche market, which in turn depended on the creation and success of advocacy coalitions supporting the photovoltaic industry. At the beginning, the photovoltaics advocacy coalition included local politicians, the Green party, researchers, environmental societies, and business associations of the infant photovoltaics industry. The successive incorporation of some of the multinational companies increased the effectiveness of political pressure against strong opposition by German utilities.

Ronit explains the development of the wind turbine energy industry through international efforts to combat climate change. He states that while national governments still remain key players in environmental and innovation policies, intergovernmental organizations have gained importance in coordinating policies of states and mitigating conflicts. Initially, wind energy was a small subdomain in national energy policy making. At that time, national and regional business associations, as well as environmental groups, had already been established, but were still exclusively linked to domestic politics and national systems of innovation. More recently, agenda-setting and policy formulation have shifted to the international level. An increasingly large number of civil society organizations take care of climate policy and wind energy.

The e-health card in Germany is the focus of Lang and Mertes' essay. The authors examine the policy and administrative coordination in this temporary, domain-boundaries-transcending policy network that was formed to implement the e-health card. They identify structural barriers to coordination and inconsistencies in goals and task settings that resulted from power asymmetries. These frictions result from different institutional logics inherent in different policy domains. The authors point out that traditional sectoral interest positions play a dominant role in the structuration of the implementation network, which explains the slow and cumbersome process of implementation. Health care providers especially have turned out to be the stalling element in the whole process.

The lessons from theory, national practice and sectoral experiences for high-tech policy and future research are synthesised in the fourth part of the book. Major implications for high-tech policy-making and the design of governance structures supporting innovation are outlined and discussed. Complexity theory offers a versatile framework not only to analyse the existing experience with innovation policy but also to draw fresh conclusions as to workable policy options. One key insight is that high-tech industries differ in their economic and technological characteristics. Thus, no panacea policy exists that will fit all domains. Rather, policies need to be congruent with the respective domain and flexible forms of multi-level coordination will often be helpful. Another insight is the importance of support for diversity, experimentation, and the potential benefits of parallel and competing policy efforts. Successful high-tech innovation policy will also have to pay attention to supporting the adoption of high-tech processes, products and services. In a globally interconnected economy, high-tech policy will also have to seek international coordination.

References

Ahrweiler, P. (Ed.). (2010). *Innovation in complex social systems.* Milton Park/New York: Routledge.

Ansell, C., & Gash, A. (2008). Collaborative governance in theory and practice. *Journal of Public Administration Research and Theory, 18*(4), 543–571.

Arthur, B. (1994). *Increasing returns and path dependence in the economy.* Ann Arbor: University of Michigan Press.

Barabási, A.-L. (2002). *Linked: The new science of networks.* Cambridge, MA: Perseus.

Barabási, A.-L. (2003). *Linked: how everything is connected to everything else and what it means.* New York: Plume.

Becker, G. S. (1983). A theory of competition among pressure groups for political influence. *The Quarterly Journal of Economics, 98*(3), 371–400.

Beinhocker, E. D. (2006). *The origin of wealth: evolution, complexity, and the radical remaking of economics.* Boston, MA: Harvard Business School Press.

Braun, D. (2008). Organising the political coordination of knowledge and innovation policies. *Science and Public Policy, 35*(4), 227–239.

Bunge, M. A. (1996). *Finding philosophy in social science.* New Haven: Yale University Press.

Butts, C. T. (2000). An axiomatic approach to network complexity. *The Journal of Mathematical Sociology, 24*(4), 273–301.

Butts, C. T. (2001). The complexity of social networks: Theoretical and empirical findings. *Social Networks, 23*(1), 31–72.

Cohen, E. (1992). *Le colbertisme "high tech": économie des Telecom et du grand projet.* Paris: Hachette.

Coutard, O. (Ed.). (1999). *The governance of large technical systems.* London/New York: Routledge.

Dodgson, M., Hughes, A., Foster, J., & Metcalfe, J. (2010). Systems thinking, market failure, and the development of innovation policy: the case of Australia. *Discussion Papers Series.*

Edler, J., & Kuhlmann, S. (2008). Coordination within fragmentation: Governance in knowledge policy in the German federal system. *Science and Public Policy, 35*(4), 265–276.

Frenken, K. (2006). *Innovation, evolution, and complexity theory*. Cheltenham/Northampton: Edward Elgar.
Godin, B. (2002). Outline for a history of science measurement. *Science, Technology, & Human Values, 27*(1), 3–27.
Gschwend, T., & Schimmelfennig, F. (2007). *Forschungsdesign in der Politikwissenschaft*. Frankfurt am Main: Campus Verlag.
Hauknes, J., & Knell, M. (2009). Embodied knowledge and sectoral linkages: An input–output approach to the interaction of high- and low-tech industries. *Research Policy, 38*(3), 459–469.
Hidalgo, C. A., Klinger, B., Barabási, A.-L., & Hausmann, R. (2007). The product space conditions the development of nations. *Science, 317*(5837), 482.
Hobday, M. (1998). Product complexity, innovation and industrial organisation. *Research Policy, 26*(6), 689–710.
Hughes, T. P. (1983). *Networks of power: Electrification in Western society, 1880–1930*. Baltimore: Johns Hopkins University Press.
Kauffman, S. A. (1993). *The origins of order: Self-organization and selection in evolution*. Oxford/New York: Oxford University Press.
Kenis, P., & Schneider, V. (1991). Policy networks and policy analysis: Scrutinizing a new analytical toolbox. In B. Marin & R. Mayntz (Eds.), *Policy networks: Empirical evidence and theoretical considerations* (pp. 25–59, Schriften des Max-Planck-Instituts für Gesellschaftsforschung Köln). Frankfurt/Main: Campus.
Kuhlmann, S. (2001). Future governance of innovation policy in Europe: Three scenarios. *Research Policy, 30*(6), 953–976.
Kuhlmann, S., Shapira, P., & Smits, R. E. (Eds.). (2010). *The theory and practice of innovation policy: An international research handbook*. Cheltenham: Edward Elgar.
Lieberman, E. S. (2005). Nested analysis as a mixed-method strategy for comparative research. *American Political Science Review, 99*(3), 435–452.
Lundvall, B.-Å. (Ed.). (1992). *National systems of innovation: Towards a theory of innovation and interactive learning* (1st ed.). London: Pinter.
Maier, M. W. (1998). Architecting principles for systems-of-systems. *Systems Engineering, 1*(4), 267–284.
Majone, G. (1994). The rise of the regulatory state in Europe. *West European Politics, 17*(3), 77–101.
Mayntz, R. (1983). The conditions of effective public policy: A new challenge for policy analysis. *Policy & Politics, 11*(2), 123–143.
Mayntz, R. (2003). New challenges to governance theory. In H. P. Bang (Ed.), *Governance as social and political communication* (pp. 27–40). Manchester/New York: Manchester University Press.
Mayntz, R., & Hughes, T. R. (Eds.). (1988). *The development of large technical systems*. Frankfurt am Main: Campus.
Mayntz, R., & Scharpf, F. W. (1995). Der Ansatz des akteurzentrierten Institutionalismus. In R. Mayntz & F. W. Scharpf (Eds.), *Gesellschaftliche Selbstregelung und politische Steuerung* (pp. 39–72). Frankfurt: Campus.
Metcalfe, J. S. (1995). Technology systems and technology policy in an evolutionary framework. *Cambridge Journal of Economics, 19*(1), 25–46.
Nelson, R. R. (Ed.). (1993). *National innovation systems: A comparative analysis*. New York: Oxford University Press.
Newman, M., Barabasi, A.-L., & Watts, D. J. (Eds.). (2006). *The structure and dynamics of networks*. New York/Cambridge: Cambridge University Press.
OECD. (1999). *Managing national innovation systems*. Paris: Organization for Economic Co-operation and Development.
OECD. (2005). *Governance of innovation systems. Volume 1: Synthesis report*. Paris: Organisation for Economic Co-operation and Development.

OECD. (2008). *Main science and technology indicators. Volume 2008, issue 1*. Paris: Organisation for Economic Co-operation and Development.
OECD. (2010a). *Measuring innovation: A new perspective*. Paris: Organisation for Economic Co-operation and Development.
OECD. (2010b). *The OECD innovation strategy: Getting a head-start on tomorrow*. Paris: Organisation for Economic Co-operation and Development.
Ostrom, E. (1999). Institutional rational choice: An assessment of the institutional analysis and development framework. In P. A. Sabatier (Ed.), *Theories of the policy process* (pp. 35–71). Boulder: Westview Press.
Pittaway, L., Robertson, M., Munir, K., Denyer, D., & Neely, A. (2004). Networking and innovation: A systematic review of the evidence. *International Journal of Management Reviews, 5*(3), 137–168.
Rhodes, R. A. W. (1996). The new governance: Governing without government. *Political Studies, 44*(4), 652–667.
Röller, L.-H., & Waverman, L. (2001). Telecommunications infrastructure and economic development: A simultaneous approach. *The American Economic Review, 91*(4), 909–923.
Saavedra, R., Earley, P. C., & Van Dyne, L. (1993). Complex interdependence in task-performing groups. *Journal of Applied Psychology, 78*(1), 61–72.
Sabatier, P. A. (1988). An advocacy coalition framework of policy change and the role of policy-oriented learning therein. *Policy Sciences, 21*(2–3), 129–168.
Sabatier, P. A., & Weible, C. M. (2007). The advocacy coalition framework: Innovations and clarifications. In P. A. Sabatier (Ed.), *Theories of the policy process* (pp. 189–222). Boulder: Westview Press.
Sawyer, P. K. (2005). *Social emergence: Societies as complex systems*. New York: Cambridge University Press.
Saxenian, A. (2007). *The new argonauts: Regional advantage in a global economy*. Cambridge, MA: Harvard University Press.
Scharpf, F. W. (1997). *Games real actors play*. Boulder: Westview Press.
Schmidt, M. G. (1993). Theorien in der international vergleichenden Staatstätigkeitsforschung. In A. Héritier (Ed.), *Policy-Analyse. Kritik und Neuorientierung* (pp. 371–393). Opladen: Westdeutscher Verlag.
Schneider, V. (2004). State theory, governance and the logic of regulation and administrative control. In A. Warntjen & A. Wonka (Eds.), *Governance in Europe* (pp. 25–41). Baden-Baden: Nomos.
Schneider, V. (2012). Governance and complexity. In D. Levi-Faur (Ed.), *Handbook on governance*. Oxford: Oxford University Press (forthcoming).
Schneider, V., & Bauer, J. (2009). Von der Governance – zur Komplexitätstheorie: Rekonstruierung gesellschaftlicher Ordnungsbildung. In I. Schultz-Schaeffer & J. Weyer (Eds.), *Management komplexer Systeme*. Oldenbourg: München.
Schneider, V., & Janning, F. (2006). *Politikfeldanalyse. Akteure, Diskurse und Netzwerke in der öffentlichen Politik*. Wiesbaden: VS-Verlag.
Schwab, K. (Ed.). (2011). *The global competitiveness report 2010–2011*. Geneva: World Economic Forum.
Shenhar, A. J. (1993). From low- to high-tech project management. *R&D Management, 23*(3), 199–214.
Steinberg, G. M. (1985). Comparing technological risks in large scale national projects. *Policy Sciences, 18*(1), 79–93.
Strogatz, S. H. (2003). *Sync: The emerging science of spontaneous order*. New York: Hyperion.
Thompson, J. D. (1967). *Organizations in action*. New York: McGraw-Hill.
United Nations Statistics Division (2011). UNdata. A world of information. Online database, available at http://data.un.org.
Williamson, O. E. (2000). The new institutional economics: Taking stock, looking ahead. *Journal of Economic Literature, 38*(3), 595–613.

Part I
Theories and Concepts of Innovation Policy

Chapter 2
Institutions and Systems: Analysing Technical Innovation Processes from an Institutional Perspective

Raymund Werle

2.1 Introduction

Institutionalism in its different facets has a long tradition in the analysis of social phenomena including the evolution and development of technical innovations. Institutional arrangements are regarded as coordinating and shaping collective action and, consequently, also influencing innovation policy. Although innovation policy addresses various kinds of innovation, this chapter will concentrate on technical innovations, product and process innovations. The studies to be reviewed examine the invention, acquisition, application, development and diffusion of new technology. They reject technological determinism, which prevailed in technology studies for a long time, and, in most cases, treat technical innovation as the dependent variable.

Innovation researchers have analysed technical innovations from various theoretical perspectives. We confine this article to studies which look at technical innovations from an institutional angle and examine what they contribute to the overall understanding of technical innovations and their repercussions. These approaches are not compared with other theories. Instead, the main focus lies on the spectrum of institutional analyses of technical innovations, including studies which primarily focus on other variables, such as economic performance, and consider the capacity to innovate only because technical innovations often enhance economic performance. These studies' suggestions or hypotheses concerning technical innovations are not less important than those developed in specialised innovation research. Thus, the studies that are of importance in our context differ gradually rather than in principle. Their conceptual understanding of institutions

R. Werle (✉)
Max Planck Institute for the Study of Societies, Cologne, Germany

Department of Organizational Sociology and Innovation Studies, University of Stuttgart, Stuttgart, Germany
e-mail: werle@mpifg.de

J.M. Bauer et al. (eds.), *Innovation Policy and Governance in High-Tech Industries*,
DOI 10.1007/978-3-642-12563-8_2, © Springer-Verlag Berlin Heidelberg 2012

and their categorization of technical innovations may differ, but all are interested in the institutional conditions under which innovations may evolve, prevail or fail.

Three groups of studies will be included: socio-economic institutionalism examining national innovation systems as its core area; politico-economic institutionalism with recent innovation-oriented research on varieties of capitalism; and techno-sociological institutionalism embracing sociological innovation research in the field of technology. The studies are institutional in the sense that they draw on particular institutions or institutional constellations as societal meso- or macro-phenomena to explain technical innovations. In their majority, they explore the effects of institutions on technology and only rarely do they touch upon processes of institutional development or change triggered by technical innovations if and when, for example, the complexity of an innovation "necessitates" a regulatory response (Feick and Werle 2010, pp. 45–47). Thus, after years of technological determinism we are now confronted with the danger that the pendulum will swing in the opposite direction towards some kind of institutional determinism. To escape the potential determinist trap, technological and institutional changes must be related to each other and their interdependence must be examined. Appropriate approaches can be found in several of these socio-economic, politico-economic and techno-sociological analyses to be discussed now.

2.2 Socio-economic Institutionalism

Since the 1980s, socio-economic research has been increasingly concerned with technical innovations. In contrast to neoclassical approaches, which treat technical innovations as exogenous variables, these studies try to endogenise innovations and to discover conducive or hindering factors. These are not necessarily always or in the first place market-related factors. Rather, the multifold institutional structures of capitalist nation states specifically determine both form and speed of technological progress (Dosi 1988, p. 1148). After several country comparisons displayed striking differences and changes in relative economic performance, the attention of researchers shifted to national institutions and their significance for the countries' innovative capabilities – thereby assuming with reference to Schumpeter that technical innovations enhance economic performance and growth.

2.2.1 National Innovation Systems

Among the socio-economic innovation studies with an institutional orientation, those focusing on national innovation systems (NIS) particularly stand out. Following Porter's (1990) groundbreaking investigation into the (particularly technological) competitiveness of ten leading industrialised countries, these studies show that varying national institutional constellations account for the divergent innovative capabilities (Edquist 1997). At the centre of these analyses lie product and process innovations within technology-based industries (Carlsson et al. 2002). Inventions and, more importantly, the development and diffusion of innovations are not considered as single acts, but instead as processes which are formed by institutional constellations and structures that vary among nations.

Prominent early studies of NIS, such as those by Freeman and Nelson, show that the prevailing understanding of institutions is rather vague and extensive, and that the concept of institutional systems remains unclear (Freeman 1987; Nelson 1988). Freeman describes NIS as "the network of institutions in the public and private sectors whose activities and interactions initiate, import, modify and diffuse new technologies" (Freeman 1987, p. 1). In his study of Japan, he alludes to the industrial structure, the education and training system, the research and development activities of businesses, and the long-term strategies of the MITI (Ministry of International Trade and Industry). Institutions encompass not only legal rules but also organizations and their activities and strategies. Such a broad understanding of institutions and the vague system concept are also typical of subsequent studies. In a more recent anthology, Nelson and colleagues define NIS as "the cluster of institutions, policies, and practices that determine an industry's or nation's capacity to generate and apply innovations" (Steil et al. 2002).

At an early stage and in collaboration with Perez, Freeman also developed a classification for innovations. Their distinction between *incremental* and *radical innovations* is used particularly frequently. Incremental innovations are seen as relatively continuous improvements of technology within one line of development. Radical innovations occur discontinuously, often as results of strategic research and development activities, and lie outside given technological trajectories (Freeman and Perez 1988, pp. 45–47).

Both Nelson and Freeman regard institutions as relatively resistant to changes. Hence, for the success of technical innovations it is decisive that they fit well into the institutional structures and that these structures have a strong absorptive capacity. Ultimately, the development and/or quick diffusion of innovations requires compatibility ("match") of new technologies and institutional constellations. As institutions have a relatively low adaptability they are treated as part of the selective external environment which ultimately determines the destiny of innovations.[1] These emerging contours of an evolutionary theory of technical innovations are elaborated on more fully by Nelson (Nelson and Winter 1982; Nelson 1987). According to Nelson, the capitalist profit motive, the competition among different sources of innovation and the market selection constitute the crucial elements of the process of technological evolution. The "selective environment" takes effect via market demand and thus determines the success or failure of innovations.

[1] Freeman shows that Japan's institutional constellation was conducive to process innovations and lead to competitive advantages in the consumer goods industry, in the automobile production and in the production of semiconductors. In other areas of technology, Japan lacked innovativeness because its institutional system was less supportive to innovations.

Corresponding to Nelson's extensive understanding of institutions, this environment also comprises numerous organizations: firms, industrial research laboratories, research universities, vocational training centres, as well as government agencies with their technology and industrial policy. These organizations, their strategies and relationships vary across countries (Nelson and Rosenberg 1993). This also applies to institutions in a narrower sense, such as the, in Nelson's view, crucial rules for appropriating and securing the profit of innovations. In the relatively simple evolutionist scheme of variation and selection, individual and corporate actors promote technological change by producing innovations (variation), while the national institutional systems in the broad sense separate the wheat from the chaff (selection).

Nelson and his colleagues conducted or inspired numerous studies in which innovative capabilities and activities are primarily measured by the expenses for research and development or by the number of patents. Another occasionally used indicator is the balance of imports and exports of high-tech industries. These indicators are differentiated according to economic sectors and whether private or public, non-military or military, as well as whether university-based or non-university research organizations are involved. Yet, no systematic country comparisons were carried out by Nelson, although he studied 15 countries in the beginning of the 1990s. Aside from the observation that strong and competent enterprises form the most important precondition for an innovative, prospering economy, no generalizable conclusions could be drawn, especially with regard to the effects of institutional constellations. But it becomes clear that a country's attempt to copy institutional factors of another, in certain areas particularly successful, country is not very promising due to the complex and multi-layered nature of innovation systems. Especially, taking cues from the U.S., complementary relationships between industry and university research on the one hand and the strongly differentiated (public) research funding on the other have been emphasised as crucial for the process of innovation (Mowery 1994, pp. 79–106; Riccaboni et al. 2003). Moreover, the availability of venture capital is the key to rapid commercialization of innovations mainly because innovative researchers are now able to leave research laboratories of universities or major enterprises and establish their own start-up companies (Mowery and Rosenberg 1993).

Lundvall developed an additional variant of the NIS-approach. According to him, the countries' historically grown economic structures, including industrial relations and their organizational and institutional structures (e.g., of research and development), are characteristic for their respective innovation systems (Edquist and Lundvall 1993). Within these frameworks, and dependent on them, innovation processes unfold as cumulative, interactive and continuous learning processes (Lundvall et al. 2002). The learning processes which ensue from the interactions among producers as well as between producers and users of technology facilitate especially the incremental development and diffusion of innovations (Lundvall 1992). This holds for technical innovations, be they process or product innovations, as much as for organizational changes of enterprises or institutional innovations. Finally, Lundvall emphasises that incremental changes resulting from learning processes are more important for a country's economic performance than the ability to create something radically new.

Within this spectrum of proposed approaches and perspectives, a large number of studies investigate the connection between institutional constellations and technical innovations. Although the list of relevant institutional factors remains enumerative, the studies successfully show that, in general, firms – securely anchored in national institutional systems – are the central agents for innovation and benefit from this situation. As a consequence this further stabilises the national institutional systems. At the same time, this has the effect that countries continue to be strong in certain technologies and weak in others. Complementarities emerge between scientific research at universities and public research organizations, research and development within enterprises, strategic networks of cooperation, public technology and industrial policy and other factors with the result of pronounced sectoral specialization patterns (Archibugi and Pianta 1992; Guerrieri 1999).

Follow-up studies in the tradition of the NIS approach reveal stagnation of the theoretical, and to a lesser extent empirical, vitality. This also holds true for more recent studies which were concerned with the excellent performance of the U.S. economy in the 1990s (Larédo and Mustar 2001; Steil et al. 2002; Block and Keller 2009). While the concept of national innovation systems can still be considered very vague, the institutional components are now more strongly emphasised than other system elements. However, there is still no clear-cut definition of what is systemic in national innovation systems (Carlsson et al. 2002).

Given these problems, it is not surprising that more recent studies in the NIS-tradition examine the connection between specific aggregates of institutional regulations and technical innovations, rather than looking at entire institutional constellations. Their main focus is regulations for the protection of intellectual property. In the past 30 years, far-reaching changes have occurred, with the consequences being discussed in these studies. Most of them concentrate on the United States. Here, particularly, the possibilities of obtaining patent protection have been extended severely since the early 1980s (Jaffe 2000; Gallini 2002). Not only was the period of validity for patents prolonged, but also the circle of organizations eligible to file patent applications was enlarged. It now includes universities and public research organizations which are allowed to patent innovations even if they were publicly-funded. More importantly, patent protection was extended to previously not patentable fields. In this context, the granting of patent protection for living organisms (such as genetically engineered bacteria), DNA sequences or other biotechnological and genetically modified innovations is particularly remarkable. But also the right to patent certain software products, which have been protected merely by copyright or not at all in the past, has raised numerous research questions. It is especially interesting to understand the effects of improved patent protection on the increase of patent activities in several countries and to check whether intensified patenting reflects an increase of successful innovation efforts (Gallini 2002, p. 133). It has been shown that strategic patenting and licensing behaviour often generates unexpected and innovation-hindering effects (Heller and Eisenberg 1998; Hall and Ziedonis 2001). Generally, the relationship between the legal opportunities to patent and innovative behaviour is more complex than frequently assumed (Bessen and Meurer 2008).

2.2.2 Repercussions of Innovations on National Innovation Systems

Most studies on the relationship between NIS and technical innovations assume that institutions change very slowly and that the diffusion of innovations depends on their compatibility with the institutions. Only in the long run and triggered by radical basis innovations are repercussions on the system of institutions expected. Freeman and Perez (1988) suggest that radical and self-accelerating innovations, which simultaneously occur in several technological fields, lead to a drastic structural shift of the entire economy. As a result, a change of the hitherto dominating techno-economic paradigm can be observed. A new "technological regime" with characteristic institutional structures is established which continues to be prevailing over centuries and only changes in long cycles. Hence, in the rare occasion of technological revolutions (and usually only then) technologically induced new institutional arrangements may evolve (Freeman and Louçã 2002).

Porter (1990) refers to possible medium-term institutional change as a reaction to technical innovations, particularly in technology-intensive industries which play a decisive role in the competition of national economies. These industries emerged under conditions originally created by the countries themselves which subsequently were influenced by the industries. Driving forces are multinational enterprises which are interested in optimally benefiting from the potential of new technologies (see Pavitt and Patel 1999). Other studies which show that institutional differences between different technologically shaped sectors within a country tend to be greater than differences between countries also point to repercussions of technology on institutional structures (Carlsson 1994; Breschi and Malerba 1997). The close examination of such sectoral innovation systems, so-called technological systems, opens better possibilities of tracing the interactions between technology and institutions, rather than only considering entire national systems (Geels 2004).

If at the same time technical innovations are analysed in more detail, it renders endogenising institutional developments possible. Institutional changes can at least partially be explained by technical innovations (see Dolata 2009). This is indicated by research on the protection of intellectual property rights in technologically innovative sectors. Graham and Mowery (2003, p. 254), for instance, characterise the relation between software innovations and legal developments towards a stronger protection of intellectual property rights in the software industry ("software patents") as "co-evolution, involving mutual causation and influence." This corresponds to rather programmatic considerations regarding the "coevolution" of technologies and institutions which Nelson (1994) coined a decade earlier.

2.2.3 *Results*

Summarizing the state of socio-economic analyses studying national innovation systems, it can be seen that the influence of institutional variables on technical innovations has been made plausible, but rarely has it been specified. Innovations comprising physical artefacts as well as technological know-how are not examined in detail. Specific attention is paid to input factors for innovation, such as public and private expenses for research and development. The output of innovation processes is often only measured by how often or how seldom innovations evolve. Some studies count the number of patents (e.g., Faber and Hesen 2004) even though changes in the number of patents do not necessarily correspond to the number of innovations. In general, technology itself is measured by simple undifferentiated categories. Most commonly, the distinction between radical and incremental technical innovations, as well as between product and process innovations, is used. It is emphasised as a general rule that innovations occur surprisingly and that their exact emergence cannot be fully explained. This does not contradict the fact that innovative technological developments tend to occur within a "technological paradigm" which constitutes a relatively stable path of development (Dosi 1982).

Regardless of the specific substance of technical innovations, the institutional conditions of countries and sectors can be judged according to whether they facilitate or impede innovations. Similar to the pioneering works, more recent studies also stress the developmental potential of technologies, the size of markets, the possibility to finance and acquire ownership rights of innovations, the structure of the respective sector, and investments in publicly available knowledge as most important factors which trigger and structure innovations. Potential repercussions of technical innovations on institutional structures have gained increasing attention. Newer studies suggest that national innovation systems tend to converge towards stronger market coordination, partly due to external pressure (see Henisz et al. 2005). Nonetheless, crucial components of national institutional constellations and, accordingly, national differences remain notably stable.

2.3 Politico-economic Institutionalism

One central shortcoming of studies pertaining to national innovation systems is their lack of a theoretical concept of institutions that could be related to and integrated with general institutional theory. Instead, institutions and institutional constellations are analysed in isolation without examining their potential relations. This is explicitly criticised by Hollingsworth, who argues that the problem is not a lack of institutional approaches, but rather their excessive supply. With a theory of innovation in mind, he contends that the specific components of a society's institutional structure and the relations between them should first be identified before any statements can be made regarding their influence on a country's innovative capacity (Hollingsworth 2000, p. 596).

In his understanding, national and sectoral arrangements of institutional governance of production constitute *social systems of production* which can differ from country to country but generally show a certain degree of internal coherence with often interdependent, complementary components (Hollingsworth and Boyer 1997; Hollingsworth 2000, pp. 613–619). Together with the structure and norms of relevant organizations (especially enterprises and public research organization), a society's social production system shapes what Hollingsworth calls its "innovative style." Accordingly societies can be more or less innovative, their innovations can be incremental or radical, and they can typically be developed in emerging high-tech sectors or in mature industries. While Germany develops successful incremental innovations in sectors such as chemistry, electrical engineering, mechanical engineering or automotive engineering, the U.S. has a rather radical innovation potential. In newer industries such as electronics or biotechnology the U.S. produces – in short time periods – completely new complex products which often have a relatively short life-span (Hollingsworth 2000, pp. 626–633). The social systems of production of both countries differ with respect to the structure of enterprises, the industrial relations, the vocational training system, the financial markets, and the university-based research system. These differences are decisive for the diverging innovation styles.

2.3.1 Varieties of Capitalism and Innovation

Hollingsworth alludes to a set of variables which play an important role in more recent politico-economic research on the *varieties of capitalism* (VoC). Similar to NIS-studies, VoC-research started in the 1980s. The goal is to explain noticeable performance differences between national economies focusing not on more or less successful public economic policies, but rather on political-institutional factors, i.e., different organizational forms or different varieties of capitalism.

Similar to studies about national innovation systems, VoC-research concentrates on countries or sectors as units of analysis, but in a more systematically comparative way. Research is directed at the global competition of social production systems and the resulting institutional change of national capitalisms. Hence, the studies not only raise the question of how institutions influence a country's economic performance, but also how institutions develop and change. However, most studies are still fixated on contrasting global convergent and national path-dependent development.

From the outset, VoC-studies have aimed at creating a typology of institutional constellations in order to classify the countries or sectors which are to be compared. A classification developed by Soskice in the last decade is generally regarded to be the most elaborated. Picking up the concept of social production systems, the author suggests that production regimes shape the rules of the institutional framework, which helps the "microagents of capitalist systems" to organise and structure their relations with each other (Soskice 1999, p. 101). Important elements of this view, which focuses on the production side of economies, include the system of corporate finance, the various models of corporate governance, the employment contract law, the industrial relations, the education and vocational training systems, and finally, the rules which govern the relations between enterprises (competition and antitrust law, technological transfer regulations, standardization guidelines, etc.). Soskice distinguishes two basic types of capitalist economies: *coordinated market economies* (CME) and *liberal market economies* (LME). In the case of CME, employers are integrated in a network of associations, coordinating wages, training and employment relations internally and together with organised labour. The network has a cooperative spirit and a long-term perspective. This includes corporate finance, which is provided in the form of long-term 'patient' capital. In the other basic type, the LME, short-term market coordinated relations prevail between enterprises, but also between enterprises and their labour force or their investors (Soskice 1994; Hall and Soskice 2001).

In the centre of attention stand enterprises and their strategies. While the enterprises can act autonomously, their actions are influenced and channelled by the afore-mentioned institutional elements. The results of such actions are hence determined by the interrelation of institutional influences and autonomous strategic interaction (Hall and Soskice 2001). They are thus never exclusively determined by preferences, resources or strategies of actors, or solely by the institutional context. This perspective is still predominantly programmatic in VoC but almost completely absent in studies on national innovation systems. Although NIS-studies regularly emphasise the importance of enterprises in the innovation process, they are merely treated as a passive "black box", influenced by "macro-social determinants" including institutions (Coriat and Weinstein 2002, p. 274).

Liberal market economy prototypes (within the OECD countries) include the United States in the first place but also the UK, Ireland, Canada and Australia. Germany is seen as the prototype of the group of countries with a coordinated market economy, to which Austria, Switzerland, Sweden, Norway, and Japan also belong. When Germany and the US are seen as two endpoints of a continuum, all other mentioned countries are very close to one of these two endpoints. Still others, including the Mediterranean countries, lie in "ambiguous positions" (Hall and Soskice 2001, p. 21). On the one hand they have a relatively liberal labour market, but on the other their governments strongly influence the economy and their agricultural sector is relatively large.

Only rarely do we find references to technical innovations in the studies about the varieties of capitalism. Generally, the studies' dependent variable is economic performance, which is shaped by the comparative institutional advantages each country has. Every institutional constellation has specific strengths and weaknesses, and no constellation is superior to others in all dimensions of economic performance. This also holds for technical innovations, which are considered in VoC-studies mainly because they influence the economic performance. Hence, innovations function as intervening variables. They are shaped by institutions and – in turn – affect performance.

Similar to the NIS-literature, the VoC-literature rather coarsely differentiates between types of technical innovations. The central distinction lies between incremental and radical innovations. It is argued that enterprises in liberal market economies exhibit a strategic alignment toward radical innovations. These innovations, mostly only patented inventions at first, emerge in the new high-tech sectors. Small start-ups especially, financed with venture capital, and – though less frequently – major enterprises introduce new products into the market or at least make the products marketable. Due to the high speed of innovation, products, which are often components of complex technological systems, have a relatively short lifespan. Under such conditions, enterprises are committed to short-term profits which they can achieve if they employ a staff with a high level of general qualifications. The enterprises must continually and flexibly position themselves vis-à-vis changing market conditions. This requires adaptable employees and involves high staff turnover. Regarding the specific institutional conditions, this is compatible with LME but not with CME. Hence, LME promote radical innovations while CME are favourable to incremental innovations (critical Streeck 2011; Soskice 1999; Hall and Soskice 2001).

In some more recent VoC-studies, the distinction between radical and incremental innovations has been developed further. This facilitates substantiating the effects of institutional constellations on technical innovations. One very interesting distinction differentiates between "discrete" and "cumulative" (or "platform") technologies, introduced in a study of enterprises in the German bio-technology and software industry (Casper et al. 1999). The authors attempt to explain why, in a coordinated market economy such as Germany, enterprises which work on radical innovations may still prosper. They attribute this to the fact that in the area of bio-technology and software one can find technologies which fit into the German institutional framework. Successful enterprises, it is argued, specialise in cumulative rather than discrete technologies. These broader platform technologies develop over a longer period of time in a comparably stable way. In information technology, it is not the standardised software, but the service segment for commercial users of software. These users ask for integrated system solutions that are regularly expanded and updated. From the perspective of technical development, this is a cumulative and long-term process. The relations of service companies to their customers are also of a long-term nature. The service providers' staff accumulates specific cumulative know-how which offers them a long-term employment perspective. Similar developments have been observed in bio-technology. German enterprises have specialised in the development and production of instruments and software which are used for pharmaceutical research or the production of pharmaceuticals. In contrast to the end products of this industry, they are applied and demanded constantly and hence must undergo continuous further development (Casper and Glimstedt 2001; Casper and Matraves 2003).[2]

[2] The view that platform technologies possess a cumulative character and can be continuously developed over years is often challenged (Dolata 2003). However, this does not alter the usefulness of distinguishing between discrete and cumulative technologies.

2.3.2 Repercussions of Innovations on National Production Regimes

One finding upon which all studies on varieties of capitalism agree is that national institutional constellations are resistant to rapid fundamental change despite the pressure of globalization. This is due to interdependencies and complementarities of institutions within the established national constellations (Amable 2000). Recently, this understanding of a close link among institutions has been criticised methodically and empirically (see Höpner 2005). The implication that there are not even niches in which innovations can develop, unless they match institutionally, had to be abandoned (Crouch 2003; Kitschelt and Streeck 2003; Lange 2009).

This could reinforce the discussion triggered by Kitschelt (1991) in the early 1990s emphasizing that technical innovations exert pressure to change on national institutions. Kitschelt criticises the fact that research on the performance of national economies tends to ignore the sectoral differentiation of national institutional arrangements and the structural features of technology. Whether or not a new technology can establish itself and develop further depends on corresponding sectoral structures of institutional governance. If these structures are missing they can be created to the extent that the encompassing national regime structures allow such a change. Within the framework of stable national institutions, sectoral structures are hence able to change under the influence of innovative technologies. Technical innovations may thus promote institutional change towards a national regime structure which shows a broader mix of sector-specific institutions and a wider array of national technologies. Kitschelt illustrates his considerations by comparing Japan's development to Western industrial countries. The author strives to describe in great detail the structures of technology and the corresponding institutions by using categories which are closely related to theories of institutional governance. These include Perrow's distinction between loosely and tightly coupled technological systems and between linear and interactive system processes (Perrow 1984), as well as the different types of governance in Williamson's transaction cost theory (Williamson 1985). Certain technology features such as "asset specificity" or "uncertainty about the causal structure of the technology" (Kitschelt 1991, p. 464), which are relevant for a transaction, co-vary with Perrow's characteristics of (more or less complex) technological systems. They each require adequate sectoral institutional environments.

Kitschelt's strategy is remarkable because he applies a differentiated, yet nonetheless rather formal concept of technology or technological systems. This allows systematically integrating technology into the analysis as an endogenous as well as exogenous variable in the process of constituting an industrial sector. The evolutionary variety of technology is reduced through institutional influences. However, technological systems are still too multifaceted to be regulated efficiently by a uniform national institutional structure. Therefore, technology-related sectoral governance structures are established. No country has general political-institutional preconditions which are equally beneficial to the development of adequate sectoral institutional structures given the diversity of technological systems. Overall, however, institutional variety at the national level benefits the development of technology-adequate sectoral structures and hence technical innovations, which, on their part, stabilise or even enhance this variety.

2.3.3 Results

Most studies on the varieties of capitalism do not focus on technical innovations. Whenever they are mentioned, they appear as dependent variables which remain under the influence of specific national institutional constellations. Hence, innovations only have good implementation opportunities if they are compatible with the national system of institutions, regardless of whether they were developed externally or within their respective country. Different institutional systems promote different types of innovations. Whereas liberal market economies tend to produce radical innovations, coordinated market economies provide favourable institutional conditions for incremental innovations.

Some enterprises acting in coordinated market economies have prospered in industries characterised by radical innovations such as biotechnology or software. This surprising fact is explained by further distinguishing certain types of radical innovations. Not all radical innovations are discrete in nature. Some are cumulative or (relatively broad) platform technologies which can be developed and improved continually over a longer period of time. This is compatible with coordinated market economies. Only rarely has it been suggested that technical innovations exert pressure on institutions to change. In Germany, the development of radical innovations has increasingly attracted venture capital – virtually incompatible with the institutions of a CME – which in turn provides incentives for further radical innovations and institutional adaptation.

2.4 Techno-sociological Institutionalism

Both socio-economic and politico-economic analyses of technical innovations generally fail to provide a detailed examination of technology and its respective stages of development. Sociological technology studies differ in that they focus on technology in more detail. But sociological technology studies often neglect the institutional arrangements in which technology evolves. The traditional rules of technology studies lack explicit reference to institutions and institutional explanations (Rammert 1997). Instead, most of the studies share the "enactment perspective" which regards the emergence and development of technology as a contingent process of social appropriation exclusively at the micro-level of individual or collective action and practice (see Schulz-Schaeffer 2000).

Institutionally oriented studies have typically been concerned with the emergence and development of technological infrastructure systems or, more general, *large technological systems* (LTS). These studies understand institutions as rule systems and focus on the problems of coordination and regulation which emerge during the process of technology development.

2.4.1 Coordinating Innovations Through Hierarchies, Markets or Networks

Based on the generic types of institutional governance – *hierarchies*, *markets* and *networks* – specific institutional arrangements, actor constellations and actor strategies are analysed regarding their effects on technology.

Research on LTS was initiated by the works of technology historian Hughes (1983) who analysed the early development of electricity supply systems in Chicago, London, and Berlin. Such systems do not simply follow technical imperatives in their development nor are they exclusively shaped by inventors and system designers with an entrepreneurial spirit. Rather, the political-institutional framework is crucial as well. Following a *phase model*, the systems develop from the stages of invention and innovation, through technology transfer, growth and competition, to the consolidated state of "momentum". Although the basic technologies for electricity supply in all three cities are similar, the systems differ with regard to the degree of centralization and integration, and also to efficiency; however, no one system outperforms the others in all respects. The decisive factors for the observed differences lie in the institutional conditions which coordinate the process of development. Whereas market factors were crucial in Chicago, corporatist networks in Berlin and administrative hierarchies in London coordinated the technical development (Hughes 1983, pp. 165–261).

The history of technological infrastructures (especially telephone, railway, electricity) shows that they unfold almost always as public or private politically tolerated and regulated territorial monopolies (see Mayntz and Hughes 1988). Large organizations internalise the system development and hierarchy has been the predominant mode of coordination for a long period of time (Chandler 1977). In general, this institutional constellation is conducive to the development of "conservative" technical innovations (Hughes 1982) and it rules out internally initiated radical system changes. Thus, far-reaching changes of technological infrastructures are typically the consequence of political-institutional changes. This has been shown by Schneider (2001) in his evolutionary analysis of the institutional transformation of telecommunications in the six most important industrial countries over the course of two centuries. The mode of coordinating telecommunications has shifted from state monopolies to more market-based structures. Deregulation and liberalization triggered a vertical de-concentration and unbundling of the systems and their architecture (see Mayntz 2009).

The politically initiated institutional change towards more market coordination alludes to its superior innovation efficiency (Baumol 2002). Hierarchical and centralised architectures are transformed into decentralised, modularised and networked architectures. The internet can be considered the most impressive example of this transformation process which released the extraordinary innovative potential of this technology (Werle 2000). Aside from some coincidental decisions, the institutional preconditions in the U.S. in the 1980s and early 1990s account for the quick national and subsequent international expansion of the internet. In contrast to Europe, U.S. telecommunications were already deregulated to a substantial degree and market principles coordinated the U.S. software industry, while European governments still supported "national champions" and protected them from competition. Moreover, the U.S. higher education sector (in which internet spread at first) was also organised competitively to a certain extent. Originally, the internet was publicly funded and developed in a protected niche. But it subsequently established itself on the market without further public support (see Mowery and Simcoe 2002). At the same time, European governments which wanted to outcompete the U.S. internet promoted national research and education networks which developed in a hierarchical institutional setting controlled by the telecommunications monopolies. But these networks failed as soon as the internet was allowed to enter Europe and the national telecommunications markets were opened for internal and external competition. The internet's advantage has been that – partly due to the heterogeneous institutional context of the U.S. – its generic protocols were designed to handle technical heterogeneity and autonomy of sub-networks and to interconnect these networks successfully. Conversely, in the hierarchical context of the European countries, the efforts of the engineers were directed towards developing rather centralised technically homogeneous networks which experienced almost un-surmountable problems when these networks were to be connected to networks with different standards (David 2001; Werle 2002).

The studies on innovations in large technological systems are generally restricted to contrasting the governance forms of market and hierarchy. Analyses of network forms of coordination which especially gained attention in the 1990s focused mainly on the level of enterprises and the organization of production (Powell 1990; Hirsch-Kreinsen 2002). The possible influence of networks on the development of technology was discovered relatively late. A special type of networks, the so called "innovation networks" which promoted the evolution and diffusion of *new* technologies moved into the centre of attention. These heterogeneous networks connect "technology-generating, technology-applying and technology-regulating social systems" with each other (Kowol and Krohn 1995, p. 78). Based on negotiations and trust, they help to manage complexity and to reduce uncertainty where markets tend to fail regarding the flow of information and where hierarchies fail with respect to flexibility (see Küppers 2002).

Rarely is the relationship between the structure of innovation networks and the technical innovation process specified. Weyer et al. (1997) take a first step in this direction, analysing in four case studies the evolution and development of the European aircraft Airbus, the Personal Computer, the high speed train Transrapid, and Satellite Television. The authors argue that successful innovations pass through the phases of emergence, stabilization and implementation as independent stages of development. From one stage to another the network of actors that are innovation-enhancing changes. In the starting phase, coincidental inventions of innovative actors are integrated into a model which includes the basic specifications of architecture, production and utilization of a technology. Potentially interested actors form a network. In the subsequent phase, strategic actors set up networks in order to promote the technology. This step absorbs uncertainty, facilitating the further development of the innovation. The innovative idea and the general model lead to a first technological prototype. In the final implementation and diffusion phase, the network opens itself to include new members such as users, affected third parties, operators, and also critics. New areas in which the innovation can be applied, as well as new patterns of utilization are invented in this phase. According to the authors, it is crucial for a successful innovation that a network is formed and socially consolidated in every single phase of development. These networks must be able to reach necessary decisions and actively participate in the construction process. Otherwise, the innovation will stagnate on the stage it has reached and not move onto further stages. For the successful implementation of a new technology in particular, a rigorous opening of the networks is essential (Weyer et al. 1997, p. 330).

Similar to this research group, other studies also restrict themselves to exclusively analyse the success or failure of innovations and of the enterprises involved in the innovation networks. The development of successful technical innovations appears to strongly depend on the formation and stabilization of networks in which actors from different institutional sectors in a certain region such as Silicon Valley work together (Castilla et al. 2000). Also, government agencies can play an important role promoting and moderating networks of innovation (Giesecke 2000). This suggests using a multilevel approach, especially if the development and transformation of large technical systems is studied (Geels 2007).

From an institutional perspective, it is important to note that innovation networks link different institutional sectors with each other. This has also been emphasised in studies which point to the close connection of institutional sectors and logics of action. Some of these studies refer to the innovation-promoting effect of symbiotic "triple helix" constellations among universities, industry and governments (Leydesdorff and Etzkowitz 1998). Others – following Gibbons and colleagues – stress the importance of these linkages and networks for the emergence of new forms of knowledge generation (Gibbons et al. 1994).

The networks' institutional character as a specific mode of coordination is nicely revealed by a study of formation, dissolution and change of networks in the field of biotechnology. Powell et al. (2005) analysed these phenomena over a period of more than 10 years in the U.S., thereby meeting the requirement to describe networks in their dynamics and to identify the underlying mechanisms (Jansen 2002). They analyse the effects of changing rules and preferences for partner selection on the population and structure of networks. These have changed remarkably over the course of time, but the networks have continued to show a high degree of heterogeneity. Taking everything together, the development of innovative bio-technology depends on the composition and structure of the networks rather than on the fate of individual hubs or organizations (Powell et al. 2005).

It is not surprising that networks are especially widespread as modes of coordination in countries with a liberal market economy, and that they are more variegated in LME than in coordinated market economies (Owen-Smith et al. 2002). The networks, in which actors connect with partners from other institutional sectors, constitute a basis for the development and diffusion of radical innovations. In purely atomistic exchange relations, on the other hand, actors working on such innovations can hardly survive. But networks as such do not guarantee an innovation's success either. If they were superior in every respect other modes of coordination would, in functionalist terms, completely disappear (Podolny and Page 1998, p. 66). Not only markets and hierarchies, but also networks can fail.

2.4.2 Repercussions of Innovations on the Modes of Institutional Governance

The majority of sociological technology studies treat innovation as the dependent variable, while institutions are seen as constant or difficult to change. To establish itself the innovation must fit in or be compatible with the institutional environment. Hence, unsuccessful innovations are not necessarily in every respect inferior to successful innovations. They merely do not match their institutional environment as suitably as successful innovations might. However, technology is not always and exclusively a dependent variable. Large technological systems are especially expected to influence institutions and to strengthen their coordinative function (Mayntz 1993). Similarly, Krücken and Meier (2003) emphasise that network structures of institutional coordination and technical innovations are recursively connected to one another.

More recent technology studies explicitly make use of the concept of co-evolution when they analyse socio-technical transformation and transition processes (Rip and Kemp 1998; Geels 2004, 2005). The studies emphasise the crucial role of technology, particularly of radical technical innovations, in such processes, but reject technological determinism. According to the studies, radical innovations are hard to predict and difficult to shape. Often, they emerge in niches. Generally, the development of technical innovations follows its own logic, resulting in pressures on the surrounding institutional structures to change. These structures tend to be inert, but some windows of opportunity are occasionally opened through which changes can be achieved. The innovation process is an interactive, co-evolutionary multilevel process, involving technological artefacts, individual actors, organizations, sectoral institutions, and finally, socio-technical regimes (Geels 2007). Technical and social factors mutually influence each other. For the transport industry, the historical process of co-evolution in ocean shipping (from sailing ships to steamships), in air traffic (from propellers to jet aircrafts) and in road traffic (from horse-drawn carriages to automobiles) has been traced by (Geels 2005). The author shows in a phase model, which is more heuristic than explanatory, that usually one technology became dominant for a certain period of time.

2.4.3 *Results*

In conclusion, sociological institutionalism understands the evolution of technical innovations as a result of coordinated efforts whereby the mode of coordination can take the institutional forms of market, network and hierarchy. These forms are often interlinked. While some studies further differentiate them, the initial typology is not advanced systematically. In some cases a correspondence can be found between the mode of coordination and the type of technical innovation. The transition to stronger market-based coordination, for instance, was accompanied by decentralization and looser coupling of telecommunication networks, facilitating the evolution and integration of radical innovations. Yet, it is still an open question how further internal differentiations of the modes of coordination affect technology.

Sociological technology studies with an institutional background are more interested in the development of large technological infrastructures than individual technical artefacts. In these studies, technology is not analysed as sophisticatedly as in other fields of the sociology of technology. But a stronger differentiation would not be useful either if the institutional concepts are not more differentiated as well. More recent research strives to overcome institutional determinism, which exclusively regards technology as a dependent variable, by interpreting the technological and institutional development as an interrelated co-evolutionary process. In such a process, the dynamics of technical innovations can exert pressure toward institutional and social changes.

2.5 Perspectives: Co-evolution and Interaction of Technology and Institution

In the institutional approaches reviewed here and summarised in the Table 2.1, technical innovations are of varying importance. Socio-economic analyses are interested in general innovative capabilities, referring to institutional and other preconditions for the emergence and diffusion of new technologies. In particular, the studies on national innovation systems argue foremost against neoclassical approaches. Technical innovations are endogenised and it is shown that different national, but also sectoral institutional systems vary according to their quantitative and qualitative innovativeness. Some countries generate innovations more often than others, and the innovations can be incremental or radical. Usually,

Table 2.1 Institutional approaches to technical innovation

	Socio-economic	Politico-economic	Techno-sociological
Main focus	National systems of innovation	Varieties of capitalism	Coordination of large technological systems
Types of innovation	Radical/incremental Frequent/seldom	Radical/incremental Discrete/cumulative	Radical/conservative Incompatible/compatible
Theoretical profile	Endogenisation of technology and innovative capacity	Correspondence of institutional variety and type of innovation	Phase models of non-linear technical development
Understanding of institutional systems	Institutions are heterogeneous elements of constellations	Systems are constellations of complementary institutions	Systems incorporate technological elements
Repercussions of innovations	Convergence of institutions; (rare) changes of techno-economic paradigms	Innovations can influence adaptive institutions	Technological momentum exerts pressure towards institutional adaptation
Theoretical perspective	Alternation of periods of social construction and technological determinism	Co-evolution of institutional and technological development	Actor-mediated interaction of institutional and technical development

socio-economic institutionalism applies a broad and rather inconsistent concept of institutions, which has hampered the theoretical development of this approach.

Politico-economic institutionalism on the other hand strives for a theoretically sound concept of institutions and institutional systems. This approach focuses not only on economic performance, and in this context on technical innovations, but at times also on institutional change. The regulatory function of institutions moves into the centre of attention. Particularly useful has been the stylised distinction of liberal and coordinated market economies. As ideal-types, both exhibit a high degree of internal complementarity of their institutional elements, which accounts for their strong stability. Technical innovations are regarded as important for politico-economic analyses because they influence a country's economic performance. It has been shown that liberal market economies are not superior to coordinated market economies in all aspects. Instead, economies prove their specific capabilities with respect to different types of technical innovations. Liberal market economies are conducive to radical innovations; coordinated market economies promote incremental innovations.

Politico-economic institutionalism further develops this distinction between radical and incremental innovations, which is predominantly used in socio-economic studies. To answer the question of why radical innovations frequently occur in coordinated market economies (contrary to all expectations), the studies further distinguish between discrete and cumulative technologies. The latter can be enhanced step by step once a technological basis has been established. In this respect, they fit in coordinated market economies, even though they emerge, as far as their fundamental basis is concerned, as radical innovations. Hence, incrementally enhanced radical innovations may also prosper in coordinated economies.

According to this perspective, innovations are not continuously regarded as dependent variables. Technologies may even at first appear as exogenous variables, when it is argued, for instance, that German enterprises face great problems with radical innovations but utilise incremental innovations successfully. Enterprises adapt their strategies to the opportunities and constraints of technology. However, they will also try to shape a technology according to their own strategic orientations. The resulting innovations may exert pressure to change on national institutional systems. Innovations are hence intervening variables. Enterprises use technologies to the extent that they can integrate them with their strategies. The integration takes place via adaptation to technology, but also via the technology's modification and change. As a result, the technological opportunity structure will change, and that, in turn, generates pressure on institutional change.

Institutionally oriented sociological technology studies regard the development and diffusion of innovations mainly as a coordination problem, with (large) technological systems as their preferred subject of research. Similar to politico-economic studies, they predominantly understand institutional systems as rule systems. However, techno-sociological institutionalism usually confines itself to reducing the modes of institutional coordination to the basic types of hierarchy, market and network. Only occasionally do the studies consider mixed types. This leads to similar problems as those confronting socio-economic institutionalism: while socio-economic institutionalism mainly gains a profile demonstrating the weaknesses of neoclassical approaches, techno-sociological institutionalism is particularly attractive where it demonstrates the dubiousness of the notion of a linear development of technology. All in all, institutional theory has only advanced rudimentarily. One promising approach is the so-called actor-centred institutionalism (Werle 1998). It helps to explain technological changes within a relatively stable institutional framework by conceptualizing actor constellations and actor strategies as varying factors of influence on technology (see Schmidt and Werle 1998).

It is remarkable that techno-sociological institutionalism examines not only success or failure of technical innovations, but also the temporal and factual sequence of the innovation process and the solution of ensuing coordination problems. The process is expected to pass several phases which may recur cyclically. The resulting dynamic momentum confronts the embedding institutions with diverging challenges. A given institutional constellation which changes only slowly or not at all may benefit or hamper the process.

The socio-economic, politico-economic and also the techno-sociological analyses usually do not scrutinise the details of technical innovations, but instead use simple descriptive categories such as radical and incremental. This 'black-boxing' – if it is not too undifferentiated – facilitates exploring the general relation or correspondence between institutional constellations and technical innovations. This holds true at least as long as institutional constellations are also characterised by means of general typologies such as hierarchy, market, network, or coordinated/liberal market economies. The alternative method to providing a more detailed account of institutions and technology has not yet yielded convincing generalizable findings, although the tendency has increased to look in more detail at hybrid or "mixed governance" modes (Weyer 2006; see also Schneider and Bauer 2009).

All three approaches treat innovations primarily as dependent variables, but apparently their development and diffusion is institutionally underdetermined. Many other factors affect innovations, too. Moreover, technology-induced factors may exert pressure on the institutions to change. But the majority of studies adhere to the assumption that institutions are relatively resistant to these pressures.

Should studies in which institutions are used to explain technical innovations not also consider the inverted causal relation? Some studies mention this possibility. The socio-economic concept of a changing techno-economic paradigm, for instance, postulates that revolutionary technological changes abolish an existing institutional regime and establish a new one (Freeman and Perez 1988). From a comparative perspective, it is argued that in countries in which new technologies are less successful than in others, reform processes are targeted at institutions which are particularly relevant for technology policy, which then may enhance the fit of technology and institutions (Giesecke 2000). Similar arguments can be found in politico-economic studies about "institutional adaptiveness" (Casper 2000). Finally, the sociology of technology repeatedly points to the dynamics and momentum of technology, which requires conceding technology's strong impact on society and its institutional structure.

Here, the concept of co-evolution plays an important role. Insofar as relevant studies using this concept are empirically oriented, they use data and ideas of the history of technology. Here the development of technology is conceived of as a process in which periods of "social construction" i.e., organizational and institutional shaping of technology, alternate with periods of "technological determinism" i.e., technology-induced changes of the organizational and institutional environment (Rosenkopf and Tushman 1994).

Particularly in its techno-sociological variant (Rip and Kemp 1998; Geels 2005), the concept of co-evolution is definitely an interesting approach, since it overcomes institutional determinism and suggests a way to explain the fact that innovations occasionally also succeed in a virtually 'incompatible' institutional environment. However, the concept is very broad. It includes not only technological and institutional but also multiple other variables. To further elaborate on the relationship between technological and institutional innovations, case studies should search for interdependencies or interactions between them. The guiding idea is that *technology and institutions change in interactive processes* which are mediated and influenced by individual and collective actors (see Werle 2007). But technological and institutional change is always also affected by other, from this perspective, exogenous factors. A research design focusing on interdependencies and interactions will require systematically relating categories and mechanisms of institutional and technological change to one another and, as a by-product, prove to be generally fruitful for institutional theorizing. It definitely shows a way out of the impasse of institutional determinism, which is puzzled by cases of successful technical innovations that do not match the institutional environment.

References

Amable, B. (2000). Institutional complementarity and diversity of social systems of innovation and production. *Review of International Political Economy, 7*(4), 645–687.

Archibugi, D., & Pianta, M. (1992). *The technological specialization of advanced countries.* Dordrecht: Kluwer.

Baumol, W. J. (2002). *The free-market innovation machine*. Princeton: Princeton University Press.

Bessen, J., & Meurer, M. J. (2008). *Patent failure: How judges, bureaucrats, and lawyers put innovators at risk.* Princeton: Princeton University Press.

Block, F., & Keller, M. R. (2009). Where do innovations come from? Transformations in the U.S. economy, 1970–2006. *Socio-economic Review, 7*(3), 459–483.

Breschi, S., & Malerba, F. (1997). Sectoral innovation systems: Technological regimes, Schumpeterian dynamics, and spatial boundaries. In C. Edquist (Ed.), *Systems of innovation: Technologies, institutions and organizations* (pp. 130–156). London: Pinter.

Carlsson, B. (1994). Technological systems and economic performance. In M. Dodgson & R. Rothwell (Eds.), *The handbook of industrial innovation* (pp. 13–24). Aldershot: Edward Elgar.

Carlsson, B., Jacobsson, S., Holmén, M., & Rickne, A. (2002). Innovation systems: Analytical and methodological issues. *Research Policy, 31*(2), 233–245.

Casper, S. (2000). Institutional adaptiveness, technology policy and the diffusion of new business models: The case of German biotechnology. *Organization Studies, 21*(5), 887–914.

Casper, S., & Glimstedt, H. (2001). Economic organization, innovation systems and the Internet. *Oxford Review of Economic Policy, 17*(2), 265–281.

Casper, S., & Matraves, C. (2003). Institutional frameworks and innovation in the German and UK pharmaceutical industry. *Research Policy, 32*(10), 1865–1897.

Casper, S., Lehrer, M., & Soskice, D. (1999). Can high-technology industries prosper in Germany? Institutional frameworks and the evolution of the German software and biotechnology industries. *Industry and Innovation, 6*(1), 5–24.

Castilla, E. J., Hwang, H., Granovetter, E., & Granovetter, M. (2000). Social networks in Silicon Valley. In C.-M. Lee, W. F. Miller, M. G. Hancock, & H. S. Rowen (Eds.), *The Silicon Valley edge: A habitat for innovation and entrepreneurship* (pp. 218–247). Stanford: Stanford University Press.

Chandler, A. D. (1977). *The visible hand: The managerial revolution in American business.* Cambridge, MA: Belknap.

Coriat, B., & Weinstein, O. (2002). Organizations, firms and institutions in the generation of innovation. *Research Policy, 31*(2), 273–290.

Crouch, C. (2003). Institutions within which real actors innovate. In R. Mayntz & W. Streeck (Eds.), *Die Reformierbarkeit der Demokratie: Innovationen und Blockaden* (pp. 71–98). Frankfurt am Main: Campus.

David, P. A. (2001). The evolving accidental information super-highway. *Oxford Review of Economic Policy, 17*(2), 159–187.

Dolata, U. (2003). *Unternehmen Technik. Akteure, Interaktionsmuster und strukturelle Kontexte der Technikentwicklung: Ein Theorierahmen.* Berlin: Edition Sigma.

Dolata, U. (2009). Technological innovations and sectoral change. Transformative capacity, adaptability, patterns of change: An analytical framework. *Research Policy, 38*(6), 1066–1077.

Dosi, G. (1982). Technological paradigms and technological trajectories: A suggested interpretation of the determinants and directions of technical change. *Research Policy, 11*(3), 147–162.

Dosi, G. (1988). Sources, procedures, and macroeconomic effects of innovation. *Journal of Economic Literature, 26*(3), 1120–1171.

Edquist, C. (1997). Systems of innovation approaches: Their emergence and characteristics. In C. Edquist (Ed.), *Systems of innovation: Technologies, institutions and organizations* (pp. 1–35). London: Pinter.

Edquist, C., & Lundvall, B.-Å. (1993). Comparing the Danish and the Swedish systems of innovations. In R. R. Nelson (Ed.), *National innovation systems: A comparative analysis* (pp. 265–298). Oxford: Oxford University Press.

Faber, J., & Hesen, B. (2004). Innovation capabilities of European nations: Cross-national analyses of patents and sales of product innovations. *Research Policy, 33*(2), 193–207.

Feick, J., & Werle, R. (2010). Regulation of cyberspace. In M. Cave, R. Baldwin, & M. Lodge (Eds.), *The Oxford handbook of regulation* (pp. 523–547). Oxford: Oxford University Press.

Freeman, C. (1987). *Technology policy and economic performance: Lessons from Japan.* London: Pinter.

Freeman, C., & Louçã, F. (2002). *As time goes by: From the industrial revolutions to the information revolution.* Oxford: Oxford University Press.

Freeman, C., & Perez, C. (1988). Structural crises of adjustment, business cycles and investment behaviour. In G. Dosi, C. Freeman, R. R. Nelson, G. Silverberg, & L. Soete (Eds.), *Technical change and economic theory* (pp. 38–66). London: Pinter.

Gallini, N. T. (2002). The economics of patents: Lessons from recent US patent reform. *Journal of Economic Perspectives, 16*(2), 131–154.

Geels, F. W. (2004). From sectoral systems of innovation to socio-technical systems. Insights about dynamics and change from sociology and institutional theory. *Research Policy, 33*(6–7), 897–920.

Geels, F. W. (2005). *Technological transitions and system innovations: A co-evolutionary and socio-technical analysis.* Cheltenham: Edward Elgar.

Geels, F. W. (2007). Transformations of large technical systems: A multilevel analysis of the Dutch highway system (1950–2000). *Science, Technology & Human Values, 32*(2), 123–149.

Gibbons, M., Limoges, C., Nowotny, H., Schwartzman, S., Scott, P., & Trow, M. (1994). *The new production of knowledge: The dynamics of science and research in contemporary societies.* London: Sage.

Giesecke, S. (2000). The contrasting roles of government in the development of biotechnology industry in the US and Germany. *Research Policy, 29*(2), 205–223.

Graham, S. J. H., & Mowery, D. C. (2003). Intellectual property in the US software industry. In W. M. Cohen & S. A. Merrill (Eds.), *Patents in the knowledge based economy* (pp. 219–258). Washington, DC: The National Academies Press.

Guerrieri, P. (1999). Patterns of national specialisation in the global competitive environment. In D. Archibugi, J. Howells, & J. Mitchie (Eds.), *Innovation policy in a global economy* (pp. 139–159). Cambridge: Cambridge University Press.

Hall, P. A., & Soskice, D. (2001). An introduction to varieties of capitalism. In P. A. Hall & D. Soskice (Eds.), *Varieties of capitalism: The institutional foundations of comparative advantage* (pp. 1–70). Oxford: Oxford University Press.

Hall, B. H., & Ziedonis, R. H. (2001). The patent paradox revisited: An empirical study of patenting in the US semiconductor industry, 1979–1995. *RAND Journal of Economics, 32*(1), 101–128.

Heller, M. A., & Eisenberg, R. S. (1998). Can patents deter innovation? The anticommons in biomedical research. *Science, 280*(5364), 698–701.

Henisz, W. J., Zelner, B. A., & Guillén, M. F. (2005). The worldwide diffusion of market-oriented infrastructure reform, 1977–1999. *American Sociological Review, 70*(6), 871–897.

Hirsch-Kreinsen, H. (2002). Unternehmensnetzwerke – revisited. *Zeitschrift für Soziologie, 31*(2), 106–124.

Hollingsworth, J. R. (2000). Doing institutional analysis: Implications for the study of innovations. *Review of International Political Economy, 7*(4), 595–644.

Hollingsworth, J. R., & Boyer, R. (1997). Coordination of economic actors and social systems of production. In J. R. Hollingsworth & R. Boyer (Eds.), *Contemporary capitalism: The embeddedness of institutions* (pp. 1–47). Cambridge: Cambridge University Press.

Höpner, M. (2005). Epilogue to 'explaining institutional complementarity': what have we learnt? Complementarity, coherence and institutional change. *Socio-Economic Review, 3*(2), 383–387.

Hughes, T. P. (1982). Conservative and radical technologies. In S. B. Lundstedt & E. W. Colglazier (Eds.), *Managing innovation* (pp. 31–44). New York: Pergamon.

Hughes, T. P. (1983). *Networks of power: Electrification in western society 1880–1930*. Baltimore: The Johns Hopkins University Press.

Jaffe, A. B. (2000). The US patent system in transition: Policy innovation and the innovation process. *Research Policy, 29*(4–5), 531–557.

Jansen, D. (2002). Netzwerkansätze in der Organisationsforschung. In J. Allmendinger & T. Hinz (Eds.), *Organisationssoziologie: Sonderheft 42 der Kölner Zeitschrift für Soziologie und Sozialpsychologie* (pp. 88–118). Wiesbaden: Westdeutscher Verlag.

Kitschelt, H. (1991). Industrial governance structures, innovation strategies and the case of Japan: Sectoral or cross-national comparative analysis? *International Organization, 45*(4), 453–493.

Kitschelt, H., & Streeck, W. (2003). From stability to stagnation: Germany at the beginning of the twenty-first century. *West European Politics, 26*(4), 1–32.

Kowol, U., & Krohn, W. (1995). Innovationsnetzwerke: Ein Modell der Technikgenese. In J. Halfmann, G. Bechmann, & W. Rammert (Eds.), *Technik und Gesellschaft. Jahrbuch 8: Theoriebausteine der Techniksoziologie* (pp. 77–105). Frankfurt am Main: Campus.

Krücken, G., & Meier, F. (2003). Wir sind alle überzeugte Netzwerktäter: Netzwerke als Formalstruktur und Mythos der Innovationsgesellschaft. *Soziale Welt, 54*, 71–92.

Küppers, G. (2002). Complexity, self-organisation and innovation networks: A new theoretical approach. In A. Pyka & G. Küppers (Eds.), *Innovation networks: Theory and practice* (pp. 22–52). Cheltenham: Edward Elgar.

Lange, K. (2009). Institutional embeddedness and the strategic leeway of actors: The case of the German therapeutical biotech industry. *Socio-economic Review, 7*(2), 181–207.

Larédo, P., & Mustar, P. (2001). *Research and innovation policies in the new global economy: An international comparative analysis*. Cheltenham: Edward Elgar.

Leydesdorff, L., & Etzkowitz, H. (1998). The triple helix as a model for innovation studies. *Science and Public Policy, 25*(3), 195–203.

Lundvall, B.-Å. (1992). User-producer relationships, national systems of innovation and internationalisation. In B.-Å. Lundvall (Ed.), *National systems of innovation: Towards a theory of innovation and interactive learning* (pp. 45–67). London: Pinter.

Lundvall, B.-Å., Johnson, B., Andersen, E. S., & Dalum, B. (2002). National systems of production, innovation and competence building. *Research Policy, 31*(2), 213–231.

Mayntz, R. (1993). Große technische Systeme und ihre gesellschaftstheoretische Bedeutung. *Kölner Zeitschrift für Soziologie und Sozialpsychologie, 45*(1), 97–108.

Mayntz, R. (2009). The changing governance of large technical infrastructure systems. In R. Mayntz (Ed.), *Über Governance: Institutionen und Prozesse Politischer Regelung* (pp. 121–150). Frankfurt am Main: Campus.

Mayntz, R., & Hughes, T. P. (1988). *The development of large technological systems*. Campus: Frankfurt a.M.

Mowery, D. C. (1994). *Science and technology policy in interdependent economies*. Boston: Kluwer.

Mowery, D. C., & Rosenberg, N. (1993). The U.S. national innovation system. In R. R. Nelson (Ed.), *National innovation systems: A comparative analysis* (pp. 29–75). Oxford: Oxford University Press.

Mowery, D. C., & Simcoe, T. (2002). The internet. In B. Steil, D. G. Victor, & R. R. Nelson (Eds.), *Technological innovation and economic performance* (pp. 229–264). Princeton: Princeton University Press.

Nelson, R. R. (1987). *Understanding technical change as an evolutionary process*. Amsterdam: Elsevier.
Nelson, R. R. (1988). Institutions supporting technical change in the United States. In G. Dosi, C. Freemann, R. R. Nelson, G. Silverberg, & L. Soete (Eds.), *Technical change and economic theory* (pp. 312–329). London: Pinter.
Nelson, R. R. (1994). The coevolution of technologies and institutions. In R. W. England (Ed.), *Evolutionary concepts in contemporary economics* (pp. 139–156). Ann Arbor: University of Michigan Press.
Nelson, R. R., & Rosenberg, N. (1993). Technical innovation and national systems. In R. R. Nelson (Ed.), *National innovation systems. A comparative analysis* (pp. 3–21). Oxford: Oxford University Press.
Nelson, R. R., & Winter, S. G. (1982). *An evolutionary theory of economic change*. Cambridge: Belknap.
Owen-Smith, J., Riccaboni, M., Pammolli, F., & Powell, W. W. (2002). A comparison of US and European university-industry relations in the life sciences. *Management Science, 48*(1), 24–43.
Pavitt, K., & Patel, P. (1999). Global corporations and national systems of innovations: Who dominates whom? In D. Archibugi, J. Howells, & J. Mitchie (Eds.), *Innovation policy in a global economy* (pp. 94–119). Cambridge: Cambridge University Press.
Perrow, C. (1984). *Normal accidents, living with high-risk technologies*. New York: Basic Books.
Podolny, J. M., & Page, K. L. (1998). Network forms of organization. *Annual Review of Sociology, 24*(1), 57–76.
Porter, M. E. (1990). *The competitive advantage of nations*. New York: Free Press.
Powell, W. W. (1990). Neither market nor hierarchy: Network forms of organization. *Research in Organizational Behavior, 12*(1), 295–336.
Powell, W. W., White, D. R., Koput, K. W., & Owen-Smith, J. (2005). Network dynamics and field evolution: The growth of interorganizational collaboration in the life sciences. *American Journal of Sociology, 110*(4), 1132–1205.
Rammert, W. (1997). New rules of sociological method: Rethinking technology studies. *The British Journal of Sociology, 48*(2), 171–191.
Riccaboni, M., Powell, W. W., Pammoli, F., & Owen-Smith, J. (2003). Public research and industrial innovation: A comparison of US and European innovation systems in the life sciences. In A. Geuna, A. J. Salter, & W. E. Steinmueller (Eds.), *Science and innovation: Rethinking the rationales for funding and governance* (pp. 169–201). Cheltenham: Edward Elgar.
Rip, A., & Kemp, R. (1998). Technological change. In S. Rayner & E. L. Malone (Eds.), *Human choice and climate change two: Resources and technology* (pp. 327–399). Columbus: Batelle Press.
Rosenkopf, L., & Tushman, M. L. (1994). The coevolution of technology and organization. In J. A. C. Baum & J. V. Singh (Eds.), *Evolutionary dynamics of organizations* (pp. 403–424). Oxford: Oxford University Press.
Schmidt, S., & Werle, R. (1998). *Coordinating technology: Studies in the international standardization of telecommunications*. Cambridge, MA: The MIT Press.
Schneider, V. (2001). *Die Transformation der Telekommunikation: Vom Staatsmonopol zum globalen Markt (1800–2000)*. Frankfurt am Main: Campus.
Schneider, V., & Bauer, J. (2009). Von der Governance- zur Komplexitätstheorie: Entwicklungen der Theorie gesellschaftlicher Ordnung. In J. Weyer & I. Schulz-Schaeffer (Eds.), *Management komplexer Systeme* (pp. 31–53). München: Oldenbourg.
Schulz-Schaeffer, I. (2000). *Sozialtheorie der Technik*. Frankfurt am Main: Campus.
Soskice, D. (1994). Innovation strategies of companies: A comparative institutional approach of some cross-country differences. In W. Zapf & M. Dierkes (Eds.), *Institutionenvergleich und Institutionendynamik: WZB-Jahrbuch 1994* (pp. 271–289). Berlin: Edition Sigma.
Soskice, D. (1999). Divergent production regimes: Coordinated and uncoordinated market economies in the 1980s and 1990s. In H. Kitschelt, P. Lange, G. Marks, & J. D. Stevens

(Eds.), *Continuity and change in contemporary capitalism* (pp. 101–134). Cambridge: Cambridge University Press.

Steil, B., Victor, D. G., & Nelson, R. R. (2002). Introduction and overview. In B. Steil, D. G. Victor, & R. R. Nelson (Eds.), *Technological innovation and economic performance* (pp. 3–22). Princeton: Princeton University Press.

Streeck, W. (2011). *Skills and politics: General and specific. MPIfG Discussion Paper*. Köln: Max-Planck-Institut für Gesellschaftsforschung.

Werle, R. (1998). An institutional approach to technology. *Science Studies, 11*(1), 3–18.

Werle, R. (2000). Innovationspotentiale im Internet: Selbstregelung auf Strukturebene. In W. Hoffmann-Riem (Ed.), *Innovation und Telekommunikation: Rechtliche Steuerung von Innovationsprozessen in der Telekommunikation* (pp. 141–160). Baden-Baden: Nomos.

Werle, R. (2002). Internet @ Europe: Overcoming institutional fragmentation and policy failure. In J. Jordana (Ed.), *Governing telecommunications and the new information society in Europe* (pp. 137–158). Cheltenham: Edward Elgar.

Werle, R. (2007). Zur Interdependenz von Innovationen. In H. Hof & U. Wengenroth (Eds.), *Innovationsforschung: Ansätze, Methoden, Grenzen und Perspektiven* (pp. 31–40). Münster/Hamburg: LIT Verlag.

Weyer, J. (2006). Modes of governance of hybrid systems: The mid-air collision at Ueberlingen and the impact of smart technology. *Science, Technology & Innovation Studies, 2*(2), 127–149.

Weyer, J., Kirchner, U., Riedl, L., & Schmidt, J. F. K. (1997). *Technik, die Gesellschaft schafft: Soziale Netzwerke als Ort der Technikgenese*. Berlin: Edition Sigma.

Williamson, O. E. (1985). *The economic institutions of capitalism: Firms, markets, relational contracting*. New York: Free Press.

Chapter 3
Ecologies of Games Shaping Large Technical Systems: Cases from Telecommunications to the Internet

William H. Dutton, Volker Schneider, and Thierry Vedel

3.1 Introduction: Social Shaping of Large Technical Systems

Large and geographically extended technical systems (LTS) are important components of modern societies and important platforms of technological innovations (Hughes 1987; Mayntz and Hughes 1988). Advances in transportation systems (automobile, railways), communication infrastructures (telecommunications, Internet), energy provision, and water management systems were important drivers for economic growth, and have become critical to all economies and societies (Joerges 1999).

Given this significance, LTS have become a focus of multi-disciplinary research. Many have studied the constitutive and functional aspects of these systems, such as their regulation and governance (Coutard 1999; La Porte 1991). Other questions concern the dynamics and performance of these social-technical configurations. These issues are focused on explaining the dynamics involved with their build-up and expansion as well as their ability to innovate and adapt (Summerton 1994). Major concepts emerging from this literature, such as a technology's "momentum", and the development of "reverse salients", such as factors that disrupt the progress of a LTS, point to the importance of such adaptive perspectives (Hughes 1983, 1987).

Authors are listed alphabetically to denote an equal contribution to the chapter.

W.H. Dutton (✉)
Oxford Internet Institute, University of Oxford, Oxford, UK
e-mail: william.dutton@oii.ox.ac.uk

V. Schneider
University of Konstanz, Konstanz, Germany

T. Vedel
Center for Political Research, Sciences-Po, Paris, France

J.M. Bauer et al. (eds.), *Innovation Policy and Governance in High-Tech Industries*,
DOI 10.1007/978-3-642-12563-8_3, © Springer-Verlag Berlin Heidelberg 2012

Large technical systems are socio-technical networks, as they combine social units and technical artefacts. Also, they are socially constructed. That is, technical systems are not just translations of physical laws into technical projects, but are inherently shaped by social processes in which multiple and heterogeneous actors interact and work together. Like all technologies, LTS are shaped by social interests and their related dynamics, but not all elements in the "construction set" are malleable, and open to be manipulated by human actions, or the actions of any single actor. While technical decisions are the outcomes of choices made by social actors, the opportunity space for choice is more or less constrained by the rules of the physical world. Nevertheless, social relations do not merely belong to the environment of a technology, but are integral components that contribute to its overall performance, whether simple tools, such as a hammer, or LTS. LTS may appear, therefore, as an "alliance of brains and electrons" (Latour 1987), but such alliances are primarily built on actor's decisions and socio-political relations.

3.1.1 Ecological Models of Socio-technical Dynamics

A problem in socio-technical analysis is to understand the complexity of interactions between socio-political and technical choices on the one hand, and the interaction between social goals and physical rules at the other. The ecological perspective provides a promising approach to model and reconstruct such complex and multi-relational settings. Ecological approaches are inspired by various branches of biology (Freeman and Audia 2006). In essence, they apply an ecosystem perspective to social processes (Pickett and Cadenasso 2002).

These models are attractive to the social sciences because they place an emphasis on: (1) the dynamic character of the interdependencies and interactions of multiple and diverse social actors; (2) the dynamics and multiplicity of relations between the actors and other components of socio-technical systems; and (3) the existence of multiple layers and levels, and emergent relations between these levels (Schneider and Bauer 2007).

3.1.2 The Ecology of Games: Combining Institutionalism with Relationalism

In the social sciences, ecological approaches have most often been applied to the sociology of organizations as variants of new institutionalist perspectives and later in studies of business and management. However, this perspective has also been applied in the political sciences, such as in studies of the local communities, policy sectors, and interest groups.

One particular variant in this perspective is the "ecology of games" (EOG), which is the focus of this paper. The idea of an EOG emphasises the complexity of social and political conflict within nested (public and private) decision-making processes that relate to complex social and technical relations and interdependencies. The EOG concept was initially developed far afield from innovation in information and communication technologies (ICTs). In the 1950s, Norton Long (1958) used the idea to provide a new perspective on pluralist versus elitist debates over who governs local communities.

However, this perspective was largely abandoned, until the 1990s, when Dutton revisited Long's notion of an ecology of games to provide a heuristically rich and useful framework for understanding the dynamics of decision-making processes in technology and innovation policy in a wide range of areas (Dutton and Mackinen 1987; Vedel and Dutton 1990; Dutton and Guthrie 1991; Dutton 1992; Dutton 1995). This reinvention of the ecology of games was joined later by scholars from other fields (such as, Cornwell et al. 2003; Firestone 1989; Shields 1995). Yet it has remained a largely under-exploited perspective on socio-technical systems in general, and LTS in particular, despite the prominence of proximate models, such as actor-network theory (Latour 1987).

The goal of this paper is to extend the EOG perspective to the study of systemic innovation processes in the communications and information sector, and to demonstrate the value of this approach within the theoretical debates on the evolution of large technical systems. The paper begins by outlining the theoretical background and the major traits of the two conceptual components in this perspective – ecological thinking and the idea of games. This is followed by a set of case studies on development and change in LTS in the area of communications – telecommunications, cable and videotext, and finally the evolution of the Internet. The paper concludes with an assessment of the strengths and limitations of the EOG as a perspective on the dynamics of technical systems.

3.2 The Foundations of an Ecology of Games Perspective

3.2.1 Origins of the Concept

The concept of an "ecology of games" was introduced by Norton Long in the study of power structures in local communities within the political and administrative sciences during the 1950s (Long 1958). At that time there was a dispute between pluralist and elitists about explaining local politics either by the interplay of a plurality of specialised, issue-based elites and actors, or by rulings of a more unified 'power elite'. Long criticised the empirical foundations of both views and used the EOG metaphor to emphasise a perspective in which both views were combined in a more structured form of pluralism.

Long recognised the participation of multiple actors in various functional circuits in the same way as he stressed the importance of interactions based on more or less important interest cleavages: His games are based on differentiated, structured, goal-oriented interactions. Community development in his perspective is a largely unplanned process with unanticipated interactions of multiple players within separate but overlapping games. In Long's view, individuals do not try to govern a local community. Instead, local communities appear as the outcome of microscopic ecosystems, in which individuals make decisions as role players in more specific, goal-oriented games, such as being council members, real estate agents, or planning commissioners fighting and negotiating over issues within a winning re-election, selling real estate, or zoning property of a new use, respectively. The unfolding history of such separate but interdependent games is then driving the evolution of local communities, like ecological relations in a "five-acre woodlot in which the owls and the field mice, the oaks and the acorns, and other flora and fauna have evolved a balanced system" (Long 1958, p. 254).

Long (1958) provides an example that also fits well in to the discussion of LTS in discussing the construction of a highway network. The overall construction process is a largely unplanned result of multiple interconnected games, including a department of public works game, a professional highway engineer game, a banking game, and a media game. The department of public works game is about a collection of contending politicians seeking to use the highways for political capital and patronage. Contractors are eager to make money by building roads. The banking game is concerned with bonds, taxes, and the effect of the highways on real estate. In the media game news reporters and publishers are interested in headlines and the effect of highways on the paper's circulation numbers. Labour leaders are interested in union contracts and their status among the community elite. Each game is played simultaneously in complex decision-making and resource mobilization processes, in which no single over-all directive authority has complete control (Long 1958).

This ensemble, indeed, looks like an ecosystem, which Alfred Kuhn once defined as follows: The core idea of ecological reasoning is that the structure of sub-systems "merely happens as an outgrowth of the interactions of [their] parts [. . .]. The essence of ecological systems [. . .] is that they consist of the uncontrolled interactions of controlled systems" (Kuhn 1974, p. 369–370).

3.3 Two Theoretical Streams Informing the Ecology of Games

Long's bringing together of the two concepts of "ecologies" and "games" followed novel approaches within the social sciences that emerged in the first half of the twentieth century, and later became important paradigms in various disciplines. The ecological approach, inspired by advances in bio-ecology (Hawley 1944), was adapted by various scholars in sociology, and became especially successful in the sociology of organizations (Astley 1985; Freeman and Audia 2006). Only very recently this perspective also has been adapted to the analysis of innovation process in business contexts (Adner 2006; Basole 2009). The second concept, the field of formal theory, game theory came into being with the seminal book by John von Neumann and Oskar Morgenstern (1944) and experienced its first growth wave during the 1950s, when the Cold War challenged traditional analyses of political conflict.

Given this reliance on two streams of theorizing, it is useful to briefly focus on each in the following sections. Before doing so, it is important to acknowledge the use of the EOG concept within a qualitative research tradition. In qualitative research, many theoretical concepts have the status of "sensitizing concepts" (Blumer 1954), in that they sensitise the reader to a more complex set of concrete empirical observations. They are different from operational concepts in the sense that they are qualitative "background concepts" that capture patterns of relationships and their interpretation remains flexible. Nor are they equivalent to theoretical concepts used in formal theories that have precise mathematical definitions, such as the concept of strategy in game theory. In this sense, we are employing the two concepts of game and ecology as "sensitizing concepts" within a qualitative perspective of social research.

3.3.1 Formal Theories of Games

The core idea in the concept of a game is that many types of social interaction, from fatal conflict to symbiotic cooperation, can be analysed by the decomposition of situations into actors (players), their action options (strategies), their goals (objectives), and the outcomes (payoffs) that each player's choices may produce (Myerson 1991). Some scholars view games to be shaped by their social environment. For instance, in an institutional perspective, games are constrained by the rules of institutions that specify players and strategies, and relate strategies to payoffs (Ostrom 1986; Ostrom and Gardner 1991). In our conception of an EOG, rules are inherent aspects of the games. In games in the real world, as opposed, for example, to parlour games, the players can change the rules, or ignore the rules. Likewise, the outcome of one game can change the rules of another game.

A game is a structured and goal-oriented interaction in which at least two actors participate – cooperate or compete. The players choose among action options by anticipating the decisions of the other players. Different combinations of strategies then result in various possible outcomes, each associated with an individual payoff that actors seek to improve or maximise. Players do not necessarily have to be conscious game players. A basic requirement is that the action unit – a human, an animal, a robot or set of players – has some choice. For example, in the LTS domain, the decision of big investors in a large infrastructure or a power station can be modelled as a "game against nature" (Milnor 1954), when outcomes are uncertain.

Empirical researchers are often sceptical of game theoretic notions, given their over-simplified and unrealistic assumptions (Bunge 1998). However, they have an advantage over approaches that are purely individualistic and actor-centred which often fail to consider the relational dimension of structured interaction in a multi-player setting. Anchored in the interactions of multiple players, game theory provides useful concepts to analyse interdependencies in decision-making that are related to various conflict constellations.[1]

A famous game in this respect is the Prisoner's Dilemma (PD). This is a game structure in which players can exploit other players in deciding for non-cooperation (defection), provided that the other actor(s) decide(s) for cooperation. The game is important because the PD structure of conflict appears in many social situations, from organizing collective action to the provision of public goods in policy-making. A classic example is the build-up of a technical infrastructure with features of a public good, which can create a "free-rider" problem.[2]

In such situations, game theoretical analyses provide tools for modelling the structures of conflicts among multiple actors. Conflicts might range from the low end of the spectrum, involving cooperation, to the high end, entailing life and death. A typical low conflict game is a pure coordination game, where all players have an interest in deciding on a common standard. Examples in the area of large technical systems range from setting standards for the allocation of radio frequencies. In such circumstances, all players could benefit if all chose a common standard, if the failure to agree resulted in incompatible solutions.[3]

---

[1] Rapoport and Guyer (1978) show that a simple game with two players, with two action options, and ordinal payoffs provides for 78 game configurations. These can be classified according to such criteria as the degree of conflict involved, or possible equilibriums. However, only a few games are theoretically interesting in the sense that they involve conflict dilemmas. Games in which actors are able to maximise their individual utility, at the same time as the collective utility is improved, are more trivial and less challenging than games in which 'motives are mixed' and strategic choices are more difficult to predict. An interesting result of this taxonomy is that about one quarter of the games are no-conflict games, and only three of the 78 games they identify imply pure conflict configurations, where interests are diametrically opposed.

[2] An illustrative example is traditional broadcast radio or television, where all persons in a given area can receive radio and TV emissions, even if they avoid sharing the costs of its provision. Such systems exhibit the basic characteristics of public goods such as non-excludability and non-rivalry in consumption. Since exclusion is technically impossible, everybody can have a "free ride", and an increasing number of receivers will not impair the quality of emission. The dilemma in this conflict structure is that if everybody free rides, system providers will find it difficult to mobilise the financial resources required for the development and maintenance of the infrastructures.

[3] A closely related low conflict game is an "assurance game", which has been applied to the collective build-up of large technical projects (Schneider 1993). The players have two strategies: to invest into the system, or abstain from contributing resources. Since the resources are specialized, the system setup only succeeds if all partners co-invest. If one player defects the whole project fails and the other system partners lose their investment. As cooperation is voluntary, strategic uncertainty is created for each of the players. However, all know that they will reap substantial profits and gain new business opportunities if the project succeeds. This game is a "no-conflict game" because the players' interests converge. Although neither player has a dominant strategy, there is an agreed best outcome when all players invest. However, if players seek to minimise maximal losses (maximin strategy), which game theory proposes as a rational strategy in such uncertain constellations, the players would end up in a Pareto-suboptimal equilibrium. But this could be avoided if each player were assured by a collective contract that the others also would invest (Sen 1969).

There are game structures in which player's interests are diametrically opposed, such as in classic zero-sum games, in which one player does better at another's expense. Even more problematic is the game of chicken, in which the pursuit of individual utility maximization leads to fatal consequences of all players.[4]

Although game theory provides powerful analytical perspectives on such general patterns of conflict, as Myerson (1991) and Schelling (1980) emphasise, its relevance to predicting the outcomes of real world situations is far more limited. Most game theoretic assumptions hold only for extremely simple and stylised situations. As Axel Leijonhufvud (1996, p. 40) put it, such models could deal with "incredibly smart people in unbelievably simple situations", whereas we rather should ask "how believably simple people cope with incredibly complex situations . . .".

Therefore, a serious restriction in game theory is that mathematical analysis of dominant strategies and equilibriums can only be applied to highly structured and (at least analytically) isolated situations, in which only a few homogeneous players interact with a limited number of strategies. Important facets of reality, such as the interplay of multiple, heterogeneous actors (with different perceptions, action orientations, and resources) that are simultaneously involved in multiple and interdependent games, becomes not only mathematically very difficult to deal with, but also empirically challenging to represent. However, as recent developments in social simulation and agent-based modelling show, a formal treatment of such complex dynamics is not impossible in principle (Cederman 2005).

An early attempt in social science to model ecological interaction by computer simulation were Axelrod's computer tournaments (Axelrod 1984) in which 63 players played the Prisoner's Dilemma repeatedly over 500 rounds, each using a different decision rule against the other. The overall approach integrated evolutionary and ecological ideas. Successful decision rules (based on simple computer programmes) proliferated, whereas less successful entries died (i.e. differential reproduction). Axelrod thus simulated the survival of the fittest decision rule and discovered the superiority of so-called "tit for tat". However, even in this context, it was a very simplified multiplayer game. An identical game was played iteratively on the basis of different decision rules. This is much less complex than the situations that the EOG seeks to explain, with multiple players involved in multiple games, using multiple decision rules.

3.3.2 *Qualitative Perspectives on Games*

While the notion of games is most often linked to formal game theory as discussed above and has often been the object of mathematical formalizations or quantitative models, it has also been used within more qualitative approaches. For instance,

[4] In biology a version of the chicken game is known as a "hawk-dove" game analyzing resource aggressive and peaceful resource sharing behaviour (Samuelson 2002).

Fritz Scharpf developed a multiple games and rules perspective with respect to economic policy making. His approach uses game concepts in a largely qualitative way to describe the essential constellations of interactions shaping economic policy in the 1970s as overlapping games in which governments play against trade unions in monetary policy and wage policy, and where governments simultaneously also have to take voters' reactions into account (Scharpf 1997).

Similar to Axelrod's varying decision rules, Scharpf applied different action orientations, such as monetarist versus Keynesian definitions of the situation, to show that when Governments tried to collectively maximise the payoff in the economic policy game, they were punished in the voting game.

Another qualitative application of the notion of games has been proposed by Crozier and Friedberg's (1977) theory of organised action.[5] For the authors, an organised action refers to the social processes leading to the structuration and stabilization of "the competitive cooperation between a set of actors who are mutually dependent for the solution of a common problem, which they cannot solve by themselves and for the solution of which they have to secure the cooperation of partners who are also potential rivals" (Friedberg 1997).

Crozier and Friedberg's contribution has been especially important in two respects. They defined games not just as goal-oriented, strategic actions but also as mechanisms of integration between actors and the system in which they operate. In other words, games embody the social interactions of actors and give these interactions a more stabilised and concrete form. This crystallization function of games is central to the EOG approach that we are developing.

Secondly, they underscored the degree to which actors attempt to control games, such as by taking advantage of the uncertainty surrounding the rules of the games. Similarly, actors attempt to move games from spaces which they do not master to other spaces in which they can impose their own rules, such as through their specific expertise or practical knowledge, control of information resources or communication channels, or organizational positions. Strategies for reconfiguring games are central to the EOG, where players have to decide between action options in nested and overlapping games.

3.3.3 The EOG and Ecological Theories in the Biological and Natural Sciences

Complex multi-actor models had been strongly influenced by bio-ecology. Ecological perspectives were first promoted by the Chicago School of Human Ecology in the 1920s, and became particularly influential during the 1950s. In an excellent

[5] While this theory was initially framed to analyse change (or absence of change) within organizations, it is perfectly applicable to the genesis and dissemination of scientific and technical discoveries and innovations, as Friedberg recognised.

review of the field Amos Hawley (1944) summarised its core idea as follows: Ecology would deal with the basic problem of how growing and multiplying beings maintain themselves in a limited but constantly changing environment. It stresses the adaptation of organisms to their environment, the interrelatedness of diverse living forms in a complex "web of life". The ecological perspective includes systemic concepts like community, society, and niche, but also a series of relational concepts such as competition, cooperation, mutualism, symbiosis, and dominance. He conceived human ecology as a special application of this ecological perspective to human behaviour, with acknowledgements to the exceptional degree of flexibility of human action.

The flourishing of this idea during the 1950s undoubtedly influenced Norton Long's (1958) idea of an ecology of games. However, whereas Long's allusion to a community's ecology was largely metaphorical, social ecological approaches have emerged since the 1950s to become a broad movement behind one of the dominant paradigms in the sociology of organizations. This is not to say that organizational ecology is a unified field. Graham Astley (1985) identified at least two perspectives in social and organizational ecology, which imply different levels of analysis, and also produce contrasting views on the mode and tempo of organizational adaptation. One is population ecology; the other is a more diversified stream of social and community ecology, which currently is better known as organizational ecology.

Hannan and Freeman (1977) applied theories of population ecology to organizations. They focused on the evolution of organizational forms, such as by explaining the birth and death of particular types of organizations through such Darwinian evolutionary mechanisms as variation, selection and retention. A major point is that environments differentially "select" organizations for survival on the basis of fit between the organizational form and characteristics of their environment. Variations occur when new organizational structures are deliberately created or when new forms accidentally happen. In the competition for scarce resources some organizational forms are more successful than others, and thus positively selected by the environment. Over time certain forms have a higher chance of survival and become more frequent – "successful" – within the population, while other forms are negatively selected and may fail to survive.

For instance, in the field of technology and technical systems, this perspective has been applied by Harvey Brooks (1980) to explain technological innovations. Brooks used the metaphor of natural selection to depict processes where the inherent logic of science leads to many technological opportunities (including some accidental discoveries), but the selection of a limited number through social processes of decision, such as market forces and political choices. As Brooks (1980, p. 66) put it, "technologies have ecological relationships with one another, and occupy ecological niches in the overall technological system, as do species in the biological world".

The second perspective is organizational or community ecology (Baum 1996; Freeman and Audia 2006). This perspective stresses organizational communities with interacting populations and their co-adaptation and co-evolution with the environment. This view implies a multi-level and multi-relational perspective which tries to understand the overall evolution of organised societies in which a broad diversity of organizations co-exist, maintaining multiple relations (resource dependencies; strategic interdependencies; communication, and not just competition) with other organizations and the natural environment.

Evolution, from this perspective, is not driven by homogeneous environmental pressure for adaptation, but by different "logics" of adaptation at different organizational levels. Organizational communities are functionally integrated systems of interacting populations that gain some autonomy from their environments. Contrary to the environmental determinism of the population ecology perspective, it rejects the premise that organizations are unable to deliberately transform their structures and behaviours (Astley 1985).

While ecological models are popular in current organizational studies, this perspective was applied by relatively few studies of technical systems. An exception is William Barnett's analysis of the early American telephone systems (Barnett 1990; Barnett and Carroll 1987). These studies concentrate on the adaptation of telephone companies to technological innovation and development in power and transmission technology, and investigate thousands of companies that proliferated or failed under conditions of technological change. The findings suggested that, in the context of systemic technologies, technological change did not necessarily favour advanced organizations. Instead, they favoured mutualism over competition. Cooperating companies survived by forming networks of complementary organizations thus increasing each other's viability.[6]

3.3.4 Combining Games and Ecologies

The two rich social science concepts of "games" and "ecologies" have some important communalities as well as differences. One common feature is that both are aimed at a dynamic reconstruction of social interactions. A second is that both are actor-centred. A third is that both assume a plural or polycentric perspective on society, where social processes are the consequences of interactions of multiple actors. A fourth common feature is that both perspectives provide a grammar for describing the structure of interactions. Finally, both insist on the dialectical interdependence of actors and their environment, with each actor being conditioned by its environment and, the environment being impacted by individual actions.

The differences between a game theoretical perspective and ecological approaches are important as well. First, formal game theory assumes that actors

[6] A study of telephone companies in Iowa had similar results. Telecommunications companies extended their systems either by "encroaching on the territories of their neighbours (competition)" or by connecting with each other (commensalism) by the creation of mutalistic organizations (Barnett and Carroll 1987).

seek to maximise their utility – act rationally. Based on this baseline assumption, it analyses conflict constellation in order to find interaction equilibriums and to make predictions about collective outcomes. Due to manifold restrictions of formal mathematical models, however, game theory largely concentrates on rather simple games with homogenous players. Formal modelling of nested and interconnected games is primarily limited to sequentially connected games. Mathematical models of multiple games with multiple, and heterogeneous, players such as suggested by evolutionary game theory, is at very early stages of development.

In contrast, ecological models place an emphasis on the dynamic interaction between multiple and heterogeneous actors in diverse interaction contexts, and changing environments, without the assumption of a single logic of action. Ecologies are permeable settings in which constantly new actors are entering and old players are leaving. Interaction is not only based on subjective preferences, but embedded in multiple functional circuits in a complex division of labour. Each of the various "species" plays their specific role.

An important difference also is a multilayered and complex perspective on the relationships between the components of these systems, where the evolutionary units do not just compete or cooperate, but are often involved in mixed competitive and cooperative relations. Because of the high compositional and relational complexity of such ecological systems, collective outcomes are largely emergent outcomes. The notion that one could predict individual and collective behaviour in such emerging interactions is impossible except at a broad level. For example, applied to a territorial ecology of local politics, "co-operation", in the EOG perspective of Norton Long, "is an unconscious affair. [. . .]. This unconscious co-operation, however, like that of the five-acre woodlot, produces results. The ecology of games in the local territorial system accomplishes unplanned but largely functional results" (Long 1958, p. 254).

Another difference between games and ecological approaches is that the former tend to focus on few and rather homogenous actors (be they individuals or social organizations), while the latter tend to focus on the larger system of action in which a multiplicity of diverse actors make decisions and choices. These comparative features make the combination of the two perspectives more valuable by enabling individual choice and systems of action to be viewed in a more integrated manner.

3.4 Applications of the EOG

The value of the EOG perspective derives from its resonance with detailed empirical studies of the development of large technical systems, such as in the area of information and communication technologies. This section briefly outlines three cases to provide a concrete sense of the meaning and value of an EOG perspective. We develop a qualitative perspective on the EOG rather than formal models in a game theoretical sense. Neither do we collect data to create operational indicators of ecological relations. Instead, we outline "explanation sketches" based on thick descriptions in which the major arenas of socio-technical interaction in various large technical communication systems such telecommunications, broadcasting, videotext, and the Internet are played.

3.4.1 Institutional Innovation in Telecommunication Systems

This paper argues that it is useful to view the development of telecommunications infrastructures and services as the outcome of an EOG – the interaction of multiple games, being played by different players, each with somewhat different stakes and interests. Each game is a local focus of interaction, partly overlapping with other simultaneously played games. This fragmentation of clusters of interacting players make innovation processes such as the introduction of new institutions and technologies in telecommunications, very difficult to predict. A key assumption of the EOG is that there is no single logic that governs the development of communication systems in a rational and comprehensive sense that the act of governing might connote. Instead of planning the development of such systems in a top-down, hierarchical, and rational manner, actors make incremental decisions about many interrelated topics such as technical design, institutional networks, rate regulation, tax policy and so forth that effect the pace and direction of development.

Telecommunication covers all technologies that make communication over distances possible – from the telegraph and telephone to email over the Internet. In analyzing the evolution of such systems, the EOG perspective can be applied to various policy and management choices which mobilise the resources for the establishment of such infrastructures and shape the organizational and the technical design of these systems (Dutton 1992). Since there is broad temporal and cross-national variation in the institutional structures governing telecommunications systems over the last 150 years (Schneider 1991), it is likely that the ecology of games shaping national telecommunication systems evolved in different ways. Whereas in Europe and Japan, public organizations and state-monopolies dominated telecommunications until the last decade of the twentieth century, private business dominated U.S. telegraphy and telephony until independent administrative commissions and agencies, such as the Federal Communications Commission (FCC), gradually regulated telecommunications. In Europe the management of telecommunications systems often was integrated in one single organization, the PTT, which led telecommunications to be shaped by intra-administrative policy-games in which infrastructural or industrial policy goals were dominant. These various national systems were separate territorial monopolies, coordinated internationally through standardization games (Genschel and Werle 1993; Schmidt and Werle 1998).

In the U.S. context of private ownership of telecommunications, the situation was more complex. The evolution of telecommunications can be seen as the outcome of a whole series of games, well beyond intra-organizational games, including a rate regulation game, an anti-trust enforcement game, an economic development game, a real estate game, a land development game, a cable franchise game, and so on. Many of these games were not aimed at the development of telecommunications per se, but affected it nonetheless. Even micro-processes such as the movement of publishing companies into the microwave communications business, which undermined the quasi-monopoly of the long-time dominant carriers, can be considered as a telecommunications sub-game (Dutton and Mackinen 1987).

Each of the games implied a specific constellation of players and conflicts. For instance, in rate regulation, the major players were telecommunication firms, public regulators and different user groups. Whereas telecom firms where interested in maximizing rates, public regulators had to balance general economic interest and users preferences. In antitrust games, the major opponents to monopolistic telecom firms were the courts, in addition to would be competitors.

A general, cross-national, trend over the last century of telecommunications has been a steady increase of players. For example, in the early twentieth century, in Europe and Japan, telecommunications policy-making was restricted to a small group of players, primarily involving the public administrations managing the telecommunication systems, the equipment manufacturers, and last, but not least, groups that represented users and influential multinational companies. Only in the 1980s and 1990s, in the process of liberalization, privatization, and re-regulation, did more and more players begin to enter the ecology of telecommunications. In fact, it was the entry of new players that initiated moves toward liberalization and privatization, which itself produced several waves of technical innovations during this time period.

For example, in Germany, institutional reform began at the end of the 1980s with moderate steps toward liberalization. The decision to privatise was made only in the early 1990s, and the telecommunications law that defined the overall transformation of the regulatory structure was not established until 1996.

Figure 3.1 displays the actor constellation of these games through a use of visualization techniques of social network analysis (Brandes et al. 1999). The network data on this liberalization policy game had been collected in a network analytic study during the early 1990s (Schneider and Werle 1991), from which a subset of the 24 most influential organizations was selected. Dotted circles in this picture represent various groups and alliances involved in different games around institutional reform, and the different forms of nodes represent a variety of players. Black lines indicate intense communication, whereas grey arrows represent major conflict relations between the various groups and alliances. The sizes of the nodes reflect varying degrees of reputed influence in the legislative process of this institutional reform.

The overall liberalization game consists of several sub-games in which various players interact. Industrial actors are often divided between the conflicting interests of the domestic telecoms industry, which prefers protection, and the computer industry, supported by foreign multinationals, which prefers drastic liberalization measures. Peak associations, such as the *Federation of German Industry* (BDI) and *Deutscher Industrie- und Handelstag* (DIHT), within German business were left

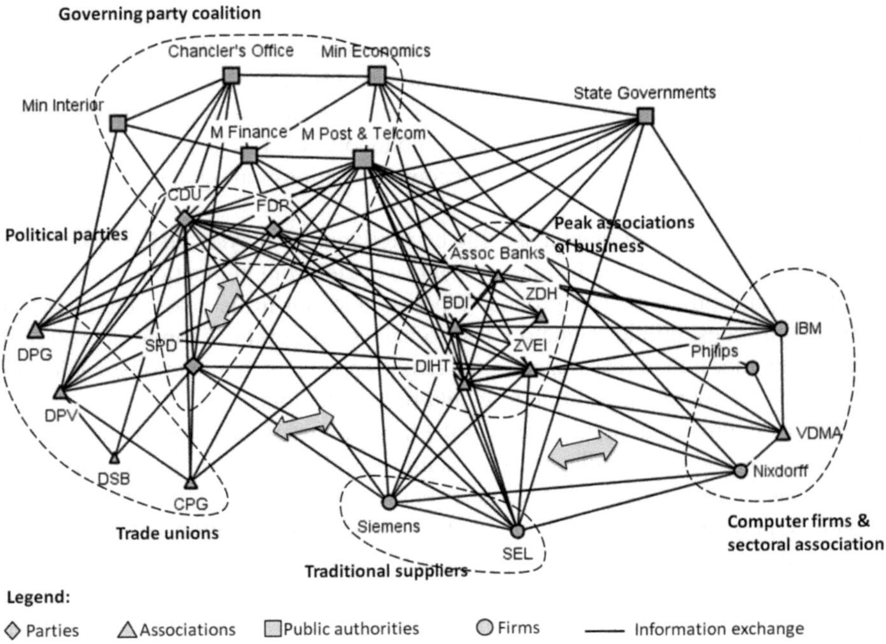

Fig. 3.1 Telecommunication liberalization in Germany (end 1980s)

with the difficult task of bargaining and coordinating a common stance for business vis-à-vis the governing coalition.

Another line of conflict existed between business and trade union interests, representing a classic capital versus labour game. A final conflict relation existed between the opposition party SPD and the governing CDU/FDP-coalition in an election game.

The interaction within and across the various games resulted in relatively more moderate liberalization measures as compared with other countries at a similar stage of telecommunications development (Schneider et al. 2005). This was a consequence of the fragmentation and the relative weakness of the "radical liberalisers" in Germany, the strength and good connections of the trade unions to the influential organizations, and the strong influence of the opposition party SPD.

3.4.2 The Emergence of New Media: From Cable TV to Videotext

Radical technological innovations in the 1970s enabled traditional telecommunications systems to expand into new areas, such as broadcasting and computer communication networks. An instructive insight into the patterns of conflicts that are created by these expansive and resource intensive technological policies is given by a case study on the introduction of cable TV in France using an EOG perspective (Vedel and Dutton 1990).

3.4.2.1 Cable Communication Games

During the 1970s, when other European countries were entering the cable television age, the French government enforced a restrictive policy on cable systems development. Cable systems were only permitted in a limited number of border areas to ensure the retransmission of foreign TV broadcasts (which could be received over-the-air as well). This limitation can be analysed as the outcome of three different games. A first game was an advertising revenues game, which opposed the French print media (notably regional dailies which enjoyed at that time a quasi-monopoly on local advertising) and advertising companies eager to take advantage of new markets.

A second game over the transmission of TV signals opposed the French telecoms operator (then known as the Directorate General for Telecommunications, DGT), which held a monopoly over the construction of telecommunications and Telediffusion of France (TDF), an organization of the public broadcast system which was in charge of the transmission of TV broadcasts.

Finally, a third partisan game was linked to the nature of President Giscard d'Estaing governing coalition. While the president himself was an advocate of the marketplace and sensitive to business circles willing to enter into a new promising communication market, he had to deal with the dominant Gaullist party within the parliamentary majority, which remained attached to a statist conception of television and feared that cable could provide a forum for challenging the existing political order.

The EOG approach illuminates the French ban on cable systems in important respects. Compared to a simple pluralistic approach which explains public decisions as the result of the competition among various actors with different resources, the EOG perspective underlines that social actors continuously operate in multiple arenas/games. This has a number of implications. First, actors may retreat in one arena if their stakes in other arenas are more important to them. Second, the balance of power in one arena is not just the result of the respective forces of the actors involved in this arena, but is also impacted by the balance of power in connected arenas. If a ban on cable development was decided in 1975, it is not just because the forces opposing cable development "won" in each of the three games that we identified. It is primarily because these games were interrelated so as to make a larger coalition against cable. The DGT was certainly stronger than TDF in the transmission game, but ran up against the overall mood swing against cable in other arenas, and preferred to focus on other strategic goals.

Similarly, the EOG perspective helps to understand how, in an outstanding reversal of previous policy, an ambitious Cable Plan was launched in 1982. This plan was designed in ways that connected it to several ongoing policy games, therefore enlarging its potential political support. Adopted just 1 year after the election of President Mitterrand, the Cable Plan was first related to a technology-led, industrial policy game, involving the DGT, the Finance and Industry ministries. Its focus on interactive fibre-optics systems, which might give French manufacturers a competitive advantage, matched well the new socialist government's willingness to implement a proactive industrial policy. The Cable Plan was also connected to the decentralization policy initiated by the socialist government in order to enlarge the role and responsibilities of local authorities. Namely, the Cable Plan required the involvement of local authorities into the wiring of cities through the establishment of specialised local companies.

Finally, cable policy ignited a cultural game, similar to that evoked in many other European nations. Its stake or meaning was the protection of French culture with the Ministry of Culture and audiovisual industries being the key players. Cable TV was perceived to permit better control over foreign programming than would satellite TV.

An important argument from the EOG perspective is that there is often a dynamic interplay of co-evolving games that shape the outcome of events. Actors within each of these separate games came to view the cable plan as a means for achieving their objectives. The Ministry of P&T was an important player in the process. It served to coordinate if not engineer these various players, but only for a time. The dynamics of the games shifted with changes in the perception of the relative strength of the cabling interests and other major players, as the costs of the Cable Plan, and the limitations of take-up, began to become clear.

3.4.2.2 Videotext as the Outcome of an Ecology of Games

The other game in which traditional telecommunications games became tied to new communication technologies was the integration of computer and telephone networks in developing videotext services. Videotext was presented as an example of a modern LTS in the early years of this debate (Mayntz and Schneider 1988). It was originally designed to enable access from private households to computer databases through the conventional telephone system. Information was displayed on traditional TV-sets. The most interesting facet in this new hybrid technology game was that the German PTT and the French DGT – taking over the idea from the British Post Office in 1975 – implemented their systems in very different ways.

In France, the DGT had learned key lessons from Plan Cable on the difficulties of coordinating a highly fragmented set of players. It therefore used a traditional telecommunications procurement strategy to establish their Minitel system in a more hierarchical manner. A small and simple display terminal was distributed for free, which was procured from a quasi-vertically integrated telecommunications industry. For data transmission the newly created packet-switched Transpac network was employed. An electronic phone book was introduced as a trigger service, which at the same time legitimated a massive resource mobilization and the free distribution of several million Minitel sets. The key strategy was to create a critical mass of users rapidly in order to enable self-sustained system growth and the further expansion of content in a virtuous cycle. Only the domain of private information providers had been left to market coordination. On the basis of this governance structure and innovation strategy, Minitel became extremely successful in the 1980s into the early 1990s (Schneider et al. 1991).

Germany introduced a different governance structure and strategy for launching its videotext system, called "Bildschirmtext". It was based on a technological alliance with a whole spectrum of industrial actors, which in the end, turned out to be very difficult to coordinate. The German PTT provided transmission networks and central storage facilities (computers to host the central database) but left the supply of information services to a branch of industry that would have to be newly created. The television industry was to supply adapters for TV sets, and banks, mail-order firms, and other large users would provide specialised data bases, such as for telebanking.

From the perspective of the EOG, this system build-up created a giant meta-coordination game in which the German PTT and its related technical agencies played the role of central coordinators. At the same time, however, this socio-technological process also incorporated a regulation game, since a number of actors felt threatened by this new medium. Media policy actors (broadcasters, the print press) feared massive negative effects on their traditional domains. Likewise, trade unions were worried about an enormous rationalization potential, which could cost jobs.

As outlined above, the coordination component can be modelled by an "assurance game structure", where all rational players choose a Pareto optimum. All actors within the technological alliance expected big profits. The German PTT expected a higher usage of its networks, TV-set manufacturers, such as Grundig, Telefunken, Blaupunkt, and Loewe, the computer industry, and information providers expected new markets. Banks anticipated new forms of electronic banking and mail-order firms expected new markets in tele-shopping. Optimistic forecasts predicted millions of users by the early 1990s. However, only a tiny fraction of the public became connected to this new communication system.

Based on data collected at the end of the 1980s (Schneider 1993), the introduction of German Videotext can be interpreted as an overlapping set of games. This can be shown by network data representing relations of information exchange and cooperation between various actors. The network graph in Fig. 3.2 shows that not all players in the coordination game actually communicated in the required way. A result, as reflected in the chart, is that the technological alliance was highly fragmented rather than coordinated or integrated. In addition, technical frictions created by a technically too advanced display standard, and opposition by various political actors, created several critical problems, i.e. "reverse salients" in the terminology of Hughes. However, an irony of history is that the development problems of BTX had forced the German PTT constantly to innovate and adapt the system to new technologies. Very early BTX began to favour pure software solutions for computers, thus enabling a smoother migration of videotext to the Internet. On the other side, because of its Minitel success, France was temporarily less flexible to adapt to the new environment of the Internet (Schneider 2000).

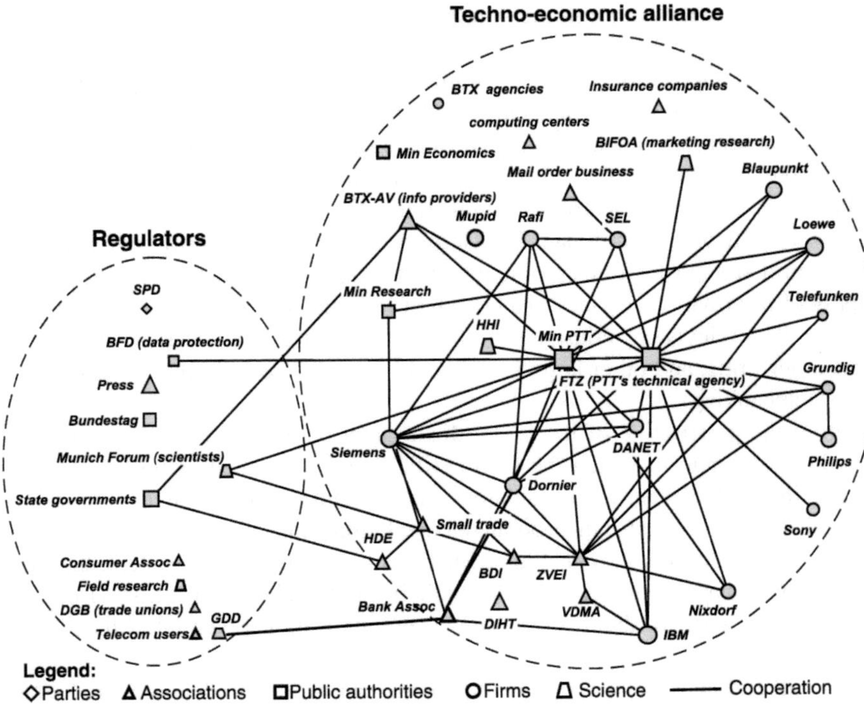

Fig. 3.2 German videotext introduction as an ecology of games

3.4.3 The Ecology of Games Shaping the Internet

The value of the ecology of games approach can also be illustrated through the evolution of the Internet. However, we face here a big challenge. So far, we have applied the EOG to different technological public policies implemented in specific countries and dealt with limited objects. Applying the EOG to the Internet raises a number of additional difficulties. It first means a change in scale. Analyzing the ecologies of games relating to telecommunications systems or cable TV systems is already a difficult task. But with the Internet it appears as a much more demanding exercise since the network of networks is now expanding in every realm of social life. Analyzing the EOG of the Internet is like analyzing the whole of society. This illustrates why the EOG of useful, but also why we need to simplify the analysis, and apply the EOG to particular dimensions of the Internet, such as the infrastructure or its governance.

A second difficulty is that the Internet is a transnational phenomenon, which has two implications. The EOG of the Internet will be more complex as it combines different levels (international and national settings) and encompasses a much larger number of more diverse actors. Moreover, interpreting the games will likely be more intricate since it would require a minimal knowledge of the local rules that govern games.

3.4.3.1 The Ecology of Games Relating to the Internet's Infrastructure

The EOG helps to explain how the growth of the Internet took place: it did not happen just because a few people turned bright ideas into practical systems, but resulted from a huge number of players in intertwined academic, commercial, technical, industrial, and other games making decisions about their own ongoing activities. These eventually had an impact on how the Internet would be designed, developed, used, and governed.

From an historical perspective, the Internet developed as a succession of different games embedded in different ecologies. In the 1960s and 1970s, the Internet grew around defence funding that supported a computational science game (in which the U.S. Department of Defense (ARPA), consulting firms and academic labs were involved), which eventually developed in specific games during the 1980s: academic cooperation, liberalization of telecommunications, communitarian utopias and revitalization of local democracies, and rivalry between computer and telecommunications groups. In the 1990's, a still more complex ecology of games developed with new, particularly commercial, players coming into the field (see Table 3.1).

Table 3.1 Illustrative broadband internet games

Game	Main players	Goals and objectives
Economic development	Governments, public agencies, investors	Players build broadband infrastructures to attract business, investment, and jobs to localities, nations, and regions
Developing country	Governments, NGOs, local activists, investors, local Internet content and service providers	Players seek to close social and economic divides in developing countries using widely available broadband infrastructures
Communitarian	Neighbourhoods, community groups, Internet enthusiasts	Individuals and groups seek free or low-cost open access to broadband Internet, including competing with commercial players
Telecommunication regulation	Telecommunications firms, regulators, investors, consumers	Regulators umpire moves of competing firms, taking account of conflicting and complementary goals of players
Broadband suppliers	Traditional telephone companies using Direct Subscriber Line (DSL) digital adaptation of existing lines, cable TV firms, wireless, and other vendors	Suppliers compete for shares in a market, where DSL and cable vendors have often been the main broadband players winning lines into homes and offices
Content provision	Media giants v. Internet entrepreneurs; media novices and non-profit content producers v. professionals	Established and emerging producers of Internet content compete to reach audiences

Source: Dutton et al. (2004), Table 4.

3.4.3.2 The Internet Governance EOG

The EOG viewpoint can also assist the reassessment and rethinking of appropriate Internet governance institutions and mechanisms. Despite the development of an increasingly informed debate on this issue, discussion on Internet governance often seems to stumble over the notion that some authority can govern it. However, the Internet community, policy makers, and the public at large are divided about what actors or authorities, if anyone, should be involved in such regulation.

Internet governance consists of a proliferating number of governance mechanisms and organizations, as well as a deepening shift in social purpose away from direct state control and toward promoting global markets, private sector control, and security. One academic participant in governance debates, William Drake, argues that Internet governance derives from six sources (Drake and Wilson 2008). The first three involve mechanisms negotiated in intergovernmental, private sector, and multi-stakeholder settings. The fourth, the unilateral imposition of mechanisms by powerful national governments, has been discussed largely with respect to U.S. authority over the Internet Corporation for Assigned Names and Numbers (ICANN), but it is a broader phenomenon. The fifth, the unilateral imposition of mechanisms by firms possessing market power in monopolistic or oligopolistic industry structures, generally has been ignored in governance discussions. So has the sixth, the coordinated convergence of policies across countries and industries.

The EOG captures well this multi-level and multi-stakeholder landscape, as illustrated by the Internet governance games summarised in Table 3.2. For instance, the management and operation of the Internet is greatly affected by the system of allocating and managing names and numbers used within the Internet's infrastructure, such as Web domain names and e-mail addresses. Key players in this game included ICANN, commercial Internet service providers (ISPs), registries, and users. Substantial commercial, personal, and national interests and convenience are at stake in these games, so great store is placed by most actors on establishing rules that create a fair playing field, without over-dominance by particular vested interests.

Interactions between players in the kinds of games shown in Table 3.2 are making it harder to reach technical agreements, which are increasing the difficulty of implementing core technical changes to Internet protocols and infrastructure. For instance, until the late 1990s, such changes could be determined largely by a relatively small group of public-spirited and mutually trusting technical experts in a relatively simple ecology of Internet-related games. The escalation since then in the number of players with an interest in the Internet is shown by more recent meetings of about 2,000 members of the IETF, the international community of network specialists concerned with the evolution and smooth operation of the Internet.

Table 3.2 A few selected games and players shaping internet governance

Game	Main players	Goals and objectives	Related games (on the internet or in real life)
Names and numbers	Individual experts, ICANN, Registries, ISPs, users	Obtain, sell, and allocate domain names, addresses, etc. to identify sites, users, etc	Intellectual property
Standards	Standards-setting bodies, World wide web consortium (W3C), Internet engineering task force (IETF)	Efforts to establish and propagate standards for the Internet	Networks interconnection Open source movement
Telecommunications operators			
Access to infrastructure	ISP, telecommunications operators, national governments	Organise networks interconnection networks, guarantee free access to the internet	Digital divide Freedom of speech
Jurisdictional 'turf struggles'	ICANN, International Telecommunication Union (ITU), UN, national governments	National actors participate in Internet governance bodies to gain or retain national control over policy (e.g. by filtering Internet traffic or gaining access to encrypted data for security reasons)	International governance game (respective roles of governments, international or intergovernmental organizations, NGOs in a globalised world)
Intellectual property	Cinema, music majors, software companies, open source movement, artists	Protect rights on the distribution of cultural contents	Freedom of information Electronic commerce
Data protection spamming, phishing	National agencies for data protection, civil liberties groups, marketing companies	How personal data can be used and which regulations should apply	Privacy electronic commerce
Political speech, freedom of expression	Media rights advocates, activists, politicians, news media, governments, writers, artists, censors	Individuals and organizations aim to facilitate or constrain flows of information, political views, and creative works	Cultural diversity

(continued)

Table 3.2 (continued)

Game	Main players	Goals and objectives	Related games (on the internet or in real life)
Harmful content (pornography, violence, sexist, hate speech)	Civil rights organizations, family groups, companies involved in pornographic contents	Very close to the previous one but with different players	Electronic commerce Freedom of speech
World information order	Developed and developing countries governments, and affiliated NGOs	Role of information in economic development	Cultural diversity, exemption for information services in WTO

Moreover, the ecology of games perspective therefore shows that not only is there unlikely to be a central source of Internet governance, but that very few people or organizations are actually seeking to govern the Internet as such. Instead, most actors try to win more focused prizes, for instance developing a market for registering names and numbers, keeping a bank's computer system secure from hackers, avoiding spam e-mails, and so on. Governance of the Internet can then be understood as the outcome of a variety of choices made by many different players involved in many separate but interdependent policy games or areas of activity. More importantly, the EOG shows that many of these games are not new but have long been played in other domains such as media, telecommunications, banking, and economic development.

3.5 Conclusions

This overview of the EOG suggests that it has key advantages as a perspective on the social shaping of LTS and the logic of large technical innovation processes.

First, the EOG provides an analytical key to better understand the multiplicity and variety of actors and relations involved in a given domain, and the related diversity of strategies and objectives. For instance, the politics of Internet governance gets clearer when one realises that many of the games that were previously played in other large technical systems continue and even cumulate in the games shaping Internet development.

Secondly, the EOG approach is a useful tool for comparative research. Not only does it help to identify games that are common to many countries and other games that are specific to some countries, it also unveils the unique social and political dynamics taking place within each country through the interrelation of games. It provides a simple way to bring the reader into an extraordinarily complex empirical reality.

Thirdly, the EOG is complementary to other approaches or concepts, such as the relational perspective in social analysis or the new institutionalism in economics and political science. Both network relations and rule systems are important elements and components of games. It is consistent with approaches that see technical systems as a dialectic interplay between technology and society. In some ways it can be described as an attempt to integrate different perspectives into a meta-approach that emphasises complex linkages among diverse policy actors and issues of conflict and cooperation.

Fourth, the EOG has the advantage of constantly reminding policy analysts of the importance of the so-called 'environment'. While policy analysts are usually aware of exogenous factors, for simplicity sake they often tend to conduct their studies within bounded or self-contained systems. When using an EOG approach, it is most apparent that large technical systems, their related policy networks and their environments are densely nested and intensely interrelated.

Fifth, the EOG is an especially valuable approach to capture ongoing innovation and developments in large technical systems. One constant problem in analyzing technical systems is that technological developments might be given different values and meanings depending on countries or sectors, or associated to different goals. As we have shown in our case studies on videotext or cable policies, the different ecologies of games in each country help to understand the specific meaning given to technology within different national cultures and political economies.

Nevertheless, the EOG also implies limitations and difficulties that must be acknowledged and tackled. The EOG is a sensitizing concept, a background theory that offers ways of seeing, organizing, and understanding complex reality (Charmaz 2003). This is not necessarily a weakness, but it does limit its usefulness for quantitative or formal mathematical approaches.

As a sensitizing concept it implies interpretive flexibility. Consequently, different researchers applying the EOG to the same object are likely to conceive different ecologies, games, actors, and relations. In this respect, it is not different from most qualitative approaches to cross-national or comparative analyses. It provides a point of view and indicates a set of methods to conduct case studies, including key variables and types of relationships to take into account. However, it is not a predictive theory in the sense that it will not be able to predict outcomes, although it does indicate the likely nature of the dynamics shaping outcomes. In addition, it can be challenged by other researchers, who can critically assess any depiction of a specific EOG.

Another problem with the EOG is that it may lead to an extremely complex mapping of social reality. Its framework or perspective can lead researchers deeper and deeper into the thicket of nested games. Partly as a consequence of this complexity, it is difficult to represent ecologies of games in a visual fashion, with graphs, charts, or pictures since the combination of games usually involves many dimensions. Also, it is necessary to arbitrarily limit the complexity of any analysis, least it become too unwieldy. While all theoretical perspectives tend to simplify reality, the need to simplify is immediately apparent to the analyst seeking to describe an EOG.

However, these weaknesses point to directions for further research. Efforts to employ new tools, such as agent-based modeling, and other approaches to modeling complex systems, could open up avenues for moving the analysis of the EOG towards more complex representations. If it were possible to represent more of the complexity of the EOG shaping LTS, it is possible that research could anticipate strategic issues early on in the process of innovation. Such representations would also enhance inter-subjectivity and replicability across researchers in ways that enable researchers to judge the value of alternative representations, and arrive at alternative explanations. At the same time, moving in the direction of more formal computer models of the EOG could also undermine the heuristic value of this sensitizing concept within the qualitative study of LTS.

References

Adner, R. (2006). Match your innovation strategy to your innovation ecosystem. *Harvard Business Review, 84*(4), 98–107.

Astley, W. G. (1985). The two ecologies: Population and community perspectives on organizational evolution. *Administrative Science Quarterly, 30*(2), 224–241.

Axelrod, R. M. (1984). *The evolution of cooperation*. New York: Basic Books.

Barnett, W. P. (1990). The organizational ecology of a technological system. *Administrative Science Quarterly, 35*(1), 31–60.

Barnett, W. P., & Carroll, G. R. (1987). Competition and mutualism among early telephone companies. *Administrative Science Quarterly, 32*(3), 400–421.

Basole, R. (2009). Visualization of interfirm relations in a converging mobile ecosystem. *Journal of Information Technology, 24*(2), 144–159.

Baum, J. A. C. (1996). Organizational ecology. In S. Clegg, C. Hardy, & W. R. Nord (Eds.), *Handbook of organization studies* (pp. 77–106). London: Sage.

Blumer, H. (1954). What is wrong with social theory? *American Sociological Review, 19*(1), 3–10.

Brandes, U., Kenis, P., Raab, J., Schneider, V., & Wagner, D. (1999). Explorations into the visualization of policy networks. *Journal of Theoretical Politics, 11*(1), 75–106.

Brooks, H. (1980). Technology, evolution and purpose. *Daedalus, 109*(1), 65–81.

Bunge, M. (1998). *Social science under discussion: A philosophical perspective*. Toronto: University of Toronto Press.

Cederman, L. (2005). Computational models of social forms: Advancing generative process theory. *American Journal of Sociology, 110*(4), 864–893.

Charmaz, K. (2003). Grounded theory: Objectivist and constructivist methods. In N. K. Denzin & Y. S. Lindcoln (Eds.), *Strategies for qualitative inquiry* (2nd ed., pp. 249–291). Thousand Oaks: Sage.

Cornwell, B., Curry, T. J., & Schwirian, K. P. (2003). Revisiting Norton Long's ecology of games: A network approach. *City and Community, 2*(2), 121–142.

Coutard, O. (1999). *The governance of large technical systems*. London: Routledge.

Crozier, M., & Friedberg, E. (1977). *L'acteur et le systѐme: Les contraintes de l'action collective*. Paris: Le Seuil.

Drake, W. J., & Wilson, E. J. (2008). *Governing global electronic networks. International perspectives on policy and power*. Cambridge, MA: MIT Press.

Dutton, W. H. (1992). The ecology of games shaping telecommunications policy. *Communication Theory, 2*(4), 303–328.

Dutton, W. H. (1995). The ecology of games and its enemies. *Communication Theory, 5*(4), 379–392.

Dutton, W. H., & Guthrie, K. (1991). An ecology of games: The political construction of Santa Monica's public electronic network. *Informatization and the Public Sector, 1*(4), 1–24.

Dutton, W. H., & Mackinen, H. (1987). The development of telecommunications: The outcome of an ecology of games. *Information and Management, 13*(5), 255–264.

Dutton, W. H., Eisner Gillett, S., McKnight, W. L., & Peltu, M. (2004). Bridging broadband Internet divides: Reconfiguring access to enhance communicative power. *Journal of Information Technology, 19*, 28–38.

Firestone, W. A. (1989). Educational policy as an ecology of games. *Educational Researcher, 18*(7), 18–24.

Freeman, J. H., & Audia, P. G. (2006). Community ecology and the sociology of organizations. *Annual Review of Sociology, 32*, 145–169.

Friedberg, E. (1997). *Local orders: Dynamics of organized action*. Greenwich: JAI Press.

Genschel, P., & Werle, R. (1993). From national hierarchies to international standardization: Modal changes in the governance of telecommunications. *Journal of Public Policy, 13*(3), 203–225.

Hannan, M. T., & Freeman, J. (1977). The population ecology of organizations. *American Journal of Sociology, 82*(5), 929–964.

Hawley, A. H. (1944). Ecology and human ecology. *Social Forces, 22*(4), 398–405.

Hughes, T. P. (1983). *Networks of power: Electrification in western society 1880–1930*. Baltimore: Johns Hopkins University Press.

Hughes, T. P. (1987). The evolution of large technological systems. In W. Bijker & T. J. Pinch (Eds.), *The social construction of technological systems* (pp. 51–82). Cambridge, MA: MIT Press.

Joerges, B. (1999). High variability discourse in the history and sociology of large technical systems. In O. Coutard (Ed.), *The governance of large technical systems* (pp. 258–290). London: Routledge.

Kuhn, A. (1974). *The logic of social systems*. San Francisco: Jossey-Bass.

La Porte, T. R. (1991). *Social responses to large technical systems: Control or anticipation*. Dordrecht: Kluwer.

Latour, B. (1987). *Science in action: How to follow scientists and engineers through society*. Cambridge, MA: Harvard University Press.

Leijonhufvud, A. (1996). Towards a not-too-rational macroeconomics. In D. Colander (Ed.), *Beyond microfoundations: Post Walrasian macroeconomics* (pp. 39–55). Cambridge, MA: Cambridge University Press.

Long, N. E. (1958). The local-community as an ecology of games. *American Journal of Sociology, 64*(3), 251–261.

Mayntz, R., & Hughes, T. P. (1988). *The development of large technical systems*. Frankfurt: Campus Verlag.

Mayntz, R., & Schneider, V. (1988). The dynamics of system development in a comparative perspective: Interactive Videotex in Germany, France, and Britain. In R. Mayntz & T. P. Hughes (Eds.), *The development of large technical systems* (pp. 263–297). Frankfurt/Main: Campus.

Milnor, J. W. (1954). Games against nature. In R. M. Thrall, C. H. Coombs, & R. L. Davis (Eds.), *Decision processes* (pp. 49–59). New York: Wiley.

Myerson, R. B. (1991). *Game theory: Analysis of conflict*. Cambridge, MA: Harvard University Press.

Ostrom, E. (1986). An agenda for the study of institutions. *Public Choice, 48*(1), 3–25.

Ostrom, E., & Gardner, R. (1991). Rules and games. *Public Choice, 70*(2), 121–149.

Pickett, S., & Cadenasso, M. (2002). The ecosystem as a multidimensional concept: Meaning, model, and metaphor. *Ecosystems, 5*(1), 1–10.

Rapoport, A., & Guyer, M. (1978). A taxonomy of 2 × 2 games. *General Systems, 23*, 125–136.

Samuelson, L. (2002). Evolution and game theory. *The Journal of Economic Perspectives, 16*(2), 47–66.

Scharpf, F. W. (1997). *Games real actors play: Actor-centered institutionalism in policy research*. Boulder: Westview Press.

Schelling, T. C. (1980). *The strategy of conflict*. Cambridge, MA: Harvard University Press.

Schmidt, S. K., & Werle, R. (1998). *Coordinating technology*. Cambridge, MA: MIT Press.

Schneider, V. (1991). The governance of large technical systems: The case of telecommunications. In T. R. La Porte (Ed.), *Responding to large technical systems: Control or anticipation* (pp. 18–40). Dordrecht: Kluwer.

Schneider, V. (1993). Networks and games in large technical systems: The case of Videotext. In F. W. Scharpf (Ed.), *Games in hierarchies and networks: Analytical and empirical approaches to the study of governance institutions* (pp. 251–286). Frankfurt/Main: Campus.

Schneider, V. (2000). Evolution in cyberspace: The adaption of national videotext systems to the Internet. *Information Society, 16*(4), 319–328.

Schneider, V., & Bauer, J. (2007). Governance: Prospects of complexity theory in revisiting system theory. *Presented at the annual meeting of the Midwest Political Science Association*, Chicago, 14 Apr 2007.

Schneider, V., & Werle, R. (1991). Policy networks in the German telecommunications domain. In B. Marin & R. Mayntz (Eds.), *Policy networks: Empirical evidence and theoretical considerations* (pp. 97–136). Frankfurt/Main: Campus.

Schneider, V., Charon, J. M., Thomas, G., & Vedel, T. (1991). The dynamics of Videotex development in Britain, France and Germany: A cross-national comparison. *European Journal of Communication, 6*(2), 187–212.

Schneider, V., Fink, S., & Tenbücken, M. (2005). Buying out the state: A comparative perspective on the privatization of infrastructures. *Comparative Political Studies, 38*(6), 704–727.

Sen, A. K. (1969). A game-theoretic analysis of theories of collectivism allocation. In M. Tapas (Ed.), *Growth and choice: Essays in honour of U. N. Ghoshal* (pp. 1–17). Oxford: Oxford University Press.

Shields, P. (1995). The ecology of games – beyond individualism and the ecology of games: Structures, institutions, and communication policy. *Communication Theory, 5*(4), 366–378.

Summerton, J. (1994). *Changing large technical systems*. Boulder: Westview Press.

Vedel, T., & Dutton, W. H. (1990). New media politics: Shaping cable television policy in France. *Media, Culture, and Society, 12*(4), 491–524.

Von Neumann, J., & Morgenstern, O. (1944). *Theory of games and economic behaviour*. Princeton: Princeton University Press.

Chapter 4
The Mass Media as Actors in Innovation Systems

Annie Waldherr

4.1 Introduction

With its recent high-tech strategy, the German government has placed a high priority on the development of advanced technologies. The performance of this sector is deemed decisive for the competitiveness of the country, a highly developed Western economy (BMBF 2006, p. 2; Prange 2006, p. 25). Modern innovation and high-tech policy pursues a systemic approach (OECD 1999; Rammer et al. 2004). The central question is how to develop lead markets where firms profit from qualified employees, excellent production and research structures, and attractive markets. Thus, the focus shifts from earlier technology-centred approaches to a broader perspective, in which the social surroundings of innovations become more significant (Meyer-Krahmer 2004).

The systemic trend in high-tech policy is based on the innovation systems framework, an approach in innovation research that has become prominent during the 1990s. Researchers extended the focus from firms as primary actors in innovation processes to the wider economic and social context of a nation (Lundvall 1992; Nelson 1993), a region (Cooke et al. 2004), or an industrial sector (Malerba 2005). The central perspective of the innovation system approach is the understanding that companies do not innovate in isolation but are interacting in networks of numerous private, as well as public, actors and institutions (Fagerberg 2005). Main participants in innovation systems are firms, public and private research organisations, government, and other public institutions (OECD 1999). The role of consultancies, professional societies, and industrial research associations as bridging institutions between industry and academic research has also been acknowledged (Metcalfe 1995). Recent research also puts a higher emphasis on demand, and therefore customers, as potential sources and drivers of innovation (Chesbrough 2004; von Hippel 2005). An understanding of the interplay between

A. Waldherr (✉)
Free University, Berlin, Germany
e-mail: annie.waldherr@fu-berlin.de

J.M. Bauer et al. (eds.), *Innovation Policy and Governance in High-Tech Industries*,
DOI 10.1007/978-3-642-12563-8_4, © Springer-Verlag Berlin Heidelberg 2012

these different actors is seen as critical for the development of efficient innovation policies (Metcalfe 1995).

Against this background, it seems astonishing that the role of the public sphere, and especially the mass media, has not yet been considered explicitly in innovation system approaches. This chapter seeks to fill this gap by theoretically exploring two major questions: (1) Do the mass media matter for innovation and high-tech policy? (2) Do the mass media have a strong interest in high-tech policy?

Since the mass media are an important factor in modern societies, they should also be seen as critical factors in innovation systems and thus, as relevant for high-tech policy. This paper argues that the mass media should be integrated conceptually into innovation system frameworks.

The need for such integration becomes even more evident when innovation systems are seen as complex adaptive systems. Many scholars agree that innovation is a highly non-linear process with numerous feedbacks, unintended consequences and often unpredictable dynamics. These dynamics foster or hinder multiple functions determining the performance of an innovation system such as knowledge creation and diffusion. Recent research, models the innovation system as consisting of numerous heterogeneous and interdependent actors that are linked together in complex innovation networks (Gilbert et al. 2001).

In such a theoretical framework, the mass media occupy an important role, which is elaborated upon in the first part of the chapter. The mass media are able to influence innovation processes and key functions, and they structurally mediate between important subsystems of the innovation system. Although they are not generic innovation actors themselves, they can influence firms, politicians and scientists by providing a public space. They are able to frame public debates, and thereby, to shape public expectations concerning innovations and innovation policy. So far, innovation systems approaches have neglected the power of these public communication processes.

However, the mass media are only relevant actors in innovation systems and influential in high-tech policy if they are interested in those issues and therefore engage in considerable coverage and debates. Therefore, the second part of the paper deals with the question of which factors are possibly influencing the activity level of media discourse. From the literature on agenda-setting and news selection, several factors are drawn and summarised in a theoretical framework at the end of the paper. News values are identified as the crucial filter mechanisms in the journalistic production process, which decide whether the mass media offer a forum for an issue. The paper develops conjectures about the news value of high-tech policy and high technologies. It concludes with hypotheses on media attention and possible influences on high-tech policy decisions.

4.2 The Mass Media as Important Actors in Innovation Systems

Basic dimensions for analysing innovation systems are the dynamic processes, the functional performance, and the structure of the system (Liu and White 2001). In the following sections, it will be shown how the mass media can be meaningfully integrated as actors in the innovation system framework when considering each of these perspectives.

4.2.1 Process Perspective

In economic innovation research, the term innovation signifies a novelty that leads to a new product, process, marketing method, or organisational method (OECD 2005b). The OECD definition of innovation is a rather broad definition that sets a minimum requirement for an innovation. It must be "new (or significantly improved) to the firm" (OECD 2005b, p. 46). Neither the idea, nor the invention, is crucial; only the successful implementation on the market or in the firm is decisive. As Fagerberg (2005, p. 4) puts it: "Invention is the first occurrence of an idea for a new product or process, while innovation is the first attempt to carry it out into practice." However, to explain the success of an idea in the firm or on the market one has to take into account the wider context of, for example, the perception of stakeholders like customers, investors, employers, and the public. Sociologists, therefore, add to the economic definition that social perception and communication processes decide whether a novelty is called an innovation or not (Braun-Thürmann 2005; Aderhold and Richter 2006). Aderhold and Richter (2006) are convinced that innovation is attributed only after a product, process or other change is successfully implemented. They suggest several characteristics of novelties that enhance the probability of success in this attribution process: market success, social acceptance and an element of surprise.

For a long time, social scientists have argued that the process of technological development is a socially constructed process (Bijker et al. 1987), a view shared by many economic researchers today. Several researchers emphasise the role of social expectations for the development and the success of new technologies (van Lente and Rip 1998; Brown and Michael 2003; Borup et al. 2006). They argue that today, in accelerated times of just-in-time markets and fetishisation of the new (Brown and Michael 2003), a new category of "strategic science" has been born (van Lente and Rip 1998, p. 221). It requires key actors to put forward promises, voice expectations and position themselves with respect to a future technology. The social shaping of expectations is an inherently discursive process (Brown and Michael 2003). As a consequence, scholars stress the role of communication processes like rhetoric (van Lente and Rip 1998) and ethical discourse (Hedgecoe and Martin 2003). Particularly for controversial technologies, discourse is regarded as "important as it both provides a negotiation space to explore the socially acceptable limits of the technology and acts as a means of enrolling support from key actors" (Hedgecoe and Martin 2003, p. 329).

With a similar focus, German sociologists concentrate largely on the role of visions in the process of technological development (Lösch 2006; Deutschmann 1997). Visions convince by showing new ways of life, and foster social acceptance of new technologies and social change. Deutschmann (1997) describes the process of shaping visions as a cyclical spiral of myths comparable to life cycle models in economics.

These theoretical suggestions are supported by empirical findings in economic research. Studies by the Gartner Group (2004) describe IT-based innovations as influenced by a "hype factor" (inflated early expectations followed by disillusionment) before a second wave of commercial application sets in. Empirical studies at the Fraunhofer Institute have described a similar pattern based on scientific publications and patent data, which is called the double-boom technology cycle (Meyer-Krahmer and Dreher 2004; Dreher et al. 2006). However, elaborated explanations for the emergence of this pattern are still missing.

Acknowledging both the economic and the social science perspective on innovation and technological change, two sub-processes have to be distinguished:

- The implementation process as the successful introduction of an invention in the market, and
- The attribution process as the collective perception of the invention as new to the firm and to the market.

While management is largely able to control the implementation process, the attribution process is heavily influenced by complex social dynamics like expectations, visions and discourses. In both processes the mass media are influential, though the impact of public communication on the attribution process is probably higher.

Implementation process: In this sub-process the mass media are important for generating awareness for the new product or service. Awareness is substantial if resources need to be mobilised or markets need to be created. Thus, in the implementation process, the mass media are mainly targets of public relations and advertising efforts. However, even in this stage, control over public communication is limited. As the mass media are independent actors the communication efforts can fail, either by not getting the attention and coverage desired or, even worse, being criticised and evaluated negatively. Thus, mass mediated societal discourses are able to influence the success and failure of new products and services on the market.

Attribution process: Since influential and sustainable attributions of innovativeness to new goods or services have to be collective (Aderhold and Richter 2006), the public sphere is the primary forum for creating such attributions. The public sphere is defined as a common space, in which members of society meet to discuss matters of common interest (Habermas 1995). Within the public sphere, the mass media today have to be considered the most important arena for societal discourse, as they reach a mass audience. They offer a public forum for different actors with competing attributions, and foster the formation of collective perceptions (Hilgartner and Bosk 1988; Gerhards and Neidhardt 1991). In the terminology of agenda-building approaches to public communication, the attribution process can be regarded as a framing process. According to Entman (1993, p. 52) to "frame is to select some aspects of a perceived reality and make them more salient in a communicating text, in such a way as to promote a particular problem definition, causal interpretation, moral evaluation, and/or treatment recommendation for the item described". Suchman and Bishop (2000, p. 331) demonstrate that attaching the label "innovation" to new products or processes nowadays generally implies an evaluation: "At least in the U.S., and to some extent within Western economies more generally, innovation is accepted without question to be a positive good. In a semiotics of bipolarised, differently valued opposites "innovation" is the preferred alternative to "stagnation" or "resistance to change". This means that framing agendas under the rubric of innovation and change is inevitably a strategic move [. . .]."

Both processes, the implementation and the attribution process, are not linear and are not independent of each other. They can be understood as mutually reinforcing processes with many feedback loops. A successful implementation of a new product or service in the market fosters the collective attribution process. Conversely, inventions collectively perceived as innovations may gain support and resources, which pushes forward the process of implementation. Framing novelties as innovations therefore becomes an important strategy to promote new products and processes. In conclusion, the mass media have to be considered important players, as they certainly influence both sub-processes of innovation: implementation and attribution.

4.2.2 Functional Perspective

Innovation systems are functionally complex systems, because they serve multiple functions. The functional view of innovation systems is a rather recent trend in innovation research (Liu and White 2001; Hekkert et al. 2007; Edquist 2005). It focuses on the activities contributing to the overall goal of an innovation system, that is, the generation and diffusion of innovations (Hekkert et al. 2007). Analysing how well an innovation system serves central functions is a way of measuring the performance of an innovation system (Jacobsson and Bergek 2004). If it can be shown that the mass media contribute to key functions of an innovation system they also have an effect on the overall performance of the system.

Currently, there is no consensus among researchers about the essential functions or activities of an innovation system (Edquist 2005). Liu and White (2001) list five functions of innovation systems. A list of seven functions has been proposed by a team of Dutch researchers, and has already been more elaborated upon and applied in empirical studies (Hekkert et al. 2007). Etzkowitz and Leydesdorff (2000) formulate the functions of an innovation system from an evolutionary perspective as *variation*, *selection*, and *retention*. Most of the specific functions proposed by the other researchers can be classified into this abstract logic. While *R&D* and *knowledge development* provide technological variation in the system, the functions *implementation*, *entrepreneurial activities*, *guidance of the search*, *resources mobilization*, and *creation of legitimacy* are part of the selection process. Providing

Table 4.1 Functions of innovation systems

	Liu and White (2001)	Hekkert et al. (2007)
Variation	R&D	Knowledge development
Selection	Implementation	Entrepreneurial activities
		Guidance of the search
		Resources mobilization
		Creation of legitimacy
Retention	End-use	Market formation
	Education	
Coordination	Linkage	Knowledge diffusion

institutional structures for *market formation*, *end-use* and *education* can be attributed to the retention function (see Table 4.1).

The functions *linkage* and *knowledge diffusion* do not fit into this classification. Both describe the need to link several actors in the innovation system to exchange complementary knowledge and coordinate the selection process. Therefore, these functions are subsumed under the term coordination. This function is important to connect the other functions with each other. While variation, selection and retention adhere to a more or less linear understanding of innovation, introducing coordination creates overlaps between different actors, their knowledge, and also the functions they serve. This results in a non-linear perspective of the innovation process, acknowledging links and feedback (Etzkowitz and Leydesdorff 2000).

Assuming that these are the central activities contributing to the performance of innovation systems, mass mediated discourse contributes to several crucial functions:

Coordination/linkage/knowledge diffusion: Until now, innovation system approaches have neglected the role of public communication for knowledge diffusion. Of course, communication in networks often takes place on smaller organisational platforms, or even along informal and personal paths. However, the different actors in the innovation system also observe each other through the mass media. Additionally, public communication leads to an exchange of information between less strongly coupled actors. Hekkert et al. (2007, p. 423) state that "Policy decisions should be consistent with the latest technological insight and, at the same time, R&D agendas should be affected by changing norms and values." Societal discourse in the mass media reflects these changes.

Selection/guidance of the search: This function describes the need of selecting between various technological options to pursue, since resources are almost always limited. While this function can be fulfilled by a variety of system components like industry, the government or the market, Hekkert et al. (2007) emphasise that this is also an important activity for society at large. Important factors in this context are changing preferences in society and expectations associated with technologies. Here again, mass mediated discourse influences the search process by articulating collective preferences and shaping expectations.

The prioritised agenda resulting from this process heavily determines the allocation of resources for the development of technologies.

Selection/creation of legitimacy: New technologies often have to overcome old technological regimes in a process of "creative destruction" (Schumpeter 1964), for which societal support and acceptance are often critical. Hekkert et al. (2007) point to the essential role of interest groups and their activities in achieving legitimacy for new technologies. Advocacy coalitions of different actors with common interests can function as catalysts. Creating legitimacy for social practices is a generic function of discourse (Foucault 1971). The mass media offer a forum for speakers of different interest groups and advocacy coalitions to reach the larger public, while at the same time acting themselves as agents of legitimation. They control which actors gain access to and credibility in the public arena of discourse on technologies (Ten Eyck and Williment 2003).

Hekkert et al. (2007) assume multiple interactions between functions, which can lead to positive, but also negative, feedback loops. In their model, they identify two functions which are common triggers for virtuous cycles in technological development and interact with other functions in "motors of change": guidance of the search and creating legitimacy. In the first case, societal and governmental agenda-setting leads to new resources, which, in turn, lead to further knowledge development and increasing expectations. In the second case, entrepreneurs lobby for resources or market formation, which increases expectations and boosts other functions, such as entrepreneurial activities or knowledge development.

Both trigger functions, guidance of the search and creating legitimacy, depend critically on societal discourse and agenda-building. They are also substantially interlinked with the coordination function of the innovation system, because they require communication and the exchange of knowledge. Since the mass media are important actors in discourse and knowledge diffusion, they can be considered critical actors in innovation systems, which are able to foster or hinder technological change.

4.2.3 Structural Perspective

It has been shown that both from a process and a functional perspective there are fundamental arguments to integrate the mass media as actors into the framework of innovation systems. But how are the mass media linked structurally to the other actors in an innovation system? What are the relationships between the mass media, the political system, the economic system, and the research system?

Several authors conceive of the public sphere as an intermediary realm, which links different subsystems of society through observation and communication (Edwards 1999; Gerhards and Neidhardt 1991). Widely accepted is the understanding that the public sphere is not one big space, but structured into different arenas (Hilgartner and Bosk 1988). Gerhards and Neidhardt (1991) distinguish three levels of the public: simple face-to-face encounters, public gatherings, and the mass media. The mass media is a highly differentiated forum with actor roles in the lay public and professional speakers. Nowadays it is also the forum where public communication is most influential.

The innovation system is structured into subsystems understood as conglomerates of interdependent actors, institutions and processes: the subsystems of economics, politics and science (Kuhlmann 1999). The mass media act as intermediaries with mutual relationships to each of these subsystems (see Fig.4.1), which are explicated in the following.

Political system: In democratic societies, the mass media fulfill essential political functions. The media exercise democratic control through information and critique. They articulate social problems and convey them to the political system, thereby enabling democratic participation (Gerhards and Neidhardt 1991). They also support social integration by communicating common norms and values (Vlasic 2004). In this context, the mass media are particularly powerful as they take the lead in the formation of public opinions (Noelle-Neumann 1984). This might be the reason why mass media communication is closely observed by political actors (Fuchs and Pfetsch 1996). On the other hand, the mass media also heavily rely on established political actors as sources for their coverage

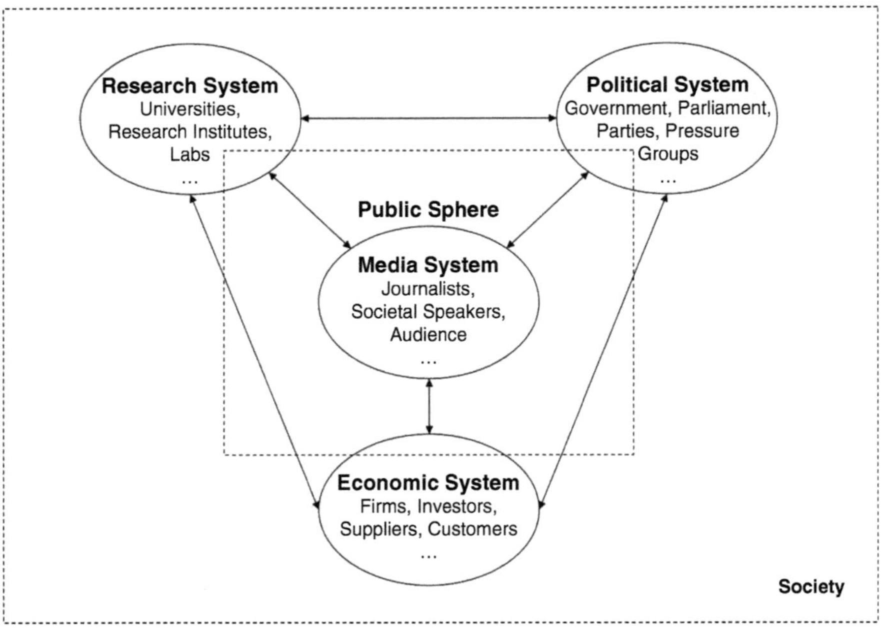

Fig. 4.1 The mass media as intermediary in the innovation system

(Berkowitz 1992). Thus, the influence of political actors on mass communication is valued relatively high (Bennett 1990).

Economic system: The relationship between economics and the mass media is characterised by the exchange of money and media attention. On the one hand, business interest in the mass media is based on the attention of its mass audiences (Davenport and Beck 2001). Media reputation is seen as a strategic resource in management literature (Deephouse 2000). Therefore, activities like advertising, marketing, and public relations are employed to profit from these resources. Economic actors also learn about changing values, norms and needs in society from the mass media. Finally, the mass media may be an indirect lobbying target to pressure political decisions by influencing public opinion. The mass media, on the other hand, depend on the financial resources of the economy for their work, which is why theorists question how independent the mass media can be from economic imperatives (Meier and Jarren 2001). The German media interest in covering business stories is increasing (Mast 2003), which also makes economic actors more important as sources.

Research system: The research system is traditionally a very autonomous system with its own established communication system of peer reviewed journals, research associations, and conferences (Weingart 2003). The interest of the mass media in science and technology issues is usually rather low (Bucchi and Mazzolini 2003; Hansen 1994). However, Weingart (2003) points out a trend towards a higher researcher interest in media coverage in order to gain public reputation and legitimacy. Like in economics, the motivation may be indirect lobbying for resources and surrounding conditions, especially in big science (Gerhards and Schäfer 2006). There exist conflicting views on the role and function of scientific journalism. The dominating view until the mid-1990s was called the popularization paradigm (Kohring 2005). In this view, the media were requested to educate the public in a way that created acceptance of new technologies. Today, scholars highlight an approach to science of social construction, where science is a matter of public negotiations (Bucchi 1998) and journalism is attributed the role of an autonomous and critical observer (Kohring 2005).

4.3 The Mass Media's Interest in High-Tech Policy

In the previous section, it was shown that from a theoretical perspective on processes, functions and structures of innovation systems, the mass media should be considered relevant actors in this framework. As a consequence, they should also be able to influence high-tech policy decisions. However, this will only be the case if the mass media is interested in offering a forum for debate of high-tech policy issues. Therefore, the following sections discuss the expected activity level of media discourse on high-tech policy and technology development.

4.3.1 Activity Levels of Media Discourse and Influencing Factors

The term discourse is used here to describe a form of reasoned debate, which includes the exchange of arguments and counter-arguments between speakers (Habermas 1995), but also has the power to legitimate or delegitimate social practices (Foucault 1971). From an agenda-building perspective, media discourse is not only produced by journalists, but develops in the interactions between different actors in the media arena (Hilgartner and Bosk 1988; Gerhards and Neidhardt 1991). On the one hand, there are media actors, which have the double role of offering a public space for other speakers while at the same time acting as speakers in the media arena themselves (Pfetsch and Adam 2008). On the other hand, there are societal actors, which have a stance on an issue and therefore are interested in actively participating in media discourse. This group of actors encompasses pressure groups, corporate strategists or public policy makers (Mahon and Waddock 1992) and has been given different names in the literature: policy entrepreneurs (Kingdon 1984), operatives (Hilgartner and Bosk 1988) or stakeholders (Mahon and Waddock 1992). In the following, the term "sponsor," introduced by Gamson and Modigliani (1989, p. 6), is used to describe actors, which are interested in pushing and framing issues on public agendas.

For both groups of actors different levels of activity can be distinguished. The boundaries between the level categories are blurred and should be understood as placed along a continuum. The levels of sponsor activity and media activity, then, determine the overall activity level of media discourse. Proposed levels of media activity, ranging from low to high activity, are (Waldherr 2008):

- *Gatekeeper*: Media actors select issues and speakers, which gain access to the arena. The need for selection is given because of the limited carrying capacity of the media arena (Hilgartner and Bosk 1988).
- *Constructing discourse*: Media actors engage in the social construction of reality, for instance, by highlighting certain facts among others (Kepplinger et al. 1991), selecting speakers, which support certain positions ("opportune witnesses"; Hagen 1993), or framing issues along specific terms (Pan and Kosicki 2003).
- *Producing discourse*: Journalists raise their own voice in media discourse by writing editorials or commentaries and take their own stance on an issue (Eilders et al. 2004). Another way of producing pieces of discourse is setting issues, actors, or frames on the agenda by using, forms of investigative journalism.

Drawing on the agenda-building model of Kingdon (1984) in public policy, and literature on issues management and social responsiveness of corporations (Carroll 1979; Mahon and Waddock 1992), three corresponding levels of sponsor activity can be distinguished:

- *Withdrawal*: No action is taken, either because sponsors deny and completely reject the existence of the issue, or because they avoid the issue in a manner of "studied ignorance" (Mahon and Waddock 1992, p. 27).
- *Shaping discourse*: Sponsors engage in interpreting problems already on the agenda and shape alternative problem solutions (Kingdon 1984). The goal is to at least position oneself in the ongoing debate (Mahon and Waddock 1992). This category includes all kinds of reactive, responsive, defensive or interactive strategies in social issues management (Carroll 1979).
- *Initiating discourse*: Sponsors proactively deploy strategies of agenda-setting and issues management by defining problems and placing them on public agendas (Mahon and Waddock 1992; Kingdon 1984).

Empirical evidence suggests an inverse relation between media and sponsor activities. The less input the media get from sponsors, the more effort they put in researching their own stories (Danielian and Reese 1989). Conversely, the more the media are inherently interested in an issue, the less successful sponsors are in dominating media discourse with their frames and interpretations (Barth and Donsbach 1992). However, different scenarios, where both media and sponsor activity are very low or very high, are also possible. The latter may be the case, when a media hype occurs, which puts sponsors under pressure to position themselves in relation to an issue (Mahon and Waddock 1992).

The agenda-building literature also proposes several factors influencing the levels of media and sponsor activity. The most important variables mentioned regularly by researchers are news values, organizational and individual orientations, and resources:

- *News values*: In professional journalism, widely accepted criteria have evolved that guide the news selection process. News values are specific issue characteristics, which are supposed to attract public attention. Important news values are conflict, drama, prominence, proximity or surprise (Galtung and Ruge 1965; Ruhrmann et al. 2003; Schulz 1990). These professional norms not only guide journalistic work, but also influence strategies of issue sponsors. Professional sponsors know the logic of mass media attention and plan their strategies accordingly.
- *Orientations*: Organizational goals and individual interests also affect the work of journalists and sponsors. For example, the political orientation of a newspaper may influence which facts are emphasised or which expert statements are dedicated more space (Hagen 1993; Kepplinger et al. 1991). The individual self-concepts and ethical convictions of journalists also play an important role (Weischenberg et al. 2006). The same considerations apply to sponsor agents.
- *Resources*: On the one hand, financial and personal resources constrain the work in media organizations and determine the "carrying capacity" of the media arena (Hilgartner and Bosk 1988, p. 56). Financial and personal resources also limit the possible actions of sponsors. For them, resources in the form of power or status are very important, because they influence how easily actors can gain access to media arenas (Kriesi 2004; Gamson and Wolfsfeld 1993).

This discussion suggests that news values are the decisive criteria that determine whether the mass media are interested in high-tech policy decisions. As universal professional norms, news values essentially guide gatekeeper decisions in the first stage of the journalistic process: the selection process. They influence the decision whether or not attention is paid to an issue or an event. The other factors, resources and orientation, are assumed to have more moderating and constraining effects. Resources restrict the time, effort and space journalists can dedicate to certain issues. However, news values may also influence how the limited resources are prioritised. Organizational and individual orientations are important in later stages of the process when journalists decide how to report on an issue. Then, questions like how to weigh facts and sources are relevant, even though the selection decision has already been made.

4.3.2 News Values of High-Tech Policy

As news values are a decisive factor in the selection of issues for media coverage, they also have to be considered in order to answer the second research question of this paper: Are the media interested in high-tech policy? High-tech policy can be seen as a subfield of the policy area of innovation policy. The term usually refers to the economic classification of industries into four different categories of technological intensity: high technology, medium-high technology, medium-low technology and low technology (OECD 2005a). In this classification, high technologies are technologies with the highest share of research and development (R&D) investment (more than 7%) in the average turnover of OECD countries (Legler and Frietsch 2007). As R&D intensities appear to be highly correlated with embodied technology intensities they are seen as reflecting an industry's "technological sophistication" (OECD 2005a, p. 170). High-tech products are most often the focus of innovation policy and therefore subject to state intervention and control through subsidies, state demand, or trade restrictions (Legler and Frietsch 2007). Developing and implementing high-tech strategies is widely seen as one of the central tasks of innovation policy.

Media are generally highly interested in politics. They observe the political system more than other societal sub-systems and are also themselves very well observed by actors of the political systems (Fuchs and Pfetsch 1996). However, the type of policy can influence politics which in turn leads to a varying degree of mass media interest. Lowi (1972) formulated this relation as policy determining politics. Many scholars have challenged this assumption during the last few decades of research. Heinelt (2003) summarises the discussion, stating that today the crucial links from policy to politics are seen in the specific arenas and actor constellations of a particular policy domain.

Nisbet and Huge (2006) show that the political arena in which an issue is negotiated has an effect on media attention. Media attention rises when issues enter overt political arenas (see Fig. 4.2). Here, chances for conflicts and polarised

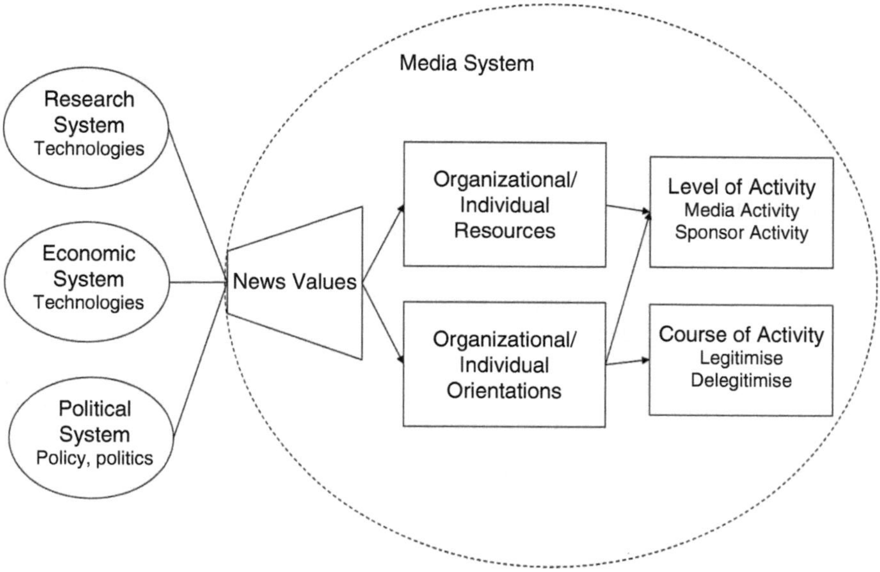

Fig. 4.2 A framework of mass mediated discourse on high-tech policy

actor constellations are high, which allows for a dramatic framing of the issue. Conflict – as mentioned above – is a traditional news selection criterion. Thus, actor constellations are not only influencing politics, but also media attention, as the mass media are particularly interested in conflict-driven political processes. Conversely, media attention is assumed to be low when issues remain in administrative arenas, where technical frames of the issues prevail. This has been the case in a study by Nisbet and Huge (2006) on media attention towards plant biotechnology. For most of the analysed time span, the issue did not leave administrative arenas and consequently, media attention was low.

Integrating Nisbet and Huge's (2006) assumptions with Lowi's (1972) perspective, the original hypothesis can be extended in the following way: Policy (precisely the specific arenas and actor constellations of a policy field) determines politics. The news values of political processes (like drama, conflict or personalization) in turn determine media attention. Thus, looking at the characteristics of the policy field should lead us to reasonable assumptions about media attention towards high-tech policy. German innovation policy shows the following characteristics (Prange 2006; Rammer et al. 2004):

- *Dominance of distributive policy*: Although elements for regulative, inhibitive and persuasive policy can be found, the majority of decisions in innovation policy are distributive. The most prominent policy instrument is providing funds for research and development.
- *High vertical integration*: Decisions have to be made in cooperation with different decision levels, which are highly integrated and interdependent.

The levels of region, federal state, and the European Union form a complicated multi-level decision system.
- *Horizontal fragmentation*: Several autonomous governmental departments have their own research sections which work on innovation policy issues including the ministry of education and research, the ministry of economy and technology, the health ministry, and the ministry of the environment. High-tech strategies need to be coordinated between these departments.
- *Blurring boundaries*: As a cross-sectional policy field, innovation policy also plays into neighbouring fields, like educational policy, regional economic policy or labour policy. The boundaries between these fields are blurred and difficult to define. The interdependencies between policy fields are high.

Similar characteristics, especially the problems of horizontal fragmentation, blurring boundaries, and vertical integration, have been identified for most OECD countries (Edler et al. 2003). The current aim of innovation policies in Europe is to integrate all relevant policy fields and decision levels into an encompassing innovation policy that uses a systemic approach. A key focus is on supporting cooperation between research and business to accelerate technology transfer (Edler et al. 2003). In Germany, another trend is the regionalization of innovation policy, where the focus is on supporting regional innovation clusters, especially small and middle-sized enterprises. Private investment in industrial and applied research is increasingly relevant. Politicians prefer indirect instruments, like setting guidelines and conditions for an innovative environment. They also introduce more and more competitive elements in funding programmes (Prange 2006). One example is the German Excellence Initiative, where universities compete for the title of elite universities and the associated financial resources.

Regarding the current high-tech strategy of the German government, the general characteristics and trends of innovation policy also apply to the field of high-tech policy. The high-tech strategy is a coordinated strategy of all German ministries involved in innovation policy (Schavan 2006). It aims at offering a concerted plan for developing lead markets in high technology (BMBF 2006). First of all, the high-tech strategy is distributive with familiar budget patterns (Beuter et al. 2007). Major funding is allocated to research institutions and specific fields of high technologies, above all space, energy, and information and communication technology (ICT). The aim is to increase research spending by up to 3% of the national gross domestic product (GDP). Newer instruments like competitions have not replaced, but have supplemented older ones, which has increased the complexity of the funding system. However, the strategy also encompasses regulative measures, which are targeted at improving the framework conditions of innovations, especially for small and medium-sized enterprises. Cooperation between science and industry is encouraged to accelerate the direct application of research findings.

What do these characteristics mean for media attention? Distributive policy decisions are mostly made in administrative arenas, often by specific project executing organizations, which organise calls for proposals and manage the funding process, for example the research centre Jülich in Germany. Media attention will be low for most funding decisions taken in such administrative arenas. The exception may be funding programmes, which are designed very competitively, like the super-cluster competitions ("Spitzenclusterwettbewerb"; BMBF 2007). Media attention can also be attracted when very large sums are distributed. Then, the sheer amount of money allocated is surprising.

Regulative policies basically have a larger chance of becoming an issue in media discourse. However, one precondition for major public interest is the emergence of conflicting and polarised actor coalitions. This may be the case with redistributive decisions like the new corporate tax law, or with issues of risk regulation like debates about genetically modified food, stem cell research or nuclear energy.

Generally, the high fragmentation and interdependence of competences in innovation and high-tech policy leads to a complex policy network with numerous actors. The attribution of responsibility to a few specific and prominent actors is difficult, if not impossible. This is problematic for media coverage as personalization is an important news selection criterion (Ruhrmann et al. 2003). Furthermore, attributing responsibility to actors is considered a central element of public discourse (Gerhards et al. 2007).

In summary, considering the characteristics of innovation policy and the resulting politics, the media attention towards high-tech policy, in general, can be expected to be rather low. Higher media attention would be exceptional. However, the policy perspective is only one aspect the media may be interested in. Instead of politics, the media may be more attentive towards the high technologies themselves, which are affected by high-tech policy decisions. Therefore, in the following section, assumptions about the news values of high technologies will also be made.

4.3.3 News Values of High Technologies

As defined in the previous section, high technologies are those with the highest share of R&D investments. Meldrum (1995) identified additional distinguishing attributes which are consequential for public communication and marketing of high-tech products:

- They are developed in a highly technical environment.
- They incorporate a new or advanced technology, which acts as a focus for their evaluation.
- They are associated with a high degree of technologically-based uncertainty on the part of both the supplier and the customer.
- They are not yet accepted as natural solutions for the problems they have been designed to address, because they are new to the market as "leading-edge" products.
- High-tech products do not yet have an associated external infrastructure, but require technological and market infrastructure to be successful.

These characteristics also determine the potential news values of high technologies. Traditional news values, which particularly apply to most high-tech innovations, are *proximity, surprise, risk,* and *benefit.*

The first criterion, *proximity*, means that the mass media are mainly interested in events or other news which are geographically, politically, culturally or economically close (Ruhrmann et al. 2003). Applied to high technologies, one can assume that the media are most attentive to high technologies which are invented and developed domestically, or which at least are relevant to the home market. This will be the case for all high technologies dealt with in national high-tech policy.

The next criterion, *surprise*, can be considered an inherent news value of technological innovations. As the terms "leading-edge," "future" or "key technologies" suggest, high technologies are often radical innovations, which trigger processes of social and economic change. However, the novelty and surprise bonus is often dampened by the fact that scientific results and technological developments take a rather long time compared to the media's need for being up-to-date. For example, scientific breakthroughs sometimes take years until they are published in peer-reviewed journals and become accessible to the wider public. In other words: "The newsworthiness of science is thus hampered by the fact that the 'event-frequency' of science (...) does not readily match the news-frequency of the press" (Hansen 1994, p. 115).

Another traditional news value is damage or failure (Ruhrmann et al. 2003). For technological innovations, this news value can be translated into *risk* of potential damage, failure or unintended consequences of a technology. As Meldrum (1995) states, high-tech products are associated with a high degree of technology-induced uncertainty. This means that all high technologies are potentially risky and associated with known or unknown dangers. However, the mass media seem to be most interested in acute risks, major accidents or disasters (Hansen 1994). Kitzinger and Reilly (1997, p. 319) found that in risk reporting, the time span of public attention is also restricted: "Far from being eager reporters of risk, the press and TV news are ill adapted for sustaining high level coverage of long-term threats. Media interest is rarely maintained in the face of uncertainty and official silence or inaction."

The counterpart to risk is the news value of *benefit* or success (Ruhrmann et al. 2003). Most high technologies promise some kind of benefit, which justifies their further development and the allocation of resources to them. Felt (1993) showed that the media are eager to create scientific success stories. However, Meldrum (1995) also points to a problem of high technologies in this respect. Compared to established technologies, high-tech products are not currently accepted as natural solutions for the problems they have been designed to address. "Thus, the chances are low that there will be a widespread understanding of how a high-tech product can provide benefits for customers" (Meldrum 1995, p. 47).

While these four news criteria apply to almost all high technologies which are relevant for national high-tech policy, other important news values depend on the specifics of the technology in question and cannot be assessed in advance. These are the criteria of *range, relevance to daily life, conflict,* and *relatedness to social problems.*

The *range of affected people*, also called relevance, is a classic news selection criterion (Ruhrmann et al. 2003). The mass media take the sheer amount of people affected by the news as a proxy for the amount of people possibly interested in the news. In the case of high technologies, the mass media ask how many people profit from an innovation or suffer from its dangers. The more people who are affected, the higher media attention will be. For example, a lot of people are generally affected by new developments in ICT technologies. The development of a drug for a very uncommon disease instead, will not attract the attention of many people.

Closely related is the selection criterion *relevance to daily life*, which describes more qualitatively the extent to which people can relate to the news (Hansen 1994). This implies that scientific and technological developments with tangible benefits for a person's daily life enjoy high attention by the mass media. The same is true for the risk or damage part of innovations. Attention to risks which affect people in their everyday life will be higher than to more abstract risks.

Depending on the technology's profile of benefits and risks, more or less controversy and *conflict* can be expected. In each case, the resulting actor constellations determine the possible media attention. As Nisbet and Huge (2006) argue, polarised actor constellations with conflicting views allow a dramatic framing of the issue, and therefore lead to more media attention. However, Bucchi and Mazzolini (2003) put into perspective the media interest in scientific controversies. In a long-term content analysis of Italian science coverage they show that the typical science story is rather positive and uncontroversial. They insist that "scientific controversy per se has little journalistic appeal because it tends to confuse both reporter and readers. Instead, media interest is aroused when it is possible to inject scientific expertise into hotly debated public issues, like the Bovine Spongiform Encephalopathy (BSE) emergency or the alleged dangers of electromagnetic fields" (Bucchi and Mazzolini 2003, p. 13).

This leads to another news value, which is particularly important for science and technology issues: the *relatedness to prevailing social problems*. Studies have shown that media attention to science and technology is continuous but generally low (Bucchi and Mazzolini 2003; Hansen 1994). Science issues are considered "the soft underbelly" which hardly competes with more mainstream issues. Exceptional coverage of technologies only occur when they are associated with major developments in the political or economic sphere: "Science becomes newsworthy, when it becomes part of wider social and political problems, or when it is linked to major accidents or disasters" (Hansen 1994, p. 116). An example is the exceptional media interest in genetics and biotechnology, which is at least partly founded in ethical problems, such as with cloning (Holliman 2004) and stem cell research (Nisbet et al. 2003). The major disaster of Chernobyl dominated the coverage of nuclear energy for years (Kepplinger 1988). Therefore, an important characteristic of high technologies is whether they are related to prevailing social problems already established on the media agenda. This relation can be positive (the technology as a solution to a problem) or negative (the technology as a cause of a problem).

Another group of news selection criteria concern the possibility of an issue to be presented in an interesting and entertaining way. These criteria are very specific to each individual case and can also be manipulated by communicators. However, they have to be viewed problematically for most high technologies. These news values are personalization, prominence, entertainment and visualization (Ruhrmann et al. 2003).

In science reporting as in other fields the mass media like to attribute success stories to specific persons, who are constructed as heroes (Felt 1993). However, the sphere of science and technology issues does not offer a lot of opportunities for such *personalization* of issues. Often, many different persons contribute to important scientific and technological breakthroughs. Furthermore, media *prominence* of researchers is often viewed critically by the scientific community and can damage scientific reputation (Weingart and Pansegrau 1999), for instance, if results are published by the media before having gone through the peer review process. An exception in this sense and also an example for media prominence is the institution of the Nobel price, which is associated with major scientific reputation *and* media prominence. A Nobel prize "bestows instant recognition, lifelong celebrity, and unrivalled authority around the globe" (Feldman 2000, p. 1). Very prominent researchers and developers of high technologies in this sense will be the exception.

As high technologies incorporate very sophisticated and often complex technologies, journalists often have problems *visualizing* them appropriately and presenting them in an *entertaining* way. Like other issues in science and technology, high technologies, in most cases, will demand a lot of "translation work" to make them accessible to the public (Hansen 1994, p. 115).

In summary, a closer look at potential news values of high technologies yields a mixed picture. Some traditional news selection criteria, such as proximity, surprise, risk, and benefit, are certainly served by most high technologies dealt with in national high-tech policy. Other news values, like range, relevance for everyday life, conflict and relatedness to social problems, are highly contingent on the specific technologies. Some news values, like personalization, prominence, entertainment, and visualization, are rather problematical and will apply less to most high technologies.

4.4 Conclusions

This article aimed at theoretically clarifying the relationship between the mass media, high-tech policy, and technology development. The first part of the paper showed from a process, a functional, and a structural perspective that the mass media are relevant actors in innovation systems and therefore also need to be considered as critical variables in political processes which lead to high-tech strategies. From a process perspective, the mass media are influential in the implementation process by creating awareness of innovations, and even more in the attribution process by labelling new technologies, goods and services as innovations. From a functional perspective, the mass media are likely to affect several key functions, which, together with other functions, encourage growth in innovation systems. These are coordination functions like knowledge diffusion, and selection functions like guidance of the search and creation of legitimacy for new technologies. In the structural sense, the mass media are, understood as a public communication system mediating between the subsystems of an innovation system: the political, the economic and the research system.

The potential relevance of mass mediated discourse for high-tech policy decisions has become clear through the previous argumentation. However, the actual relevance of the mass media for high-tech policy depends on the mass media's attention towards high-tech issues and the resulting activity level of media discourse. Therefore, the second part of the paper developed theoretical assumptions about the mass media's interest in high-tech policy and high technologies. In the agenda-building literature, important factors influencing the activity level of media actors and issue sponsors are news values, organizational and individual orientations, and resources.

News values are common news selection criteria and therefore the central filtering mechanism in news production processes. They are considered the critical variable for the activity level of media discourse, while the other variables are interpreted as intermediary variables. Figure 4.2 shows the interplay of factors in a theoretical framework.

News values of high-tech policy, related political processes, and high technologies influence the decision whether the media system is going to pay attention to the political, the economic and the research subsystems of the innovation system. Organizational and individual interests, as well as resources, mediate between news values and the levels of media and sponsor activity in discourse. Organizational and individual interests also determine whether actors work on legitimizing or delegitimizing a technology in discourse. Considering the characteristics of German innovation policy in general, and high-tech policy in particular, one can conclude that media attention to high-tech policy will usually be low, because administrative arenas and technical frames are dominating, and scattered competences inhibit personalization of issues.

However, for high technologies higher media attention can be expected. They satisfy several traditional news selection criteria like proximity, surprise, risk and benefit. Although attention to science and technology is relatively low in routine coverage, studies have shown that technologies can make issue careers in the mass media if they are shown to be relevant for the daily life of a range of people, and if they are related to major social or economic problems which are already prominent in coverage and allow for polarizing and conflicting debates (Bucchi and Mazzolini 2003; Hansen 1994). These characteristics will differ between technologies. Therefore, mass media attention to high technologies is hard to predict, but nevertheless influential on high-tech policy. If such issue careers occur, media discourse is likely to contribute decisively to knowledge diffusion, guidance of future research, legitimation of technologies and related policy decisions. Thus, the mass media may influence innovation policy more through debating high technologies than by discussing policy instruments and political processes.

This chapter showed that the mass media, in principle, has a strong potential to influence high-tech policy through media discourse. Whether this potential is exploited depends on the specific characteristics of the high technologies in question, the prominence of related problems, and the nature of arenas and actor constellations in the political process. Nevertheless, the identified potential of mass media influence calls for integrating mass media actors into the framework of innovation systems. Future studies should empirically analyse the role of the mass media in innovation systems and their actual influence on innovation processes and innovation policy.

References

Aderhold, J., & Richter, G. (2006). Paradoxe Innovationsstrukturen – Orientierungs- und Lernerfordernisse für KMU. In R. Abel, H. H. Bass, & R. Ernst-Siebert (Eds.), *Kleine und mittelgroße Unternehmen im globalen Innovationswettbewerb. Technikgestaltung, Internationalisierungsstrategien, Beschäftigungswirkungen* (pp. 9–43). Münster: Lit-Verlag.

Barth, H., & Donsbach, W. (1992). Aktivität und Passivität gegenüber Public Relations: Fallstudie am Beispiel von Pressekonferenzen zu Umweltthemen. *Publizistik, 2*, 151–165.

Bennett, L. W. (1990). Toward a theory of press-state relations in the United States. *Journal of Communication, 40*(2), 103–125.

Berkowitz, D. (1992). Who sets the media agenda? The ability of policymakers to determine news decisions. In J. D. Kennamer (Ed.), *Public opinion, the press, and public policy* (pp. 81–102). Westport: Praeger.

Beuter, K., Hufnagel, S., Orlowski, M., Peng, K., & Schneider, V. Germany's high-tech strategy: Background, risks and opportunities. *High-Tech Policy – Innovation Policy for High Technology in a Comparative Perspective. Spring meeting of the working group "Politics and Technology" of the DVPW*, Meersburg, 22–23 June 2007.

Bijker, W. E., Hughes, T. P., & Pinch, T. J. (1987). *The social construction of technological systems: New directions in the sociology and history of technology*. Cambridge: MIT Press.

BMBF. (2006). *The high-tech strategy for Germany*. Berlin: Bundesministerium für Bildung und Forschung.

BMBF. (2007). *Deutschlands Spitzencluster: Mehr Innovation. Mehr Wachstum. Mehr Beschäftigung*. Berlin: BMBF.

Borup, M., Brown, N., Konrad, K., & Van Lente, H. (2006). The sociology of expectations in science and technology. *Technology Analysis & Strategic Management, 18*(3/4), 285–298.

Braun-Thürmann, H. (2005). *Innovation* (Einsichten. Soziologische Themen – Themen der Soziologie). Bielefeld: transcript.

Brown, N., & Michael, M. (2003). A sociology of expectations: Retrospecting prospects and prospecting retrospects. *Technology Analysis & Strategic Management, 15*(1), 3–18.

Bucchi, M. (1998). *Science and the media: Alternative routes in scientific communication*. London: Routledge.

Bucchi, M., & Mazzolini, R. G. (2003). Big science, little news: Science coverage in the Italian daily press, 1946–1997. *Public Understanding of Science, 12*(1), 7–24.

Carroll, A. B. (1979). A three-dimensional conceptual model of corporate performance. *The Academy of Management Review, 4*(4), 497–505.

Chesbrough, H. W. (2004). *Open innovation: The new imperative for creating and profiting from technology*. Boston: Harvard Business School Press.

Cooke, P., Heidenreich, M., & Braczyk, H.-J. (Eds.). (2004). *Regional innovation systems: The role of governance in a globalized world* (2nd ed.). London: Routledge.

Danielian, L. H., & Reese, S. D. (1989). A closer look at intermedia influences on agenda setting: The cocaine issue of 1986. In P. F. Shoemaker (Ed.), *Communication campaigns about drugs: Government, media, and the public* (pp. 47–66). Hillsdale: Lawrence Erlbaum.

Davenport, T. H., & Beck, J. C. (2001). *The attention economy: Understanding the new currency of business*. Boston: Harvard Business School Press.

Deephouse, D. L. (2000). Media reputation as a strategic resource: An integration of mass communication and resource-based theories. *Journal of Management, 26*(6), 1091–1112.

Deutschmann, C. (1997). Die Mythenspirale. Eine wissenssoziologische Interpretation industrieller Rationalisierung. *Soziale Welt, 47*, 55–70.

Dreher, C., Frietsch, R., Hemer, J., & Schmoch, U. (2006). Die Beschleunigung von Innovationszyklen und die Rolle der Fraunhofer-Gesellschaft. In H.-J. Bullinger (Ed.), *Fokus Innovation. Kräfte bündeln – Prozesse beschleunigen* (pp. 275–306). München: Hauser.

Edler, J., Kuhlmann, S., & Smits, R. (2003). New governance for innovation: The need for horizontal and systemic policy co-ordination. *Fraunhofer ISI Discussion Papers/Innovation System and Policy Analysis, 2*.

Edquist, C. (2005). Systems of innovation: Perspectives and challenges. In J. Fagerberg, D. Mowery, & R. Nelson (Eds.), *The Oxford handbook of innovation* (pp. 181–208). Oxford: Oxford University Press.

Edwards, A. (1999). Scientific expertise and policy-making: The intermediary role of the public sphere. *Science and Public Policy, 26*(3), 163–170.

Eilders, C., Neidhardt, F., & Pfetsch, B. (2004). *Die Stimme der Medien: Pressekommentare und politische Öffentlichkeit in der Bundesrepublik*. Wiesbaden: VS-Verlag.

Entman, R. M. (1993). Framing: Toward clarification of a fractured paradigm. *Journal of Communication, 43*(4), 51–58.

Etzkowitz, H., & Leydesdorff, L. (2000). The dynamics of innovation: From national systems and "Mode 2" to a Triple Helix of university-industry-government relations. *Research Policy, 29*(2), 109–123.

Fagerberg, J. (2005). Innovation: A guide to the literature. In J. Fagerberg, D. Mowery, & R. Nelson (Eds.), *The Oxford handbook of innovation* (pp. 1–27). Oxford: Oxford University Press.

Feldman, B. (2000). *The Nobel prize: A history of genius, controversy and prestige*. New York: Arcade.

Felt, U. (1993). Fabricating scientific success stories. *Public Understanding of Science, 2*, 375–390.

Foucault, M. (1971). Orders of discourse. *Social Science Information, 10*(2), 7–30.

Fuchs, D., & Pfetsch, B. (1996). Die Beobachtung der öffentlichen Meinung durch das Regierungssystem. In W. van den Daele & F. Neidhardt (Eds.), *Kommunikation und Entscheidung. Politische Funktionen öffentlicher Meinungsbildung und diskursiver Verfahren, WZB-Jahrbuch* (pp. 103–138). Berlin: Edition sigma.

Galtung, J., & Ruge, M. (1965). The structure of foreign news: The presentation of the Congo, Cuba and Cyprus crises in four Norwegian newspapers. *Journal of International Peace Research, 1*, 64–91.

Gamson, W. A., & Modigliani, A. (1989). Media discourse and public opinion on nuclear power: A constructionist approach. *American Journal of Sociology, 95*(1), 1–37.

Gamson, W. A., & Wolfsfeld, G. (1993). Movements and media as interacting systems. *The Annals of the American Academy of Political and Social Science, 528*(1), 114–125.

Gartner. (2004). *Hype cycle for retail technologies: Strategic analysis report*. New York: Gartner Group.

Gerhards, J., & Neidhardt, F. (1991). Strukturen und Funktionen moderner Öffentlichkeit: Fragestellungen und Ansätze. In S. Müller-Dohm & K. Neumann-Braun (Eds.), *Öffentlichkeit, Kultur, Massenkommunikation: Beiträge zur Medien- und Kommunikationssoziologie* (pp. 31–89). Oldenburg: BIS.

Gerhards, J., & Schäfer, M. S. (2006). *Die Herstellung einer öffentlichen Hegemonie: Humangenomforschung in der deutschen und der US-amerikanischen Presse*. Wiesbaden: VS-Verlag.

Gerhards, J., Offerhaus, A., & Roose, J. (2007). Die öffentliche Zuschreibung von Verantwortung. Zur Entwicklung eines inhaltsanalytischen Instrumentariums. *Kölner Zeitschrift für Soziologie und Sozialpsychologie, 59*, 105–124.

Gilbert, N., Pyka, A., & Ahrweiler, P. (2001). Innovation networks: A simulation approach. *Journal of Artificial Societies and Social Simulation, 4*(3), 8.

Habermas, J. (1995). *Strukturwandel der Öffentlichkeit. Untersuchungen zu einer Kategorie der bürgerlichen Gesellschaft*. Frankfurt am Main: Suhrkamp.

Hagen, L. M. (1993). Opportune witnesses: An analysis of balance in the selection of sources and arguments in the leading German newspapers' coverage of the census issue. *European Journal of Communication, 8*(3), 317–343.

Hansen, A. (1994). Journalistic practices and science reporting in the British press. *Public Understanding of Science, 3*(2), 111–134.

Hedgecoe, A., & Martin, P. (2003). The drugs don't work: Expectations and the shaping of pharmacogenetics. *Social Studies of Science, 33*(3), 327–364.

Heinelt, H. (2003). Politikfelder: Machen Besonderheiten von Policies einen Unterschied? In K. Schubert & N. C. Bandelow (Eds.), *Lehrbuch der Politikfeldanalyse* (pp. 239–255). München/Wien: Oldenbourg.

Hekkert, M. P., Suurs, R. A. A., Negro, S. O., Kuhlmann, S., & Smits, R. E. H. M. (2007). Functions of innovation systems: A new approach for analysing technological change. *Technological Forecasting and Social Change, 74*, 413–432.

Hilgartner, S., & Bosk, C. L. (1988). The rise and fall of social problems: A public arenas model. *The American Journal of Sociology, 94*, 53–78.

Holliman, R. (2004). Media coverage of cloning: A study of media content, production and reception. *Public Understanding of Science, 13*, 107–130.

Jacobsson, S., & Bergek, A. (2004). Transforming the energy sector: The evolution of technological systems in renewable energy technology. *Industrial and Corporate Change, 13*(5), 815–849.

Kepplinger, H. M. (1988). Die Kernenergie in der Presse: Eine Analyse zum Einfluss subjektiver Faktoren auf die Konstruktion von Realität. *Kölner Zeitschrift für Soziologie und Sozialpsychologie, 4*, 640–658.

Kepplinger, H. M., Brosius, H.-B., & Staab, J. F. (1991). Instrumental actualization: A theory of mediated conflicts. *European Journal of Communication, 6*(3), 263–290.

Kingdon, J. W. (1984). *Agendas, alternatives and public policies*. Boston: Harper Collins.

Kitzinger, J., & Reilly, J. (1997). The rise and fall of risk reporting: Media coverage of human genetics research, 'false memory syndrome' and 'mad cow disease'. *European Journal of Communication, 12*, 319–350.

Kohring, M. (2005). *Wissenschaftsjournalismus: Forschungsüberblick und Theorieentwurf*. Konstanz: UVK.

Kriesi, H. (2004). Strategic political communication: Mobilizing public opinion in "audience democracies". In F. Esser & B. Pfetsch (Eds.), *Comparing political communication: Theories, cases, and challenges*. Cambridge: Cambridge University Press.

Kuhlmann, S. (1999). Politisches System und Innovationssystem in "postnationalen" Arenen. In K. Grimmer, S. Kuhlmann, & F. Meyer-Krahmer (Eds.), *Innovationspolitik in globalisierten Arenen. Neue Aufgaben für Forschung und Lehre: Forschungs-, Technologie- und Innovationspolitik im Wandel* (pp. 11–39). Opladen: Leske und Budrich.

Legler, H., & Frietsch, R. (2007). Neuabgrenzung der Wissenswirtschaft – Forschungsintensive Industrien und wissensintensive Dienstleistungen (NIW/ISI-Listen 2006). In BMBF (Ed.), *Studien zum deutschen Innovationssystem* (Vol. 22). Berlin: BMBF.

Liu, X., & White, S. (2001). Comparing innovation systems: A framework and application to China's transitional context. *Research Policy, 30*(7), 1091–1114.

Lösch, A. (2006). Means of communicating innovations: A case study for the analysis and assessment of nanotechnology's futuristic visions. *Science, Technology & Innovation Studies, 2*(2), 103–125.

Lowi, T. J. (1972). Four systems of policy, politics, and choice. *Public Administration Review, 32*, 298–310.

Lundvall, B.-A. (1992). *National systems of innovation*. London: Pinter.

Mahon, J. F., & Waddock, S. A. (1992). Strategic issues management: An integration of issue life cycle perspectives. *Business & Society, 31*(1), 19–32.

Malerba, F. (2005). Sectoral systems of innovation: A framework for linking innovation to the knowledge base, structure and dynamics of sectors. *Economics of Information and New Technology, 14*, 63–82.

Mast, C. (2003). *Wirtschaftsjournalismus: Grundlagen und neue Konzepte für die Presse* (2nd ed.). Wiesbaden: Westdeutscher Verlag.

Meier, W. A., & Jarren, O. (2001). Ökonomisierung und Kommerzialisierung von Medien und Mediensystem: Einleitende Bemerkungen zu einer (notwendigen) Debatte. *Medien & Kommunikationswissenschaft, 49*(2), 145–158.

Meldrum, M. J. (1995). Marketing high-tech products: The emerging themes. *European Journal of Marketing, 29*(10), 45–58.

Metcalfe, J. S. (1995). Technology systems and technology policy in an evolutionary framework. *Cambridge Journal of Economics, 19*(1), 25–46.

Meyer-Krahmer, F. (2004). Vorreiter-Märkte und Innovation. Ein neuer Ansatz der Technologie- und Innovationspolitik. In F.-W. Steinmeier & M. Machnig (Eds.), *Made in Germany '21* (pp. 97–109). Hamburg: Hoffmann und Campe.

Meyer-Krahmer, F., & Dreher, C. (2004). Neuere Betrachtungen zu Technikzyklen und Implikationen für die Fraunhofer-Gesellschaft. In D. Spath (Ed.), *Forschungs- und Technologiemanagement. Potentiale nutzen – Zukunft gestalten*. München: Hauser.

Nelson, R. R. (Ed.). (1993). *National innovation systems: A comparative analysis*. New York: Oxford University Press.

Nisbet, M. C., & Huge, M. (2006). Attention cycles and frames in the plant biotechnology debate: Managing power and participation through the press/policy connection. *Press/Politics, 11*(2), 3–40.

Nisbet, M. C., Brossard, D., & Kroepsch, A. (2003). Framing science. The stem cell controversy in an age of press/politics. *Press/Politics, 8*, 36–70.

Noelle-Neumann, E. (1984). *Spiral of silence: Our social skin*. Chicago: University of Chicago Press.

OECD. (1999). *Managing national innovation systems*. Paris: Organisation for Economic Co-operation and Development.

OECD. (2005a). *OECD science, technology and industry scoreboard 2005*. Paris: Organisation for Economic Co-operation and Development.

OECD. (2005b). *Oslo manual: Guidelines for collecting and interpreting innovation data* (3rd ed.). Paris: Organisation for Economic Co-operation and Development.

Pan, Z., & Kosicki, G. M. (2003). Framing as a strategic action in public deliberation. In S. D. Reese, O. H. Gandy, & A. E. Grant (Eds.), *Framing public life: Perspectives on media and our understanding of the social world* (pp. 35–65). Mahwah: Lawrence Erlbaum.

Pfetsch, B., & Adam, S. (2008). Die Akteursperspektive in der politischen Kommunikationsforschung: Fragestellungen, Forschungsparadigmen und Problemlagen. In B. Pfetsch & S. Adam (Eds.), *Massenmedien als politische Akteure* (pp. 9–26). Wiesbaden: VS-Verlag.

Prange, H. (2006). *Wege zum Innovationsstaat – Globalisierung und der Wandel nationaler Forschungs- und Technologiepolitiken*. Baden-Baden: Nomos.

Rammer, C., Polt, W., Egeln, J., Licht, G., & Schibany, A. (2004). *Internationale Trends der Forschungs- und Innovationspolitik – Fällt Deutschland zurück?* Baden-Baden: Nomos.

Ruhrmann, G., Woelke, J., Maier, M., & Diehlmann, N. (2003). *Der Wert von Nachrichten im deutschen Fernsehen: Ein Modell zur Validierung von Nachrichtenfaktoren*. Opladen: Leske + Budrich.

Schavan, A. (2006). Hightech-Strategie der Bundesregierung: Neue Grundlage für Deutschlands Innovationspolitik. Regierungserklärung der Bundesminsterin für Bildung und Forschung vor dem Deutschen Bundestag am 21. September, Berlin.

Schulz, W. (1990). *Die Konstruktion von Realität in den Nachrichtenmedien: Analyse der aktuellen Berichterstattung* (2nd ed.). Freiburg/München: Karl Alber.

Schumpeter, J. A. (1964). *Theorie der wirtschaftlichen Entwicklung* (6th ed.). Berlin: Duncker & Humblot.

Suchman, L., & Bishop, L. (2000). Problematizing 'innovation' as a critical project. *Technology Analysis & Strategic Management, 12*(3), 327–333.

Ten Eyck, T. A., & Williment, M. (2003). The national media and things genetic: Coverage in the New York Times (1971–2001) and the Washington Post (1977–2001). *Science Communication, 25*(2), 129–152.

van Lente, H., & Rip, A. (1998). The rise of membrane technology: From rhetorics to social reality. *Social Studies of Science, 28*(2), 221–254.

Vlasic, A. (2004). *Die Integrationsfunktion der Massenmedien: Begriffsgeschichte, Modelle, Operationalisierung*. Wiesbaden: VS-Verlag.

von Hippel, E. (2005). *Democratizing innovation*. Cambridge, MA: MIT Press.

Waldherr, A. (2008). Gatekeeper, Diskursproduzenten und Agenda-Setter – Akteursrollen von Massenmedien in Innovationsprozessen. In B. Pfetsch & S. Adam (Eds.), *Massenmedien als politische Akteure – Konzepte und Analysen*. VS-Verlag: Wiesbaden.

Weingart, P. (2003). *Wissenschaftssoziologie*. Bielefeld: Transcript.

Weingart, P., & Pansegrau, P. (1999). Reputation in science and prominence in the media: The Goldhagen debate. *Public Understanding of Science, 8*(1), 1–16.

Weischenberg, S., Malik, M., & Scholl, A. (2006). Journalismus in Deutschland 2005. *Media Perspektiven, 7*, 346–361.

Part II
National Systems of Innovation and High-Tech Policies

Chapter 5
Entrepreneurship and Government in U.S. High-Tech Policy

Johannes M. Bauer

5.1 Introduction

The U.S. has long been considered a leader in high-tech industries. Among the factors identified as creating favourable circumstances for innovation are: a culture supportive of risk-taking and entrepreneurship, abundant availability of venture capital, low costs of starting a business, and diverse forms of government support despite the absence of an overarching high-tech policy. Comparative studies in the national innovation systems (NIS) and the varieties of capitalism (VoC) literature (see Werle, Chap. 2) often describe the U.S. as the prototype of a laissez-faire economy that is particularly apt to spawn radical innovations. A closer look reveals, however, historical variations in the relative importance of market forces and government intervention. This flexible and differentiated interaction is seen as one of the distinct features of the U.S. innovation system and its success.

The decisive American lead over other industrialised nations in the post-World War II era has weakened recently. Not only has the U.S. lost a large number of manufacturing jobs during the past decade, it also suffers from a trade deficit in advanced technology products, amounting to $81.8 billion in 2010 (U.S. Census Bureau 2011). The deficit was particularly high in information and communications, as well as opto-electronics and life sciences. In 2008, six OECD member states spent a higher share of their GDP on R&D than the U.S., whose gross expenditures on research and development were 2.8% of its GDP (OECD 2010). And in a recent report benchmarking the global innovation-based competitiveness of the EU and the U.S., the Information Technology and Innovation Foundation (ITIF) asserted that the U.S., while still ahead of the EU overall, was ranked sixth behind Singapore, Sweden, Luxembourg, Denmark and South Korea. Alas, during

J.M. Bauer (✉)
Michigan State University, East Lansing, MI, USA
e-mail: bauerj@msu.edu

J.M. Bauer et al. (eds.), *Innovation Policy and Governance in High-Tech Industries*,
DOI 10.1007/978-3-642-12563-8_5,

the last decade, the U.S. had made the least progress in improving its competitiveness (R. D. Anderson and Andes 2009).

This slipping performance is the outcome of several domestic and international developments. Other countries have increased their efforts and developed considerable presences in high-tech industries, as illustrated, for example, by the emergence of a European aircraft industry, the strength of European and Asian countries in high-speed trains, and the emergence of Asian nations as formidable players in semiconductor manufacturing, consumer electronics and communication technologies. At the same time, U.S. multinational firms have shifted some of their production, including related research and development, to developing and emerging nations. With an increasing share of services in GDP, the U.S. is facing new challenges in sustaining innovation. Some services, particularly highly personalised services, may offer only limited innovation opportunities; some, such as advanced electronic commerce services, require new forms of cooperation and coordination among players, and therefore may not thrive in a laissez-faire environment; and others, such as many internet-based services, face daunting challenges in finding a sustainable revenue model. As international markets for services are less open than those for goods, the U.S. economy, with its high share of services, is in a difficult position (for which the structural current account deficit is but one indication). U.S. policy has created additional domestic hurdles for high-tech industries, for example, by limiting stem cell research. Although several other factors were at work, the economic crisis of 2008 can also be interpreted as a sign of the potential difficulties faced by a large, service-based economy in sustaining innovation, especially when mixed with a laissez-faire market model. Complicated derivatives, once described by Warren Buffett as "financial weapons of mass destruction" (Berkshire Hathaway 2002), can be seen as financial high-tech innovations whose potential risks and effects on the economy are poorly understood.

In response to lingering concerns about lackluster innovation performance, aggravated by the economic crisis of 2008, the Obama Administration released a Strategy for American Innovation (White House 2011). It is the first attempt at formulating a comprehensive high-tech and innovation policy agenda in decades, combining funding for strategic research areas, promotion of markets and entrepreneurship, and measures to catalyse national priorities in areas such as renewable energy, energy efficiency and health IT. Although many implementation details and actual effects remain to be seen, the strategy reflects increasing disappointment with the innovation performance of the U.S. economy and a backlash against laissez-faire policy. However, given the number of rivaling interests and possible veto players, the U.S. will most likely not embrace an interventionist industrial policy, but rather steer a middle course by recalibrating the relative roles of the public and the private sectors.

The first three sections of this chapter discuss the basic elements of U.S. high-tech policy, the strengths and weaknesses of the historical system, and current forces and initiatives that are transforming it. Sections 4–7 focus on the information and communication technology (ICT) sector, which has historically been central to high-tech policy and continues to be one of its cornerstones. Section 8 discusses the challenges faced by U.S. high-tech policy and the complexity of effective coordination in a changing global environment for technology industries. Main points of the chapter are synthesised in the conclusions.

5.2 The U.S. Innovation System

Notwithstanding strong beliefs in unfettered markets, government plays an important role in U.S. innovation policy (Block and Keller 2011). This role has changed over time, but it continues to be multi-faceted and distributed. It is this differentiated, parallel and sometimes redundant nature of efforts that has historically fuelled the dynamism of the U.S. innovation system (Alic et al. 2003). For reasons external and internal to the U.S., this approach currently faces considerable stress. Some of these weaknesses, such as the imbalance of military and civilian research funding, were already visible during the 1970s. Others emerged more recently with the changes in the global economy and the political repositioning of the U.S.. A brief synopsis of the main features and the evolution of this system will help in putting these challenges into context.

Prior to World War II, the states played a proactive role in innovation policy by funding public higher education and extension activities (i.e., knowledge transfer to practice) of many public universities, prominent among them the "land grant" universities (Mowery and Rosenberg 1993). During that time period, research facilities were established in universities and first informal linkages between universities and industry were developed. The federal government assumed only a secondary role in non-agricultural research support. World War II and the subsequent Cold War changed this picture quite dramatically. Largely motivated by national security concerns rather than an economic strategy, the role of the federal government greatly expanded and superseded state involvement. Table 5.1 shows that spending on R&D increased from 1.51% of GDP in 1955 to a peak of 2.82% in 1965, thereafter fluctuating between a low of 2.18% in 1975 and an estimated 2.75% in 2008, the most recent year for which data was available. The data also shows a fundamental shift in the relative importance of various sources of funds. In 1955, government spent $3.6 billion (57.4%), industry $2.5 billion (40.2%) and others, including universities, only $0.2 billion (2.5%). Government's share increased to 65.1% by 1965, when it started a steady decline. Total R&D expenditures for all sources were $398 billion in 2008, to which government contributed 26.1%, industry 67.4% and other sources 6.6%.

During the height of the Cold War, a vast share of more than 80% of federal spending on R&D was defense-related. Although military R&D funding declined somewhat, its share remains high. After dropping to below 50% in the late 1970s, it increased again to nearly 70% during the Reagan Administration. In 2008, the share of Department of Defense research spending stood at 57.3%, which constitutes a lower boundary, as other agencies also contribute to military R&D funding (NSF

Table 5.1 Sources of U.S. R&D funds

Year	Federal US$ 10^6	Industry US$ 10^6	Other US$ 10^6	Federal (%)	Industry (%)	Other (%)	% of GDP	Done by federal	Done by industry	Done by other
1955	3,603	2,522	156	57.4	40.2	2.5	1.5	15.5	73.9	10.6
1960	8,915	4,516	280	65.0	32.9	2.0	2.6	13.1	76.6	10.2
1965	13,194	6,549	511	65.1	32.3	2.5	2.8	15.6	70.0	14.4
1970	14,984	10,449	839	57.0	39.8	3.2	2.5	15.8	68.8	15.4
1975	18,533	15,824	1,314	52.0	44.4	3.7	2.2	15.6	67.8	16.6
1980	29,986	30,929	2,310	47.4	48.9	3.7	2.3	12.4	70.4	17.2
1985	52,641	57,962	4,068	45.9	50.5	3.5	2.7	11.4	73.5	15.1
1990	61,610	83,208	7,175	40.5	54.7	4.7	2.6	10.3	72.2	17.5
1995	62,969	110,871	9,786	34.3	60.4	5.3	2.5	9.2	71.9	18.9
2000	66,417	186,136	14,746	24.8	69.6	5.5	2.7	6.7	75.6	17.7
2005	93,817	207,826	20,461	29.1	64.5	6.4	2.6	7.6	71.0	21.4
2008[a]	103,696	267,847	26,073	26.1	67.4	6.6	2.8	6.8	74.0	19.2

Source: NSF (2010), Appendix Tables 4–3 and 4–7
[a]2008 data preliminary

2010, pp. 4–22). Mowery and Rosenberg (1993) were concerned that this high share of defense R&D put the U.S. economy at a disadvantage compared to industrialised peer nations such as Japan and Germany, countries that dedicated a much higher share of research resources to civilian projects. Although it funded a large portion of research, the U.S. government was not the main locus where it took place. The last three columns of Table 5.1 show that most of the research was performed by private industry and universities. The share of federally performed R&D has declined steadily since the 1970s. It was picked up by universities and projects carried out jointly with federal labs, which are included in the "other" column (NSF 2010).

Since the late 1940s, several distinct periods can be distinguished. In their concise overview of the contribution of government to U.S. innovation activity, Block and Keller (2011) identify three turning points: (1) the major expansion in the late 1940s and 1950s, (2) the decentralization of the federal research support system until the 1980s, and (3) a continued proactive role of government initiatives during the era of market fundamentalism until the economic crisis of 2008. A first major expansion of resources dedicated to R&D happened during and immediately after World War II. This was an extraordinarily productive time with significant breakthroughs both in military and civilian technology (e.g., the computer, the transistor, nuclear power, radar and semiconductors), albeit with very high government funding. During this time, government capacity to pursue scientific work was expanded with the creation of a network of federal laboratories, such as Los Alamos, Lawrence Berkeley, Oak Ridge and Sandia, which had their roots in the Manhattan Project's atomic research programme. Many of the initiatives funding basic research were entrusted to the National Science Foundation (NSF) in 1950.

In the late 1950s, this system was further differentiated and decentralised (Block and Keller 2011, p. 8). Early Russian successes in the space race were answered with the creation of several new agencies, including NASA (National Aeronautics and Space Administration) and DARPA (the Defense Advanced Research Projects Agency). President Kennedy formed the Office of Science and Technology to provide analysis and recommendations mostly related to space technology. In 1976, Congress renamed the office to the Office of Science and Technology Policy (OSTP) and gave it a broader mandate. At the same time, spin-offs from large corporations and universities emerged as a new model for commercializing research. DARPA took advantage of this new organization of the research and innovation process by instigating competition between start-up firms, which stood in contrast to the large and expensive corporate research labs that had dominated earlier. Established firms and their research activities were forced to respond to these challenges with increased efforts of their own. Unlike NSF, DARPA did not rely on elaborate peer review of grant applications, which allowed the agency to fund projects in an agile and flexible fashion (Alic et al. 2003).

A third turning point occurred in the 1980s. Despite the emerging market fundamentalist political attitude, Block and Keller (2011) argue government continued to take a proactive, if not as visible role. The increasing trade deficit provided the backdrop to a discussion of how an active technology and science policy could be used to narrow the deficit. Rival nations such as Japan and Germany had been able to catch up with American firms in several areas, which had negative effects on the U.S.'s foreign trade balance. After the Keynesian consensus had crumbled during the stagflation period of the 1970s, innovation promotion had become rooted in a strong trust in market forces. Two legislative initiatives in 1980 sought to improve technology transfer and commercialization. The Stevenson-Wyler Technology Innovation Act of 1980[1] intended to encourage research collaboration between federal and other institutions and required the federal laboratories to strengthen technology transfer efforts. The Bayh-Dole Patent and Trademark Amendments Act of 1980[2] permitted recipients of federal grants to file patents on the results of such grants and to award exclusive licenses to other parties. An underlying assumption was that such stronger protection would accelerate technology commercialization. Whereas other countries have attempted to emulate the Act, the empirical evidence assessing its effects is weak. Some analysts claim it facilitated the breakthrough of the biotechnology industry,[3] but empirical research has found only small or even negative impacts (Mowery et al. 2004; Rafferty 2008).

Among the most successful programmes established during the 1980s were the Small Business Innovation Research (SBIR) Program and the Small Business Technology Transfer (SBTT) Program. Signed into law by President Reagan in 1982, SBIR made permanent an NSF pilot programme started under Jimmy Carter. Administered by the U.S. Small Business Administration (SBA) Office of Technology, the programme required that government agencies with large research budgets

[1] Stevenson-Wyler Technology Innovation Act of 1980, Public Law 96–480, October 21, 1980. .

[2] Bayh-Dole Patent and Trademark Amendments Act of 1980, Public Law 96–517, December 12, 1980.

[3] E.g., "Innovation's Golden Goose," The Economist, December 14, 2002, p. 3.

dedicate a percentage of their funds to small for-profit start-up companies. In 2010, 11 federal government departments participated in SBIR and five in SBTT, awarding approximately $2 billion to small high-tech companies. Funding of up to $150,000 is available during the initial Phase I of projects. If successful, up to $1 million may be awarded during Phase II. The vast majority of empirical studies confirm that the programme has been successful in stimulating innovation in high-tech start-up businesses (Link and Ruhm 2009; Link and Scott 2010; Audretsch 2003), although a few authors reach more skeptical conclusions (Wallsten 2000). Ambitious programmes, such as the Advanced Technologies Program (ATP) failed, often due to agency conflicts (Negoita 2011). Others, such as the decentralised Manufacturing Extension Program, did not reach their intended profile.

5.3 The Rejuvenation of High-Tech Policy

Faced with a widening trade deficit, President Bush signed the America Creating Opportunities to Meaningfully Promote Excellence in Technology, Education, and Science Act of 2007 (short America COMPETES Act), a comprehensive set of initiatives to improve the international competitiveness of the U.S. The Act was reauthorised in 2010 and signed into law in January 2011.[4] Its provisions strengthen the role of the federal agencies supporting research into advanced technologies. This includes increased funding, new organizational units to better promote innovation, prizes for innovation, and improved dissemination strategies for scientific information. For example, Section 601(a) directs the Secretary of Commerce to establish an Office of Innovation and Entrepreneurship "to foster the innovation and commercialization of new technologies, products, processes, and services to promote productivity and economic growth in the United States". Moreover, the Act establishes to better coordinate activities that support education in science, technology, engineering and mathematics (STEM).

A further major rejuvenation of high-tech policy was launched by the Obama administration in its Strategy for American Innovation (White House 2011). A $780 billion stimulus package was passed in response to the 2008 economic crisis.[5] As part of this package, funding for the national institutes, including NSF and NIH, was substantially increased, infrastructure projects were funded, and large amounts of money were channelled into strategic initiatives. However, the Administration also initiated a coherent overhaul of the fragmented system of high-tech policy. Orchestrated by the overarching motto, expressed by President Obama in his

[4] America COMPETES Reauthorization Act of 2010 or America Creating Opportunities to Meaningfully Promote Excellence in Technology, Education, and Science Reauthorization Act of 2010, Public Law No: 111–358, January 4, 2011.

[5] Authorised in the American Recovery and Reinvestment Act (ARRA) of 2009, Pub. L. No. 111–115, Feb. 19, 2009.

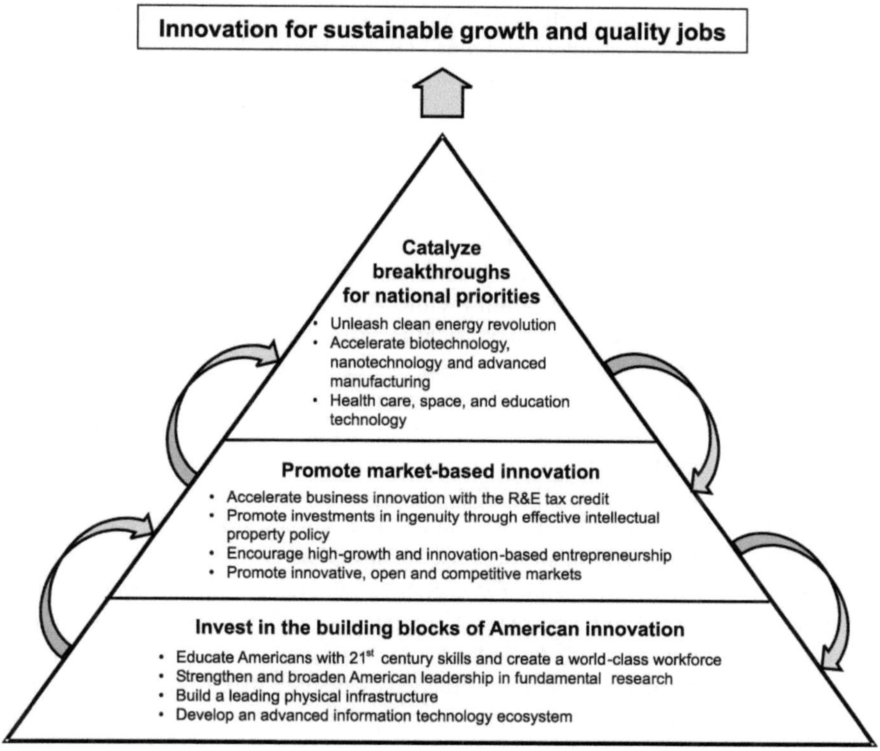

Fig. 5.1 Strategy for American innovation. Source: White House (2011), http://www.whitehouse.gov/innovation/strategy/executive-summary

January 2011 State of the Union address, to "out-innovate, out-build, out-compete, and out-educate" the rest of the world, a broad range of initiatives was designed. The Innovation Strategy contains three major components (see Fig. 5.1): (1) investment in the building blocks of American innovation, (2) the promotion of competitive markets and entrepreneurship, and (3) measures that catalyse breakthroughs for national priorities. The strategy recognises that the bubble economies of the 1990s (built on highly inflated information technology stocks) and of the early 2000s (driven by inflated housing prices and an associated financial services industry) were unsustainable forms of economic growth that need to be replaced with more lasting achievements. It walks a middle ground between massive expansion of government intervention and reliance on unfettered markets, casting government as a convener and a facilitator of robust economic growth.

Investment in the building blocks of American innovation was boosted with $18.3 billion in research funding. In accordance with longer-term goals of increased funding, a doubling of the budgets of the National Science Foundation (NSF), the Department of Energy's Office of Science, and the National Institutes of Standards and Technologies (NIST) was set. Investment in these agencies is seen as a critical input to the development of "industries and jobs of the future, such as the convergence of bio, info, and nanotechnologies" (White House 2011, p. 10). A second long-term goal is to increase economy-wide public and private investment in R&D to more than three percent of GDP. Lastly, the strategy proposes to make the research and experimentation tax credit permanent, a $75 billion relief to businesses. Additional research funding is combined with several initiatives to enhance American education, including the goal to reclaim the global top position in the number of college graduates by 2020, an increase of $200 billion in scholarships over a 10-year period, and the utilization of online learning in continuing, post-secondary education (White House 2009). A third key aspect of this building block is investment in physical infrastructure. Roads, rail, the electricity grid and air traffic systems are in need of massive modernization, a goal that had not been supported well by the liberalization policies since the 1970s.The last component, which will be discussed in more detail in the next section of this chapter, is the development of an advanced information technology ecosystem (White House 2009).

Measures to promote market-based innovation are a second major element of the strategy. Here the administration lists a broad set of aims, ranging from an extension of the Research and Experimentation (R&E) Tax Credit to trade agreements that intend to help double U.S. exports by 2014. Measures to support innovative entrepreneurs include better access to credit, faster patenting, and the Startup America Partnership, which is branded as an "initiative to celebrate, inspire, and accelerate high-growth entrepreneurship throughout the nation."[6] Four measures are packaged to assist start-up companies in crossing the "valley of death", which separates the early phases of business development from a sustainable business model: (1) additional access to venture capital in the form of government matching funds ($2 billion), (2) a simplification of regulations affecting small businesses, (3) mentoring for founders of start-up companies, and (4) tax relief for small businesses. Complementing these drivers of high-tech growth is additional support for innovation hubs, "large, multi-disciplinary, highly-collaborative teams of scientists and engineers working to achieve a specific high priority goal" (White House 2009, Appendix C). For example, Department of Energy innovation hubs have been established for efficient building technology, liquid fuels from sunlight, and the modeling and simulation of nuclear reactors.

The final layer of the national innovation strategy seeks to advance five national priorities: unleashing of a clean energy revolution; acceleration of biotechnology, nanotechnology and advanced manufacturing; breakthrough space capabilities and applications; breakthroughs in health technology; and significant advances in educational technology (White House 2011). Several of the initiatives are modelled after DARPA, widely seen as one of the most successful federal agencies in stimulating high-tech industries. The Advanced Research Projects Agency-Energy

[6] See http://www.whitehouse.gov/issues/startup-america.

(ARPA-E) is the Department of Energy's vehicle to promote clean, secure and independent energy technologies, whereas the Advanced Research Projects Agency-Education (ARPA-ED) is supposed to spearhead similar high-tech projects in the education field. In the health area, a new National Center for Advancing Translational Sciences designed to foster new forms of knowledge transfer between research labs and clinics, was proposed by the National Institutes of Health. A National Nanotechnology Initiative (NNI) carries hopes of enabling revolutionary breakthroughs in nano-electronics. At first glance, these priority areas resemble traditional forms of industrial policy. However, the role of government is conceptualised differently, often as the catalyst, the convener of networks of researchers and of public-private partnerships. The hope is that many of the regulatory and governance issues that arise, for example, in the sharing of medical data, will be addressed by new forms of networked governance (Berejka 2011).

The next four sections of the chapter discuss the ICT sector, which occupies a central position in U.S. high-tech policy. It can be used to illustrate key elements of U.S. policy, particularly, the diversity of agents, their interaction, and the expected and unexpected emergent properties of the entire system.

5.4 The Evolution of U.S. ICT Policy

The structure and performance of the U.S. ICT sector are the outcome of the co-evolution of technology, institutions (including public policy), business strategy, and user decisions. Decisions in each of these areas influenced the further development of coupled realms. Technological decisions affected subsequent policy options and choices; business strategy was enabled by technology developments and public policies; and public policy shaped technology and business strategy. This interaction had sometimes beneficial and sometimes undesirable or unintended consequences. Decisions in each of the realms constrained subsequent options in the others, thus generating forms of path dependency. These processes unfolded in the various segments of the ICT sector, most importantly in components (e.g., semiconductors, microelectronics and optoelectronics), computing and devices (e.g., mainframes, PCs, tablet computers, mobile handsets, and TV sets), networks and network services (e.g., fiber optical networks, mobile communications networks), software to operate the physical networks and devices (e.g., TCP/IP, DOCSIS, Android), as well as applications and end-user services. Not only are these activities high-tech industries in their own right, with a high share of R&D expenditures and high knowledge intensity, they are also important components of high-tech policy directed towards other sectors. ICTs are general purpose technologies that permeate an increasing number of other economic and social activities. By enabling smart appliances and smart grid technology, they are critical for increased energy efficiency. ICTs can reduce the environmental impact of transportation. They facilitate health services and are an indispensible element of advanced manufacturing technologies. At the same time, ICT components are embedded in products and services that are, per se, not considered "high-tech," such as networked refrigerators and other appliances.

High-tech policy has affected and continues to affect all these segments, but it does so in different ways. The configuration of this system has significantly changed since World War II. Historically, ICTs were engineered to provide a narrow range of services. For example, the telephone system was designed to provide high-quality, voice-grade service. Early data communication networks were engineered to support communication between computers. Even the predecessors of computers were designed for relatively specific uses. This gradually changed with the diffusion of digital technology from computing to devices and communication networks. As analog technology was replaced by digital technology, former industry boundaries started to blur. Modern ICT can be seen as layered systems in which physical networks of nodes and communication links enable the configuration and delivery of services and applications (Fransman 2010). Convergence has eliminated many historical boundaries between ICT subsectors with ambiguous effects on industry structure. On the one hand it resulted in considerable industry consolidation. On the other hand, it also coincided with new forms of differentiation and specialization, as illustrated by the vibrant and diverse application market segment.

A deeper understanding of U.S. ICT policy requires examining three aspects of this system: (1) policies toward components and hardware, (2) policies towards networks and basic services, and (3) policies affecting enhanced services and applications. These areas are often complementary to each other. Periods during which these areas were aligned with each other and periods when they were misaligned are visible during the past decades. Tensions have risen because components and hardware, data communications, as well as software and applications, are generally organised as competitive, unregulated industries, whereas networks and basic services have historically been regulated. The internet straddles these areas. Evolving from the unregulated data communications environment, it benefitted from the regulations governing traditional telecommunication networks. The U.S. ICT innovation system has repeatedly proven adaptive to respond to forms of misalignment but often only with considerable friction and delay.

5.5 Computing, Components, and Networking Protocols

Technical advances in components, hardware and protocols were more directly influenced by government actions than any other ICT market segments. During the formative years of these industries and through the height of the Cold War, military public procurement and research funding were primary instruments to promote innovation. More recently, indirect measures such as tax incentives, the facilitation of collaboration, and a strengthening of intellectual property have gained in importance. Government policy had a most decisive effect on the development of computing, was a catalyst in the development of software and the design of protocols to link computers, and was least influential in industries producing components such as semiconductors. The co-evolution of policy, technology and business strategy until the 1990s is documented elsewhere (e.g., Langlois 2002; Ruttan 2006; Flamm 1988). This section draws on this earlier work and supplements it with a brief look at recent developments.

Computing benefited directly from military procurement. During World War II, the military needed technology to aid in the calculation of artillery range tables. The Army Ballistics Research Laboratory (BRL) risked a bet on a new technology by commissioning the development of the first digital computer, the Electronic Numerical Integrator and Calculator (ENIAC), by researchers at the University of Pennsylvania (Ruttan 2006, p. 92). A new computing architecture, anticipated by several researchers and more fully developed by John von Neumann, who was involved in the computer-intensive Manhattan Project, logically separated instructions from the processing of operations, thus laying the groundwork for a separate software industry. A big boost for the commercial computing industry resulted from the funding by the U.S. Air Force of the Semi-Automatic Ground Environment (SAGE) system, a computerised air defense system (Ruttan 2006, p, 95), which in turn was an outgrowth of projects funded by the U.S. Department of Defense. With the emergence of minicomputers and eventually the microcomputer or personal computer (PC), commercial development and demand by the private sector became the main driving force behind further innovation. The rapid price decline in computing is an outcome of the open, modular design of the PC, which allowed many firms to pursue solutions to problems simultaneously, "leading to a rapid trial-and-error learning" (Langlois 2002). The shift to private sector activity was further accelerated by the migration to notebook computers, tablets, mobile devices, and advanced entertainment devices, such as internet-enabled TV sets and game consoles.

Beginning in the 1950s, transistors, and later integrated circuits, replaced vacuum tubes and contributed to the rapid decline in costs of digital technology. The transistor was invented by Bell Laboratories, whose researchers were working on a solution to handle the anticipated volume of telephone calls, for which the prevailing electromechanical switching technology was insufficient. The government was minimally involved in funding research leading to advancements in semiconductor technology. However, as a large buyer it contributed to and stabilised demand. Most importantly, the government bought from new specialised suppliers that could meet its needs, supporting new entrants such as Texas Instruments (TI) and Fairchild. These two companies were the birthplaces of the planar process of manufacturing integrated circuits. In part to avoid antitrust action, TI and Fairchild made their technology available to all new firms at relatively favourable licensing terms. The resulting open environment allowed other firms that were able to successfully commercialise the planar process to capture part of the innovation rents (Langlois 2002). U.S. leadership was challenged by Japan during the 1980s, in particular in the VSLI initiative. In response, SEMATECH (Semiconductor Manufacturing Technology) was established as a public-private partnership between the government and 14 private sector companies. Until the mid-1990s the public sector provided $100 million in annual matching funds to help develop processes and technologies beyond the capabilities of each individual participant. Experts are divided in their assessment of the success of SEMATECH, with some authors suggesting positive effects (Link et al. 1996; Grindley et al. 1994) but others arguing that public funding reduced private R&D expenditures (Irwin and Klenow 1996). In the mid-1990s the government withdrew funding and SEMATECH broadened its scope to international members, hence diluting the effect it might have on U.S. manufacturers.

Government also served as a catalyst for software and protocols needed to network computers, which eventually led to the emergence of the internet. Early funding of networking research was provided by the Department of Defense, which was interested in dual-use technology that could also serve civilian purposes (Langlois and Mowery 1996). Other federal agencies, such as the Department of Energy, NASA and NSF, were also involved. Of key importance in these early efforts was DARPA, an agency with relatively flexible spending rules that could fund projects bypassing lengthy peer reviews. The multitude of agencies and research centers led to duplicate and parallel efforts that allowed for simultaneous search in different directions, thus enhancing the chances of successful innovation (Langlois 2002). A standardised open networking protocol – TCP/IP – emerged from these initiatives and was deployed in ARPANET. In 1981, NSF started a parallel effort to network computer science departments in its Computer Science Network (CSNET). The agency's efforts to provide access to NSF's supercomputing centers and other computing-intensive research centers led to the deployment of NSFNET, an advanced network linking these centres, in 1985. NSFNET also used TCP/IP and eventually, beginning in 1990, ARPANET was subsumed into NSFNET. In the early 1990s, a gradual transition to private operation started, leading to the decommissioning and full privatization of NSFNET in 1995. Concern by researchers about the implications of privatization for higher education and research led to the formation of the Internet2 consortium. By 2011, the consortium had grown to include several hundred universities, corporations, and government agencies in the U.S. and in 50 other countries.[7] Since the 1960s, the role of the U.S. government in this realm of ICT has therefore changed from early catalyst to participant (and indirect funder) in broad-based joint private-public efforts.

5.6 Telecommunication Networks and Services

Much of high-tech policy seeks to shape the unregulated activities in the ICT system. Nonetheless, the older, more regulated telecommunications networks and services are an important aspect and help to understand some of the successes and

[7] See http://www.internet2.edu/about, retrieved 5 April 2011.

failures of U.S. ICT policy. Until the gradual migration to flatter IP-based architectures beginning in the 1990s, telecommunication networks were highly centralised and hierarchically organised. The mix of nodes and links varied with the relative cost of these two components. Much of the functionality (the "intelligence") of the services was embedded in the networks. Some of the finest research laboratories, such as Bell Labs, were operated by telecommunications companies. The centralised organization of the industry was supported by its monopolistic legal and regulatory framework. Based in a shared believe that telecommunications constituted a "natural monopoly," a situation in which technological and economic conditions are such that the lowest cost of serving demand is to pool all supply in one organization, the sector was operated by privately owned, government regulated firms. Eventually, however, technological and policy change heralded a new era in which the monopolistic organization gave way to more open markets and diverse technological platforms. In the U.S., this change unfolded over the course of several decades during which significant tensions and conflicts between the principles governing highly regulated telecommunications services and those governing essentially unregulated data communication services, as well as the boundaries of the organizations providing these services, had to be resolved (Brock 1981, 1994, 2003; Schneider 2001).

Telecommunications services, particularly voice communications, were historically treated as common carriers. Firms that are so designated have certain obligations (e.g., to provide services at non-discriminatory conditions) and rights (e.g., an opportunity to earn a fair return on the invested capital) that go beyond those of commercial firms. When data communications services became available, these common carriers were often slow in making the inputs needed for the new services, such as leased lines and modems, available. Some incumbents abused their strong market position to thwart competition. Consequently, the largely unregulated computing industry and users of the new services sought to address these problems in the public policy arena. Both antitrust and regulatory action were undertaken to mitigate the problems. In 1956, AT&T signed a Consent Decree with the U.S. Department of Justice in which it agreed to focus on common carrier voice services and to license technology related to data communications freely to other companies including the Unix operating system and the C programming language. Both technologies facilitated innovation in the emerging nascent ICT industries (Mowery and Simcoe 2002).

By the late 1970s, the telephone companies recognised the growth opportunities in data communications. At the same time, regulatory agencies were looking for ways to reduce the realm of regulated monopoly activities. Problems related to the presence of the dominant AT&T in both competitive and non-competitive markets were addressed with the break-up of the company in 1984. Again the agreement created a beneficial, but largely unanticipated, consequence by facilitating competition for leased lines and other data communications services and allowing companies providing these services to connect to local networks at regulated, reasonable conditions. The Telecommunications Act of 1996 furthered these causes by adopting a broad range of measures intended to expand competition to the last monopoly bastion – local exchange markets – and to support the further growth of data communications services and the internet in a market environment "unfettered by state and federal regulation."[8] As part of the Act's implementation, the Federal Communications Commission (FCC) continued the exemption of information service providers from per-minute access charges to local networks, which allowed the budding dial-up Internet Service Providers (ISPs) to connect modem banks to local networks at low, unmetered rates. As local phone services typically also were available at flat rates to end-users, dial-up internet service became available at relatively low, flat prices. Supported by these fortuitous conditions, dial-up internet diffused in the U.S. much faster than in other regions (Greenstein 2005).

In the early 1990s, there was widespread concern among U.S. policymakers that the country was losing the race toward modernizing communication networks. The National Telecommunications and Information Administration (NTIA) published a comprehensive report and strategy to improve sector performance and infrastructure investment (NTIA 1991). One of the key suggestions was to expand the realm of competition. Although not solely a response to this concern, the Telecommunications Act of 1996 followed that vision and established conditions to facilitate competition in local markets. This goal was initially implemented with stringent regulatory measures allowing resale of existing services and by requiring incumbent local exchange carriers (ILECs) to make network and service elements available on an unbundled basis at low, regulated prices. By the early 2000s, empirical evidence suggested that these measures had attracted new service-based competitors but had depressed incentives for new infrastructure investment. Confronted with the need to upgrade network infrastructure to broadband capacity and to roll-out next-generation networks, the FCC, in part prodded by the courts, reclassified the technologies capable of providing broadband access (cable modem service, Digital Subscriber Line service, broadband over powerline (BPL), and wireless broadband) from common carrier status to essentially unregulated information services (Bauer 2005; Bauer and Bohlin 2008).

These measures were designed to unleash the private sector's ability to invest and innovate in network infrastructure. They did increase investment but some of the hopes were not met. Rather, the weaknesses of a purely market-driven approach became more visible. For example, the expansion of networks to rural areas has been progressing slower than anticipated. Network upgrades are initially targeted to urban centers and market segments, such as mobile broadband, for which demand is high. One could argue that these policies have generated undesirable and unexpected consequences for ICT development by constraining further innovation dependent on more advanced network infrastructure. As in earlier cases, the

[8] The history of U.S. telecommunications policy reform is multifaceted and evolved in a highly complicated policy-making environment with many checks and balances between the main actors. This system facilitates challenges to existing policies but makes broad and sweeping overhauls more complicated. Compared to the European Union (EU), U.S. reform started much sooner but unfolded in a much more gradual and slower pattern.

policy-making system has responded to the perceived shortcomings. The Obama Administration has dedicated $7.2 billion of stimulus funding to support network upgrades and has launched additional initiatives to promote infrastructure modernization. For the first time since 1992, the U.S. has put together an ambitious National Broadband Plan, including a plan to expand wireless broadband. In contrast to the early 1990s, the measures proposed to overcome the perceived innovation shortfall entail an increase, rather than a decrease, in government involvement, including subsidies to roll-out networks to rural areas and public-private partnerships to accelerate adoption of new services (FCC 2010).

5.7 Software, Applications and Services

Hardware and the networks linking nodes and devices are intermediate technologies that enable applications and services, which are the ultimate sources of value generated by ICTs. In this vast and sprawling subsector, government policy was least present and other aspects of the U.S. innovation system are more critical in shaping the overall direction of innovation. The early investment in computer science departments probably had a positive effect on the development of a knowledge base in software development. The close relation between universities and business as well as the ease with which new firms could be established facilitated a vibrant culture of entrepreneurship. Many new firms were spun-off from universities and became the cores of entrepreneurial clusters such as Silicon Valley (California), Route 128 (Massachusetts), and the Research Triangle (North Carolina). Until the late 1960s, most software was bundled by computer manufacturers with the hardware and given away for free. This situation started to change slowly in the 1960s as more specialised software needs were met by products developed by software companies. When IBM announced in 1969 that it would start to unbundle and separately sell software, the industry received a further boost that sent it on a fast growth path. By the 1980s, software was on a steep growth path, but further change was in the make. The digitization of communication networks, and particularly the broad diffusion of the internet beginning in the mid-1990s, when the first browsers started to simplify its use, supported the emergence of new types of software-based services and applications on the edge of the network, epitomised in the concept of cloud computing. Growth in web applications was further propelled by the fast adoption of social media and wireless data communications and the mobile internet.

Government policy played a fairly limited direct role in these developments. At present, research funding by the military and federal agencies supports the development of advanced knowledge in computer science, simulation software, video games, and specialised areas such as health informatics or advanced graphics. In 2007, all federal agencies contributed slightly less than 2.5% of total R&D funding in the area of software and computer-related services (and only 0.5% to computer and electronic products) (NSF 2010). The remainder of R&D funds was generated by the private sector. The situation is similar in application markets, where the share of privately financed R&D may be even higher. Web applications use browsers to access tools such as online calendars, webmail, document processing, and electronic commerce functions. They first emerged in the fixed internet, where a number of open development frameworks were available. Like advanced communications systems, web applications consist of multiple tiers, often a presentation tier (the browser), an application tier (the functions, programmed in languages such as Java, PHP, or Ruby on Rails), and a storage tier (database).

Public policy has had an indirect – and most likely inadvertent – stimulating effect by promoting an open internet platform. Network and development platform openness has facilitated modular types of innovation and contributed to the vigorous growth of the application market. Open standards and open networks allow "permission-less" innovations at the edge of the internet, enabling entrepreneurs to design new applications without having to incur the potentially high transaction costs of negotiating with multiple network operators and service providers (Van Schewick 2010). The introduction of user-friendly mobile internet devices, such as the iPhone and other smart phones, has similarly boosted the market for mobile web applications. However, as the example of the iPhone and the rivaling open Android platform illustrate, openness is not a necessary condition for innovation. Developers have created numerous applications for both the walled-garden iPhone environment, controlled by Apple and based on its proprietary iOS mobile operating system, and the open Android environment, supported by Google. A key question is whether public policy makers ought to mandate network and platform openness. In the U.S., as in other countries, this has fueled an intense controversy conducted under the heading of network neutrality. Although this is primarily a discussion of communications policy principles, it also often refers to the effects of regulatory choices on innovation and U.S. competitiveness. In December 2010, the FCC adopted rules that proscribe certain minimal safeguards for open networks, but it promulgated less stringent rules for mobile network operators given the more urgent concerns about capacity constraints of mobile networks (FCC 2010).[9] However, this area continues to be in flux, with challenges to the FCC Order pending in appeals courts.

[9] Specifically, the FCC adopted three principles for fixed broadband access network operators: (1) no blocking of traffic and applications, (2) transparency of rules governing network management, and (3) non-discrimination, except in relatively limited scenarios, such as security concerns or binding capacity shortages, where network management is required. Mobile operators are subject to no blocking and transparency obligations but not the third one (FCC 2010).

5.8 Prospects and limits of high-tech policy

During the second half of the twentieth century, the U.S. ICT sector followed a strong performance trajectory, although considerable variations existed over time in the three subsectors. Toward the end of the millennium, competition from firms in other regions had considerably intensified and other nation's innovation policies had narrowed the gap between them and the U.S. In all three areas discussed above, global leadership became more contested and changed among the regions toward the end of the millennium. The time pattern varied between the subsectors. Until the 1990s, the U.S. could boast one of the most widely available and efficient network infrastructures, a considerable achievement given the size of its territory and the presence of vast rural areas with very low population density.[10] This strong performance became more lackluster during the 1990s. As other countries and regions revamped their historically monopoly-centric systems, they were able to unleash substantial efficiency improvements. In fixed broadband networks, South Korea and Japan pursued aggressive roll-out programmes, often orchestrated in public-private partnerships, followed by strict regulation, which put them in a global leadership position (Frieden 2005). Several smaller European countries, including the Netherlands, Sweden and Denmark, also adopted policies in support of rapid broadband deployment.

During the 1990s, as a consequence of enlightened policy choices, such as a Europe-wide standard for mobile communications (GSM), the European Union took the global lead in mobile voice communications. However, this position did not last and as mobile communications moved from voice toward a first generation of mobile data communications and mobile internet access, Japan and South Korea once again offered the most conducive environment. In the ongoing transition toward next-generation fixed and mobile broadband communications, the U.S. has regained a strong position (Falch et al. 2010). This position is stronger in mobile communications, where the country is arguably leading the charge toward new services and applications, and is somewhat more modest in next-generation fixed networks, where the weaknesses of a purely market-driven model have become visible. For example, while investment in urban areas has been strong, network upgrades in rural and sparsely populated areas are slower than anticipated. Furthermore, during the past two decades, a significant shift in the locus of R&D has occurred. Whereas network providers operated important research laboratories that contributed major inventions, since the 1980s the R&D intensity of network operators has diminished.[11] Much of the R&D has migrated to device manufacturers and software vendors (Fransman 2010).

---

[10] Throughout most of the twentieth century, Sweden held the top position with regard to ICT infrastructure availability and efficiency.

[11] Researchers at Bell Laboratories invented the transistor and made major contributions in a number of fields, including laser technology (used in fiber communication networks), mobile cellular and wireless local area technology, the C programming language, and sensor and imaging

The U.S. position in computing and components was somewhat more volatile and was successfully challenged, at least temporarily, in some areas. Japan made inroads into the semiconductor and consumer electronics business during the 1980s and other Asian countries followed in the 2000s. The U.S. regained some of its prowess in semiconductors when companies such as Motorola, Intel, and Texas Instruments accelerated efforts with Dynamic Random Access Memory (DRAM) technology. Through 2000, the U.S. share in global semiconductor sales recovered but manufacturers in the Asia-Pacific region have been able to increase their global share steadily ever since. America's position was probably least contested in software and in web applications, although even in some of these segments the competitive landscape shifted in favour of other nations. The three global leading software vendors are U.S. firms (Microsoft, IBM, Oracle) and U.S. firms hold dominant positions in search markets (Google), browsers (Microsoft, Mozilla, Google) and cloud computing. In some of these markets, open source software has captured a significant market share and foreign vendors have gained in selected national and regional markets (e.g., search engine Baidu in China). In applications and services, many innovative foreign competitors have emerged in the past 10 years. For example, European firms have been able to capture noticeable market shares in business software (e.g., SAP), browsing (e.g., Opera) and in online music markets (e.g., Deezer).

Public policy influenced the historical innovation performance of the U.S. ICT sector in multiple ways. However, as the technology and its applications become increasingly complex, future policies will be exceedingly difficult to design and implement. Software-based innovation at the edges of the network has greatly increased the combinatorial space for technological advances (Arthur 2009). Complex adaptive systems theory would suggest that diversity of public policy interventions and an institutional environment that supports public policy and commercial experimentation would stimulate innovation (Beinhocker 2006). The diversity and multifaceted nature of U.S. innovation policy, therefore, seems to be an important contributing factor. One could argue that the U.S. innovation system, in part by deliberate design, in part by serendipity, was structured in non-linear fashion long before this became recognised by the mainstream R&D policy (OECD 2005). The parallel and duplicative activities by different government agencies, the interaction between federal research laboratories, universities, and private sector

technology. Seven Physics Nobel Prizes were awarded to researchers at Bell Labs. After the break-up of the Bell System in 1984, Bellcore (now Telcordia) was spun-off from Bell Labs to produce similar research services for the newly established Regional Bell Operating Companies (RBOCs) that focused on the provision of local exchange carriers. In 1996, AT&T divested AT&T Bell Labs, integrating it into a new company, Lucent Technologies. Lucent Technologies became part of Alcatel-Lucent in 2006.As a consequence of the merger, Lucent Bell Laboratories and Alcatel Research and Innovation were integrated into a new Bell Laboratories. However, in 2008, only four physicists remained employed in basic research functions and the company announced that it would withdraw from basic research to focus on more immediately marketable areas such as networking technology and high-speed electronics.

researchers, and a culture of entrepreneurship and risk-taking generated a highly dynamic co-evolutionary system. Furthermore, during several post WWII decades, national security interests, a powerful motivating force of public policy, and high-tech innovation in ICT were mutually reinforcing. This congruence facilitated support for R&D in computing, semiconductors, networking, and software. As the Cold War receded, this parallelism also weakened. In other areas of R&D policy, for example, environmental policy, energy efficiency, and health IT, the nexus between national interests and R&D is more difficult to establish in a compelling fashion. This is illustrated in the latest fiscal policy documents by the House leadership that propose a return to purely market-driven energy policy without regard for the innovation opportunities that a more aggressive policy might open.

Whereas the very successful interaction and collaboration of government, universities and the private sector continues to show considerable dynamic, it also exhibits signs of strain. Significantly reduced military R&D spending has only partially been compensated by higher funding for the National Science Foundation and other federal institutes. Given current fiscal pressures, significantly higher levels of government spending seem out of reach for the near future. Private sector spending can be stimulated with tax incentives such as the extended F&E tax credit and simplifications of regulations affecting innovation. The goal of increasing U.S. R&D spending to above 3% of GDP seems ambitious but within reach. However, one has to keep in mind that other countries also envision increases and that the set mark is not international best practice. Another problem for the U.S. innovation system might be that the low-hanging fruit have been harvested, as Cowan (2010) argues. If indeed technical change has reached a plateau with lower rates of change and fewer innovation opportunities, higher R&D efforts may not translate into more innovation.

In ICT industries, these concerns are aggravated by three factors: (1) the shifting of production and some of the R&D to overseas locations, (2) the ease with which some software-based innovations can be imitated by competitors, and (3) the difficulty of finding sustainable business models in markets with very low incremental costs but high costs of generating a first copy of a product or service. Globalization of supply chains has weakened the domestic effects of knowledge generation. On the one hand, the production activities and jobs associated with knowledge-intensive products are often located abroad. For example, whereas Apple has manufacturing plants in the U.S. and in Europe, a very large share of its products is produced in Asia. On the other hand, after an initial design is created, many subsequent product and service innovations emerge from the practical knowledge generated in the production process (Stoneman 2010). Offshore production limits these learning effects, which may weaken longer-term innovation performance of a nation.

Innovation is in part fuelled by the ability of companies to recoup temporary innovation rents. In industries where imitation is easy, innovations may not have the same multiplier effect on production and jobs compared to industries where imitation is more cumbersome. This effect is aggravated if production is internationally mobile. In the long run the innovation rate in such industries may even be lowered. Products and services in several ICT market segments, such as software or software-based applications, can be imitated relatively easily. This is one of the reasons why intellectual property right protection has become an important issue. However, the optimal scope and duration of protection is difficult to determine and subject to considerable controversy (Jaffe and Lerner 2004). In areas where complementary skills and knowledge are important, proprietary knowledge may even diminish innovation. An increasing number of firms are therefore making some of their intellectual property available freely to others in the hope of benefiting from indirect spill-backs. In other areas, the fragmentation of intellectual property rights may create barriers to further innovation built on such knowledge. Patent thickets are often generated by firms who want to fend off competitors, but they also reduce the set of available innovation opportunities by narrowing "adjacent possibilities" for other players.

Lastly, many segments of ICT have peculiar cost structures: high costs to generate a first copy and low or even zero incremental costs. Competitive pressure in such markets will relentlessly drive market prices to incremental costs. Some firms may be able to build sustainable business models around giving away a software package or a service for free. This can be done by deriving an income stream from advertising, by bundling a free product with one that is more difficult to imitate, or by various models of price and product differentiation (Shapiro and Varian 1999; Anderson 2009). However, it must be doubted whether all firms in the ICT sector can pursue such a strategy, especially because digital technology allows unlimited copying, whether legal or illicit, that may further undermine the ability to derive a revenue stream. Thus, compared to manufacturing industries, technology and cost characteristics inherent to ICT challenge the notion that the generation of advanced knowledge and innovation will subsequently improve economic growth and employment.

5.9 Conclusions

After two decades of strong belief in unfettered markets, during which the U.S. has slipped in terms of high-tech performance, the country has begun to reshape its approach to high-tech policy. This process started under the previous Republican administration but it has become a much more concerted effort under the Obama Administration. Even during the period of high trust in unregulated markets, the government shaped private sector innovation policy in many ways. Nonetheless, the massive public funding for R&D associated with the Cold War has declined substantially, although the private sector has picked up a large part of the share. It is unlikely that similar levels of overall public funding will be forthcoming, despite strong commitments to increase resources channeled through the major federal research agencies, such as the NSF, NIH and NIST. Given the current absence of bipartisan politics and the country's fiscal problems, it is even questionable whether the Administration's innovation strategy will be funded at the desired level. The country does not lack the instruments to pursue a forward-looking high-tech innovation strategy, but rather the political consensus necessary to find feasible solutions.

Despite these handicaps, the country's entrepreneurial culture, the willingness of individuals and organizations to take risk, continues to be a strong force of economic dynamics. However, market forces alone do not channel this energy into the full range of projects that have a high payoff for the welfare of future generations, such as clean energy. In international comparison, the U.S. has not necessarily slipped back, but other nations and regions have gained – some, such as the Asian countries, with a policy mix quite different from the U.S. approach. Whereas the Anglo-Saxon model may have lost in credibility and the performance of other approaches has been respectable and occasionally even better, gradual rather than radical reform of that model with more coherently designed interaction between government, universities, and industry may be the only way forward for the U.S. in the near future. According to emerging policy visions, an important role of government is to serve as a convener and facilitator of private sector innovation networks. Although this seems to be a step in the right direction, it will likely not suffice to propel U.S. high-tech performance back to the pole position. The complexity of high-tech industries will require additional measures, such as continued research funding, support for the adoption of advanced technologies, and reform of the educational system. Most importantly, the country would benefit from an overarching framework for the continuous refinement, monitoring, and adaptation of high-tech policy.

References

Alic, J. A., Mowery, D. C., & Rubin, E. S. (2003). *U.S. technology and innovation policy: Lessons for climate change*. Washington, DC: Pew Center on Global Climate Change.

Anderson, C. (2009). *Free: The future of a radical price*. New York: Hyperion.

Anderson, R. D., & Andes, S. M. (2009). *The Atlantic century: Benchmarking EU and U.S. innovation and competitiveness*. Washington, DC: The Information Technology and Innovation Foundation.

Arthur, W. B. (2009). *The nature of technology: What it is and how it evolves*. New York: Free Press.

Audretsch, D. B. (2003). Standing on the shoulders of midgets: The U.S. Small business innovation research program (SBIR). *Small Business Economics, 20*(2), 129–135.

Bauer, J. (2005). Unbundling policy in the United States: Players, outcomes and effects. *Communications & Strategies, 57*, 59–82.

Bauer, J., & Bohlin, E. (2008). From static to dynamic regulation: Recent developments in U.S. telecommunications policy. *Intereconomics: Review of European Economic Policy, 43*(1), 38–50.

Beinhocker, E. D. (2006). *The origin of wealth: Evolution, complexity, and the radical remaking of economics*. Boston: Harvard Business School Press.

Berejka, M. (2011). *Multi-stakeholder-ism: Making it fly*. Paper presented at the Silicon Flatirons Conference The Digital Broadband Migration: The Dynamics of Disruptive Innovation, February 13–14, 2011, Boulder, Colorado.

Berkshire Hathaway (2002). Annual report. Omaha, NE.

Block, F., & Keller, M. R. (Eds.). (2011). *State of innovation: The U.S. government's role in technology development*. Boulder/London: Paradigm Publishers.

Brock, G. W. (1981). *The telecommunications industry: The dynamics of market structure*. Cambridge, MA: Harvard University Press.

Brock, G. W. (1994). *Telecommunications policy for the information age: From monopoly to competition*. Cambridge, MA: Harvard Business School Press.

Brock, G. W. (2003). *The second information revolution*. Cambridge, MA: Harvard University Press.

Cowan, T. (2010). *The great stagnation: How America ate all the low-hanging fruit of modern history, got sick, and will (eventually) feel better*. New York: Dutton.

Falch, M., Henten, A., & Vandrup, K. (2010). Mobile internet developments in Europe, East Asia, and the US. In A. Gentzoglanis & A. Henten (Eds.), *Regulation and the evolution of the global telecommunications industry* (pp. 317–340). Cheltenham/Northampton: Edward Elgar.

FCC (2010). In the matter of preserving the open internet; broadband industry practices. GN Docket No. 09–19; WC Docket No. 07–52, Report and Order, December 23, 2010 (Preserving the Open Internet Order). Washington, DC: Federal Communications Commission.

Flamm, K. E. (1988). *Creating the computer: government, industry and high technology*. Washington, DC: Brookings Institution Press.

Fransman, M. (2010). *The new ICT ecosystem: Implications for policy and regulation*. Cambridge: Cambridge University Press.

Frieden, R. (2005). Lessons from broadband development in Canada, Japan, Korea and the United States. *Telecommunications Policy, 29*(8), 595–613.

Greenstein, S. (2005). The economic geography of Internet infrastructure in the United States. In S. Majumdar, I. Vogelsang, & M. Cave (Eds.), *Handbook of telecommunications economics, volume 2: Technology evolution and the Internet* (pp. 287–372). Amsterdam: Elsevier.

Grindley, P., Mowery, D. C., & Silverman, B. (1994). SEMATECH and collaborative research: Lessons in the design of high-technology consortia. *Journal of Policy Analysis and Management, 13*(4), 723–758.

White House. (2009). *A strategy for American innovation: Driving towards sustainable growth and quality jobs*. Washington, DC: Executive Office of the President, National Economic Council, Office of Science and Technology Policy, The White House.

White House. (2011). *A strategy for American innovation: Securing our economic growth and prosperity*. Washington, DC: Executive Office of the President, National Economic Council, Office of Science and Technology Policy, The White House.

Irwin, D. A., & Klenow, P. J. (1996). High-tech R&D subsidies: Estimating the effects of Sematech. *Journal of International Economics, 40*(3–4), 323–344.

Jaffe, A. B., & Lerner, J. (2004). *Innovation and its discontents: How our broken patent system is endangering innovation and progress and what to do about it*. Princeton: Princeton University Press.

Langlois, R. N. (2002). Computers and semiconductors. In B. Steill, D. G. Victor, & R. R. Nelson (Eds.), *Technological innovation and economic performance* (pp. 265–284). Princeton: Princeton University Press.

Link, A. N., & Ruhm, C. J. (2009). Bringing science to market: Commercializing from NIH SBIR awards. *Economics of Innovation and New Technology, 18*(3–4), 381–402.

Link, A. N., & Scott, J. T. (2010). Government as entrepreneur: Evaluating the commercialization success of SBIR projects. *Research Policy, 39*(5), 589–601.

Link, A. N., Teece, D. J., & Finan, W. F. (1996). Estimating the benefits from collaboration: The case of SEMATECH. *Review of Industrial Organization, 11*(5), 737–751.

Mowery, D. C., & Rosenberg, N. (1993). The U.S. National Innovation System. In R. R. Nelson (Ed.), *National Innovation Systems: A comparative analysis* (pp. 29–75). New York/Oxford: Oxford University Press.

Mowery, D. C., & Simcoe, T. (2002). The Internet. In B. Steil, D. G. Victor, & R. R. Nelson (Eds.), *Technological innovation and economic performance* (pp. 229–264). Princeton: Princeton University Press.

Mowery, D. C., Nelson, R. R., Sampat, B. N., & Ziedonis, A. A. (2004). *Ivory tower and industrial innovation: University–industry technology transfer before and after the Bayh–Dole Act in the United States*. Stanford: Stanford University Press.

Negoita, M. (2011). To hide or not to hide? The advanced technology program and the future of U.S. civilian technology policy. In F. Block & M. R. Keller (Eds.), *State of innovation: The U.S. government's role in technology development* (pp. 77–95). Boulder/London: Paradigm Publishers.

NSF. (2010). *Science and engineering indicators 2010*. Washington, DC: National Science Foundation, Division of Science Resources Statistics, Survey of Industrial Research and Development.

NTIA. (1991). *The NTIA infrastructure report: Telecommunications in the age of information*. Washington, DC: U.S. Dept. of Commerce, National Telecommunications and Information Administration.

OECD. (2005). *Governance of innovation systems*. Paris: Organisation for Economic Co-operation and Development.

OECD. (2010). *Main science and technology indicators, OECD science, technology and R&D statistics (database). doi: 10.1787/data-00182-en, (Retrieved 24 April 2011)*. Paris: Organisation for Economic Co-operation and Development.

Rafferty, M. (2008). The Bayh-Dole Act and university research and development. *Research Policy, 37*(1), 29–40.

Ruttan, V. W. (2006). *Is war necessary for economic growth? Military procurement and technology development*. Oxford/New York: Oxford University Press.

Schneider, V. (2001). *Die Transformation der Telekommunikation. Vom Staatsmonopol zum globalen Markt (1800–2000)*. Frankfurt am Main: Campus.

Shapiro, C., & Varian, H. R. (1999). *Information rules: A strategic guide to the network economy*. Boston: Harvard Business School Press.

Stoneman, P. (2010). *Soft innovation: Economics, product aesthetics, and the creative industries*. Oxford: Oxford University Press.

U.S. Census Bureau. (2011). *U.S. bureau of economic analysis news, U.S. international trade in goods and services, December 2010*. Washington, DC: U.S. Department of Commerce.

Van Schewick, B. (2010). *Internet architecture and innovation*. Cambridge, MA: MIT Press.

Wallsten, S. J. (2000). The effects of government-industry R&D programs on private R&D: the case of the small business innovation research program. *RAND Journal of Economics, 31*(1), 82–100.

Chapter 6
Innovation Switzerland: A Particular Kind of Excellence

Beat Hotz-Hart

6.1 Introduction

The internationalisation and globalisation of business implies increased transparency of national economic activities and improved comparability in terms of costs and market prices, which therefore challenges nations and national firms in several ways. Opening up of markets leads to greater mobility but requires faster adjustment and restructuring. Businesses increasingly combine and control value-added beyond national borders, organising value chains transnationally. In the process, traditional exports are beginning to be side-lined by new forms of international cooperation, such as direct investments, joint ventures, and strategic alliances.

For national economies, this causes a shift from competition between (domestic) companies and products to rivalry between business locations. The international competitiveness of a location is therefore increasingly important for employment and prosperity, determining the ability to provide attractive work and income for its inhabitants. Companies at a given location have to position their business activities more consistently on world markets and in transnational value chains (Hotz-Hart et al. 2006, p. 15).

In the case of a highly developed economy such as Switzerland, boasting a high income level and a hard currency, this cannot, or only to a very limited extent, be achieved by means of cost competition. It is far more likely to be accomplished through competition in terms of quality and innovation. Firms in a location have to be able to offer products and services to an international clientele interested in innovative and unique products for which the buyer is willing to pay a premium price. This presupposes a high capacity for innovation, encompassing the implementation of each new, useful idea from its inception to successful application in

B. Hotz-Hart (✉)
University of Zurich, Zurich, Switzerland
e-mail: beat.hotz-hart@ethrat.ch

J.M. Bauer et al. (eds.), *Innovation Policy and Governance in High-Tech Industries*,
DOI 10.1007/978-3-642-12563-8_6, © Springer-Verlag Berlin Heidelberg 2012

the market. It also requires an ability to generate innovations quicker and better than competitors, and to maximise potential in international markets.

As the situation in highly developed economies is quite similar, there is much common ground, but also fierce competition between locations. This is where national policy plays a pivotal role. The main competing nations and locations in Europe and elsewhere are increasing their efforts to improve their capacity for innovation. This has given rise to a race between countries to gain at least a temporary innovation advantage.[1] However, the best policy approach and the role of the state are hotly debated (see Ergas 1987; Grande 2000; Dolata 2004). Four models can be distinguished. In some countries an active industrial policy is advocated and pursued. This policy aims at helping new businesses get off the ground, thereby creating jobs that are at least partially compensating for the jobs lost from declining industries. In other cases, the state tends not to get involved in the steering of technological development at all. It is left to the market to generate innovations. If necessary, the business environment is shaped so as to create a climate that is as conducive to innovation as possible.

During the 1990s, a "third way" between the market and the state emerged, with Germany being a case in point (Weyer et al. 1997; Dohse 2001). The government serves as a facilitator of networks in which actors develop technology meeting market requirements. In this model, the state steers clear of the thematic fine-tuning of technology by means of direct project promotion without, however, entirely relinquishing its task as a driving force and co-financier of innovation processes. An example of this third paradigm are contests organised by the German Federal government (e.g., Bio-Regio, Inno-Regio, Mobility in Agglomerations, Initiative for Excellence, and Spitzencluster Competition), in which locations compete for funds to assist in developing innovative industries. The aim of these programmes is to link actors from business, science and politics within a region to collaborate on innovative solutions by drawing on local strengths.

A fourth variation is an independently organised innovation process primarily driven by the activities of non-governmental actors. In this mode, the state plays a somewhat more active role than in the pure laissez-faire model, but restricts itself to the tasks of coordination and regulation, such as the creation of the conditions necessary for the successful self-coordination of the relevant actors. Innovation activities are supported by protecting competition, risk regulation, standardisation efforts, and the facilitation of service interoperability (Werle 2001; Dolata 2004).

Because nations want to use their capacity for innovation to achieve competitive advantage and international success, there is intensifying competition in the relevant policy areas. As a result of both scientific work and actual experience, an

---

[1] Major recent policy efforts to strengthen innovation are, for example, EU: Lisbon Strategy (2000) and Digital Agenda for Europe (2010); Canada: Achieving Excellence, Knowledge Matters (2002); UK: new positioning of science and innovation policy (since 2004); Sweden: Innovative Sweden – A Strategy for Growth through Renewal (2004); U.S.: National Broadband Plan (2010); Germany: High-Tech Strategy (2006); Denmark: Progress, Innovation, and Cohesion (2006); Ireland: Strategy for Science, Technology, and Innovation (2006); and Finland: Finland's National Innovation Strategy (2008).

increasing understanding has emerged in many countries that innovations have to be considered in a systemic perspective. It is necessary to go beyond political considerations. A comprehensive view incorporates all the factors that have an impact on education, research, and innovation in an economy, taking into account the global environment. This systemic view was developed in the analytical concept of the National Innovation System (NIS) (Nelson 1993; Lundvall 1988, 2007; Edquist 1997).

A NIS is composed of various institutions interacting closely to promote the development and dissemination of innovations. What is vital in this regard is the systemic consideration of interaction in a dynamic perspective. Over time, innovations and the system co-evolve and can be seen as a social learning process. If successful, they adapt to changing circumstances in a timely and appropriate manner. Furthermore, if growth should be achieved through innovation, then the emphasis must shift from a narrow focus on academic teaching, research, and high-tech, to the further development and use of a comprehensive, high-quality education and scientific system and its efficient funding. This broader approach seeks to particularly support close collaboration between this dynamic system and the private sector for joint market-driven problem solving.

The role and function of technology and innovation policy therefore needs to be seen in the context of the respective NIS. In the Swiss context, confidence in the reliability and pre-eminence of free markets and competition is at the heart of the approach. This is complemented by a guarantee and promotion of suitable economic frameworks, as well as good governance of the NIS. Therefore, the particularities of Switzerland's NIS have to be examined before innovation policy can be described and analysed.

6.2 Strengths, Weaknesses and Risk Profile of Switzerland's NIS

Throughout its history, Switzerland has repeatedly used the image of being a special case among European states to emphasise its uniqueness in terms of a singular, historically-evolved situation that is unlikely to be copied. From an economic point of view, a combination of factors distinguishes Switzerland from other countries, providing it with a comparative location advantage. These special characteristics include:

- The small size of the country and the economy render it relatively easy to understand. The high degree of decentralisation and geographical proximity result in a wide range of effective formal and informal networks.
- Diversity both in the size of businesses and in the sectors in which they operate. Switzerland has a good mix of several large global companies and a large number of SMEs with a high degree of independence and proclivity to innovate.

The interplay between innovators, marketing people, and major investors works well in the majority of cases.

- High employment rate and good work discipline (i.e., low level of absenteeism) combined with high productivity. In several sectors, this results in internationally competitive unit labour costs despite high salaries.
- A high level of international economic integration in education and science. This was stepped up further with the introduction of the free movement of labour with the EU.[2]
- High political stability with moderate but continuing reforms, high ability to integrate various cultures and political affiliations[3]; pronounced federalist structure.

A successful and efficient NIS in a small country like Switzerland shows characteristics that differ from those of a large country. As it is more difficult to achieve the necessary critical mass, a small country only in exceptional cases will be able to make its mark internationally as a pioneer or prime mover with radical innovations at the cutting edge of scientific and technological development. Possibilities for radical innovations based on systematic R&D and their developmental impacts in the national economy are limited. A far greater incentive for economic development comes from incremental innovations and from the absorption and intelligent, effective use of new technologies that have been invented elsewhere. The ability to absorb such technologies as a fast follower and the rapid diffusion of state-of-the-art applications across the economy are more important than breakthroughs achieved at the cutting edge of technologies.

Thus, experience-based learning on the job or while providing services, the use of advanced technologies, as well as learning by doing, using and interacting, make an important contribution. These processes may even be more important than the static knowledge-based culture of the private sector. Of course, it is particularly advantageous to have both: a good basic knowledge for radical innovations, as well as rapid absorption and effective experience-based learning for incremental innovations. Before the backdrop of these contextual factors, the strengths and weaknesses of the Swiss innovation system can be assessed.

---

[2] Examples are the high level of internationalisation in Swiss tertiary level institutions (in 2006, 44% of professors and 52% of other teaching staff were foreign nationals); the establishment of R&D centres of international companies such as IBM or Google; and the high percentage of foreign nationals of all those in gainful employment. As a result of the lifting of quotas for labour immigration for the EU15/EFTA from 1 June 2007, there was an influx in 2008 alone of around 25,000 persons qualified on secondary level and 44,000 on tertiary level, all together 1.5% of the workforce who found jobs in Switzerland.

[3] The four linguistic regions in Switzerland (German, French, Italian, and Romansh) favour multilingualism and the above-average incidence, in international comparison, of a knowledge of foreign languages. The co-existence of various cultures means that people have always had a high capability to integrate and adapt culturally.

6.2.1 Strengths

For years, Switzerland has been in the top European and world rankings for various aspects of innovative performance and competitiveness Volery et al. 2007 and multiple other years; Arvanitis et al. 2007; Bundesamt für Statistik 2006). For example, in the EU's European Innovation Scoreboard, Switzerland was considered the most innovative European economy in 2008 and 2009 (European Union 2010 and multiple previous years). It ranked second behind Sweden in 2006 and 2007. The country improved in the World Economic Forum's (WEF) Global Competitiveness Reports from eighth in 2004 to first in 2009 (World Economic Forum 2008, 2010).[4] It is worth noting that the WEF's report shows that Switzerland's performance in tertiary level education, research and innovation is excellent precisely in those areas that are increasingly important for growth and prosperity in a knowledge-based economy.

Switzerland's economy offers basic conditions which are particularly business-friendly and which encourage innovation. This is especially true in terms of the high level of flexibility in the labour market and the business-friendly regulation of intellectual property generated from public sector-supported R&D. The flexibility afforded to private sector initiative is relatively large in comparison to, for example, Germany or France. According to periodic surveys between 1994 and 2005, regulatory restrictions on innovation (such as on construction and environmental protection legislation) have become less prohibitive (Arvanitis et al. 2007). A gene protection initiative that would have greatly restricted research on the subject was rejected in a referendum in 1998 with 66.6% of the votes.

Switzerland also has a well-developed and high-quality education system which permits a high degree of vertical and horizontal permeability from vocational education and training to the top universities. The majority of young people (around 67%) receive practical vocational training; around 23% receive an academic education; and a little over 10% do not enrol in any post-compulsory education (Bundesamt für Berufsbildung und Technologie 2010). This makes for a good mix of vocational and academic qualifications, benefiting the economy and society as a whole.

Related to this is the high-quality of teaching and research and development. Measured in terms of bibliometric indicators, the Swiss science system has a number of excellent universities and is among the world's best in some disciplines (Staatssekretariat für Bildung und Forschung 2007; ETH Board 2010). In addition, the private sector, headed by several major companies, also spends large sums of money on R&D. These factors taken together, Switzerland's NIS can deliver first-class performances in several areas of science and technology which can also lead to breakthroughs and radical innovations. The result is an advantageous mix of

[4] Switzerland also ranked first in 2006.

companies with breakthrough and adaptation strategies performing as leaders and fast followers.

In addition to Switzerland's highly developed knowledge and education infrastructure, the country also has a well-functioning transport and communications system. The per capita density of the provision and use of information and communications technologies has long been one of the highest worldwide. For example, in 2009, Switzerland ranked number three in the OECD's comparison of fixed broadband access lines per 100 inhabitants (OECD 2009).

An analysis of patent applications shows that the technological foundations for the renewed dynamism of innovative activity in Switzerland are excellent and increasingly put to good use. Since the late 1990s, companies have been quick to focus their skills on areas in which the growth of know-how has been above average internationally and value added is being generated in new markets where technical advances are closely linked to science. Swiss firms achieve this, for the most part, by means of ever greater integration in cross-border networks of knowledge production and technology development.[5]

It is a sign of strength that the companies use their own funds to carry out a large amount of R&D in other locations, some outside Switzerland. Thanks to such cooperation projects, Switzerland has acquired substantial skills and know-how. To develop their technological basis, Swiss firms are sourcing technology worldwide with great success.[6] Knowledge components are adopted from abroad and combined with home-made components, or are transferred to expand the knowledge base in Switzerland. This model of innovative behaviour – away from mostly in-house R&D to internationally networked development of innovations from a large number of external knowledge sources, termed "open innovation" by Chesbrough (2003) – has helped turn SMEs into export and innovation-driven units.

This broadening of the knowledge base makes Switzerland more attractive as a location for business. In the future, it will be even more important for universities and R&D in Switzerland to be attractive partners for Swiss and foreign companies in knowledge and technology transfer. Even in years when Switzerland did not rank at the very top, it featured among the leaders in almost every respect and therefore achieved an excellent overall result. The resulting balance and complementarities in diversity are among Switzerland's great strengths.

[5] According to a survey, for the period 2003–2004 some 39% of companies and 23% of the tertiary education institutions reported international cooperation for their innovation activities. Both figures are well above those for Germany or the UK. For tertiary education institutions it is even the second best figure in Europe after Finland (Eurostat multiple years).

[6] The level of R&D outsourcing by companies in Switzerland has greatly increased in the last few years. Contracts worth about CHF 4bn were awarded in 2004 although mostly to private companies (26%) or abroad (60%) and only a small share to Swiss tertiary education institutions (6.4%). At the same time 74% of Swiss tertiary education institutions reported KTT activities with foreign companies (Arvanitis et al. 2005).

6.2.2 Weaknesses

Despite this impressive record, weaknesses in Switzerland's capacity for innovation also need to be identified. Historical data on innovation performance show that in the 1990s some countries caught up with Switzerland. While Swiss industry was able to stay in the top European rankings, important competitors closed the gap and occasionally overtook it. The average growth rate of Switzerland's Summary Innovation Index (SSI) for the period 2003–2007 was below the EU average, behind Germany and the UK, but ahead of Sweden, Finland and the U.S., but it has improved again in subsequent years as reflected in the data collected for the European Innovation Scoreboard (see European Union, 2010). A convergence and closing of the gap between countries is predicted by development theory: the higher the level of prosperity, the harder it is to sustain an above average growth rate. Nonetheless, Switzerland is under pressure to improve its innovation dynamic to hold on to its relative ranking. Various surveys examining the factors influencing Switzerland's innovation performance have revealed weaknesses (see Arvanitis et al. (2007) and Eurostat (2010) and multiple earlier years):

- High costs of innovation and financial constraints during certain stages of the innovation process. Despite a general abundance of capital, actors have complained of financial difficulties during the seed and venture capital phase. The fiscal environment is generally favourable to businesses but tax disadvantages for employee stock options cause problems for start-up activities.
- Risk aversion. There is a low inclination to take risks, a lack of willingness to try new products or processes (basically a feature of a wealthy and prosperous society), and a penchant to cling to what has been tried and tested. This is not just an outcome of individual values but also of institutional arrangements, such as restrictive insolvency laws with long liability periods.
- Sceptical attitude not only towards new products and technologies, but also a one-sided insistence on technical aspects to the detriment of a demand and client-based approach with innovations.
- Human capital limits to growth. Due to demographic trends, Switzerland faces a looming shortage of talent in the near future. Total reliance on hiring personnel from abroad is not a long-term solution.
- Low participation of women in innovation processes and a generally sceptical attitude toward women's participation in the economy, resulting in underutilisation of labour potential.
- Sluggish deregulation in areas important for innovative performance, such as telecommunications, energy, public and freight transport. Progress has been made, but remains well behind developments in the rest of Europe. Due to limited competition, price levels are substantially above the EU average in certain areas of the domestic market.
- Public spending on R&D has stagnated in real terms for years and, according to some calculations, may even have declined.

The picture resulting from this profile of strengths and weaknesses and the development trajectory is that Switzerland's NIS is facing several risks in the future.

6.2.3 Risks

Small countries such as Switzerland are more exposed to international competition than larger ones. That they have to adapt faster is generally a positive incentive, but greater exposure to international competition also comes with the risk of rapidly falling behind. Globalisation and new technological possibilities attract new competitors and increase (international) competitive pressures, especially for SMEs. To sustain economic success, the economy has to remain flexible and agile and display an above-average ability to respond quickly to and assimilate related conflicts.

The growing pressure to reform public finances, particularly in the course of restructuring the social security system, diminishes financial resources for innovation. Consequently, owing to a shortage of funds and the dominance of policy areas other than education and R&D, in the medium term, the publicly financed basis of science and research runs the risk of only modest growth or of being crowded out by higher-priority items.

Inherent in Switzerland's political system is the necessity to seek consensus combined with complex mechanisms for reaching such consensus in economic affairs. This institutional arrangement allows only slow reforms in policy areas relating to innovation promotion. It sometimes may create temporary stalemates, as is the case, for example, in energy policy during the 1990s. This can be a serious disadvantage in relation to competing nations and locations.

On account of demographic challenges and structural changes in the economy, Switzerland needs to attract and employ more foreign researchers and experts than it has thus far. However, rising immigration intensifies the challenges of social integration and increasingly arouses political opposition.

6.3 Features of Switzerland's Innovation-Oriented Economic Policy

Switzerland's innovation policy differs in many respects from other OECD countries. This is particularly visible in its regulatory policy principles and the philosophy guiding the promotion of innovation. Much of this distinctiveness is contextual. It is shaped by the features described above, such as the country's smallness, its political culture, federalism and direct democracy with a large number of popular votes on specific matters. It is also influenced by the assignment of responsibilities to the various political actors, such as the Confederation, the cantons, the communes, public and private institutions, and the range of private partners and interest groups in policy areas involving innovation.

For the most part, policies to promote innovation are implicit. No overarching, explicit innovation policy based on a concept consented to by all parties, with comprehensive programmes and transparent mechanisms of coordination, exists; there are only general guidelines.[7] Thus, when actors are setting policy objectives, innovation is not considered the main goal. For example, promotion of innovative performance is subsumed in goals such as growth, research excellence, energy efficiency, or sustainability, and is instrumental to achieving these goals. Innovation policy is operated from a fragmented system in which various actors pursue their own agendas. There is no support or integration, for instance, by means of an advisory body or an innovation council. It is almost impossible to establish a cross-sectional reference inherent in innovation policy.[8]

However, efforts in various policy areas relevant to innovative performance indirectly contribute to the high level of innovative performance. They create the right conditions in the form of solid foundations and a favourable framework. In light of the difficulties of achieving across-the-board coordination of the various policy areas due to the specific features of the Swiss system of government, this pragmatic approach might be the best way forward.[9] The following sections outline the key features of Switzerland's policy in relation to the promotion of innovative capacity.

6.3.1 Innovation-Oriented Legal and Regulatory Policy

Swiss regulatory policy aims at creating a favourable environment for the successful positioning of Switzerland in the international innovation competition. Its emphasis is on competition and private initiative, and keeping government regulation and intervention at a low level. The various markets for goods and services,

[7] The documents which most explicitly outline innovation policy of Switzerland are from the Federal Council as the government of Switzerland (Bundesrat 1992) and the Federal Department of Economic Affairs (Eidgenössisches Volkswirtschaftsdepartement 2003).

[8] It is perhaps possible to talk of an *informal coordination* of the individual policy areas. Actors aim at generally recognised goals such as "supporting innovation", "being innovative" with a view to their own political success.

[9] The 2008 growth report by the *Staatssekretär für Wirtschaft* (SECO), describes innovation policy's relation to growth policy as follows: "Innovation policy is not considered as a field of action of growth policy in its own right because innovation derives at least as much from the intensity of competition and international opening as from expenditure on education, research and technology transfer" (Staatssekretariat für Wirtschaft SECO 2008). Innovation policy is therefore subordinated to growth policy.

labour and capital should be competition-driven, and access to international markets should be as free and open as possible.

Swiss labour market regulation allows a high degree of flexibility and mobility in international comparison. In markets that particularly lend themselves to innovative performance and where the state is heavily involved, such as telecommunications, energy and public transport, the goal is deregulation and privatisation. However, there are deficits, for example, in the slow and late liberalisation of these markets compared with other countries. The domestic market is protected from international competition by the rejection of the Cassis de Dijon principle in trade relations with the EU member states and also the application of the national exhaustion of patent law (ban on parallel imports). The resulting higher innovation costs receive, as mentioned, frequent criticism from companies in surveys on capacity for innovation.

There should be clear and simple rules for regulating intellectual property stemming from R&D that has been supported with public funds and whose results are also suitable for SMEs. The rules governing intellectual property for universities vary because they are codified in relevant cantonal legislation. In most cases, private sector interests are generously taken into account. SMEs are offered public coaching to deal with intellectual property issues. Special solutions are sought for the regulation of intellectual property regarding services.

In the area of tax laws, the aim is to create a fiscal environment favourable for businesses. In Switzerland as in practically all countries, R&D expenditures are allowed to be expensed, thereby reducing the tax burden. Attempted tax incentives for venture capital companies have so far had little success. However, new businesses, start-ups, and small, fast growing businesses all benefit from the fact that Switzerland has no capital gains tax. More extensive tax incentives in support of innovation only exist in environmental and energy policy. Switzerland generally has a low level of taxation in international comparison, which contributes to a business – and investment-friendly environment while offering few additional incentives to promote innovations by means of tax measures. The current debate on tax reform focuses on tax relief and simplification, such as a single VAT rate for all or, in connection with social deductions, simplified returns, more generous tax thresholds, and flat-rate assessments.

In contrast to many other countries, public procurement has little bearing on the promotion of innovation in Switzerland. As public procurement is controlled politically and political actors tend to be conservative in terms of risk-taking ("politicians don't take risks"), in the practical application of public procurement rules there is hardly any impulse for innovation.[10]

---

[10] The proposal for a new Federal Law on Public Procurement (Bundesgesetz über das öffentliche Beschaffungswesen BöB), which is currently under consultation, mentions that innovative content should range among the non-monetary criteria for successful bids. Variants including features that were not part of the call for tender are expressively welcome, thus promoting innovative solutions.

6.3.2 Education Policy

Education policy is instrumental in improving innovative performance. Responsibility for regulatory matters in the education sector lies mainly with the cantons. "Education policy in Switzerland is a complex interchange between various bodies and authorities. There is no main player; everything is based on agreement between the Confederation, the cantons and the communes. There is a special federal policy only in certain areas, mainly in vocational education and training (VET), but even this policy cannot be enforced by ordinance, which is what centralist Austria is continually striving for" (Oelkers 2005). Under this policy of educational federalism there is a decentralised adaptation to regional and local circumstances and needs, which creates great diversity. The necessary coordination and mutual agreement between the federal levels that is needed for a successful education policy are to be reinforced by a new constitutional provision that has been in force since 2006, and legislation yet to be created governing the tertiary sector. The new law will not come into effect until at least 2013. The new framework consists of an intensified cooperative federalism with the development of responsibilities delegated to a body jointly constituted by the Confederation and the cantons.

Education policy rests mainly on two pillars: vocational education and training, which covers around 65–70% of young people, and academically-oriented education (*Gymnasien* – grammar schools), covering 20–25%. A similar division is also found in tertiary education, in which Tertiary B (professional colleges) and Tertiary A (universities of applied sciences, universities, and the Federal Institutes of Technology) are distinguished. From an economic perspective, this results in a good mix of qualifications of different types (practical and application-oriented and scientifically oriented/academic), reflecting the lines and strengths of Switzerland's business activities. For innovation policy, it is important that the universities of applied sciences have an independent and strong profile in the development and implementation of innovations in direct contact with private sector businesses.

Despite the fact that education is largely a cantonal responsibility, the Confederation has been drawn into funding a number of areas (vocational education and training, universities). This has led to relatively complex financing mechanisms whose incentives and steering effects are not always favourable for innovation policy. The current debate on reform of the tertiary sector revolves to a large extent around these mechanisms and the steering of the education system, such as the choice of criteria on which the Confederation's funding is based.

The Swiss education system as a whole is well funded[11] in international comparison, especially given the size of the country. Expenditure on education per

---

[11] In 2007 the 5.5% share of public education expenditure on GDP was slightly below the average for OECD countries (5.7%) and well below the top group led by the U.S. with 7.6% and Denmark with 7.1% (OECD 2010, p. 217).

apprentice or student from primary to tertiary level (including private VET institutions, upper secondary level) is one of the highest in international comparison (OECD 2010). A special feature of education policy in Switzerland is the promotion of VET qualifications by means of a dual system of vocational education and training. Alongside 2 or 3 days a week of schooling, the apprentices also spend 2 or 3 days in a business known as a "host company". This model thus combines a high-level of practical educational orientation, which is very valuable for employability, but also a significant material training contribution on the part of the business community. For a long time, human capital presented a barrier limiting the growth of Switzerland's economy. Recently, many adjustments were made in all qualification areas due, to a large extent, to the introduction of the free movement of persons within the EU.

Triggered by the results of the OECD's PISA study, questions are now being raised about the quality of the education system. Switzerland is positioned in the top middle of the PISA rankings, which, given some of the highest education expenditures per student, provokes questions with regard to efficiency and effectiveness. The high level of reproduction of existing social structures and the minimal contribution of the education system to social mobility hinder potential innovation resources. The debate on higher education reform, which has been ongoing since 1997, strives to improve efficiency and quality so that Switzerland can hold its own in international competition. Governance in the tertiary sector should be improved particularly with regard to greater national coordination. This should be achieved by means of greater transparency in the financial matters of universities, more autonomy for and competition between them, more reliance on performance-oriented funding criteria, proof of quality (accreditation), and greater mobility of students between schools.

6.3.3 Research Policy

Clear differences exist between Switzerland and other countries in research policy and performance. Most importantly, expenditure on R&D as a percentage of GDP in Switzerland is high at 3.0% (2008) (Federal Office for Statistics 2010). A large portion is generated by the high involvement of the private sector in R&D, especially large companies in the pharmaceuticals sector. The private sector provides around 74% of Switzerland's total R&D expenditure (or CHF 12 billion, equivalent to 2.2% of GDP), whereas the state finances 26% (or CHF 4.3 billion, about 0.8% of GDP). Very little R&D expenditure is included in the defence budget. In recent years, public involvement in R&D has even fallen in real terms. This has a particularly negative impact as basic research is becoming more important, but SMEs, if they engage in R&D at all, focus on the development stage.

The focus of the state's research promotion is on basic research, mainly in form of the Swiss National Science Foundation (SNSF) budget of CHF 700 million

(2010) (Swiss National Science Foundation 2010).[12] In contrast, applied research and development is funded by the Innovation Promotion Agency (CTI) with a mere CHF 100 million (2010). Different from many other countries, the public sector spends four to five times more on the promotion of pure research than on applied research (in Finland, for example, the ratio is reversed).

Applied research is promoted by the CTI. It supports innovation processes in two ways: first, by co-financing projects that can be jointly funded by university institutions and businesses (main target group SMEs); second, by the promotion of entrepreneurship and high-tech start-up companies. In the 1990s, various promotion programmes in areas such as microelectronics, machine tools, production technology, nanotechnology, software, and medical technology also came about.

Private companies in Switzerland receive practically no funding for R&D directly from the state. The relatively small share of public funding for R&D goes exclusively to public university institutions (universities, Swiss Federal Institutes of Technology (FITs), and universities of applied sciences (UASs)). This is in stark contrast to the promotion policy of most comparable countries and to the EU. While not being a member of the EU, Switzerland, through bilateral agreements, is nonetheless a fully paid up member of the EU's R&D framework programme. This situation results in tension between national and EU funding practices: whilst a company from Switzerland can receive direct support within an EU-project, it is not possible for the same company to receive such support under the national research promotion.

National funds are distributed on the basis of decentralised initiatives of directly interested parties ("bottom-up principle"). Expert opinions (peer reviews) are the decisive factor. Consequently, in Switzerland R&D promotional funds are only awarded to politically determined and targeted priority themes in exceptional circumstances. Programme-oriented funding is the exception and is, in absolute terms, of little significance. If it does exist, then it is largely conducted in the context of sectoral policies, in particular on the environment and energy, but not via research policies. In as far as there are promotional priorities, their themes for the most part are generated based on the initiative of directly interested parties. Thus, even projects within these priorities are usually generated from the bottom up. This also includes university institutions which in some cases have considerable research funds of their own, which, in the case of the FITs, for example, amount to around 50% of their budget.

All the above-mentioned attributes and intentions on the part of the responsible authorities mean that Switzerland's R&D resources, to a far greater extent than in most competing countries, are allocated in line with market economy principles and competition in the target markets. Risks are borne directly by those responsible for R&D resources. Companies' behaviour regarding innovation is therefore to a large extent a consequence of market incentives. The positive effect of these

[12] Some 81% of this amount is for basic research, 19% for targeted research, programmes, national competence centres, and the promotion of young talent.

particularities is further enhanced by the smallness of the country and the good mix of large and small companies.

Cooperation between university institutions and business, and hence, knowledge and technology transfer, tend to take place by means of "head hunting", providing jobs for young graduates, and through strong informal networks.[13] Training and teaching are the most important transfer channels, but a special role is played by joint R&D projects of university institutions and business. These make up the core promotion of the Innovation Promotion Agency CTI. An explicit promotional policy by the Confederation to establish institutionalised knowledge and technology transfer (KTT) via transfer points at university institutions and research institutes in Switzerland, is stuck at the beginning.

6.3.4 Large Scale and Further Internationalisation of the NIS

Empirical research shows that the R&D activities of Swiss public university institutions and those of private companies are largely internationalised. In 2008, 57% of the R&D budget of Switzerland's private sector was spent abroad – some CHF 15.8 billion in comparison to CHF 12 billion that was spent domestically. Companies seek out the top centres of excellence worldwide. Empirical studies show that, for the economy as a whole, the effects of the complementary nature of R&D activities in Swiss and foreign locations are stronger than substitution effects (Arvanitis et al. 2005). In addition, the faculty of university institutions shows a high degree of internationalisation. Decisions of public authorities support and encourage this development. Elements of Switzerland's policy towards the internationalisation of innovation are:

- The government is actively making efforts to use R&D bodies to facilitate access to international programmes. Partnerships with other countries are fostered multilaterally and bilaterally: from research cooperation projects to common vocational education and training policy. Switzerland is a full member of the seventh EU Framework Programme, a founding member of the pan-European R&D initiative EUREKA as well as the global initiative Intelligent Manufacturing System (IMS), and a participant in joint research activities as defined in Article 185 of the Treaty on European Union as amended (e.g., Ambient Assisted Living (AAL) and Eurostars[14]). The ultimate initiative and responsibility

[13] This is seen in the KOF (Konjunkturforschungsstelle at the ETH Zurich) survey on the channels of knowledge and technology transfer (KTT) (Arvanitis et al. 2006).

[14] Eurostars is a EUREKA programme dedicated exclusively to SMEs, see http://www.eurostars-eureka.eu/ for more details (retrieved August 4, 2010). With the coming into force of the Lisbon Treaty in 2009, former Article 169 was renumbered Article 185.

for international scientific cooperation lie with scientists and universities with a direct interest.

- Switzerland is actively involved in the planning and organisation of international programmes in connection with, for example, the EU, EUREKA or IMS. The degree of participation of Swiss universities and companies in international R&D cooperation projects is high. The applications submitted by Swiss researchers in EU programmes often have an above-average success in EU selections. Evaluations of Swiss participation generally show a positive return on investment (e.g., Bieri et al. 2005).
- Successes in international cooperation require a national basis that is strong and attractive to international partners. The fostering and development of corresponding domestic competencies are therefore an explicit aim of government policy. It is a matter of internationally recognised achievements in education and R&D: infrastructure, content/quality, and *modus operandi* for education and R&D.

6.3.5 *Entrepreneurship*

No national public promotion programmes for financing SMEs or starting up companies exist in Switzerland. Governance of money and capital markets that generates financing options for start-up companies, whether for setting them up or for their initial growth, is considered sufficient. In the last 10 years, the area of venture capital has been significantly expanded, but criticism continues to be articulated. A focal point of policy efforts is the improvement of the general conditions for setting up companies and targeted service offerings.

Entrepreneurship is promoted by the CTI entrepreneurship initiative (under the brand name "venturelab") across Switzerland in all linguistic regions. This includes the establishment of targeted education and training opportunities for young people at universities. "venturelab" has been very well received by the target groups addressed.[15] The aims and implementation plan also received good marks in comparison to other internationally comparable promotion measures. The programme is expected to be developed further with the entrepreneurship theme developing into a nationwide education standard. These measures are carried out in close collaboration with professional organisations, vocational schools and universities.

[15] Over 24,000 young people were taught and motivated over the period from 2004 to 2007 in information events to consider entrepreneurial independence as a meaningful alternative to a dependent employment relationship and to act as founders of companies. In addition 7,500 people took part in courses for setting up businesses companies.

Under the national CTI start-up development programme launched in 1996, support is given for the setting up and development of technology-based companies with high growth potential. Coaching is offered to start-up projects that passed jury selection after the submission of an application. After coaching is successfully completed, high-quality start-up companies can receive a seal of approval based on an examination by a jury of independent, distinguished experts. This seal qualifies the companies for venture capital funding and provides the opportunity to present their case to and seek funding from a club of private investors (CTI-Invest). Each member of CTI-Invest is committed to invest a minimum amount of CHF 150,000 per year into projects that are part of the initiative.

This programme helps reduce the information and transaction costs of potential investors and increases the chances of obtaining funding for projects. It works as a public-private partnership: the public partner helps to build up projects and improve their quality; the private partners are financiers who invest seed and venture capital. The CTI start-up label is widely recognised in business circles as a mark of quality for innovative companies with great growth potential. With this programme Switzerland possesses an innovative plan for the promotion of entrepreneurship and start-up companies.

Start-up promotion helps complete the last link of the innovation chain. It contributes to the successful application of an innovative idea in the market and hence further develops the creation of customer benefits. As many start-up projects emanate from publicly funded R&D projects, these promotional measures also make a contribution to the valorisation of their results.[16]

While the number of companies started by women in Switzerland is above the international average, the general conditions for female entrepreneurs must be considered as poor. According to experts consulted in the 2006 Global Entrepreneurship Monitor (GEM) project, in a European comparison, the conditions for women to set up companies in Switzerland are among the worst. Switzerland, the GEM Report states, has a social infrastructure that does not sufficiently support the independence and flexibility of women. For example, women could continue to work only under difficult conditions after they had started a family. In the Fetz report, the *Bundesrat* (Federal Council) (2006) identified three action areas to improve the compatibility of work and family life: (1) development of child-care facilities; (2) development of family-friendly working conditions in corporations; and (3) elimination or reduction of negative incentives that impede employment of parents.

[16] Companies with a CTI start-up label show a survival rate that is well above average, a much higher profit and turnover volume and double the size of employee growth. They have also raised seven times more debt capital than a corresponding comparison group which was not involved in the CTI start-up coaching process (ITM 2006).

6.3.6 *Funding of Education, Research and Innovation*

The financing of the federal share of spending on education, research and innovation (ERI) of the NIS is decided in legislative periods of 4 years. As mentioned above (see Sect. 2.2), federal government expenditure fell in real terms during the 1990s as a result of a lengthy period of stagnation and since then has risen only moderately, following efforts to restructure federal finances. In the period from 2008 to 2011, a 6% annual growth rate is targeted, higher than other items in the federal budget. In this respect, government and parliament have clearly signalled that they are making ERI a priority. However, growth rates are below those that experts consider to be necessary to ensure the position of Switzerland's NIS; in the parliamentary debate, growth rates of up to 10% were considered necessary. There is a risk that Switzerland's innovation policy therefore will be underfunded.

The main part of the budget consists of transfer payments (so-called "subsidies") which are guaranteed to the beneficiaries (in most cases the cantons) in federal legislation. Only a small part consists of freely available funds (so-called "financial assistance") that are allocated on request. The financial room for the manoeuvring of discretionary innovation policy is therefore small. New promotion measures need to be planned before the start of the financing period. No funds are normally available for measures that are launched at short notice during a legislative period. If such initiatives are introduced, they require an absolute majority in the parliamentary vote.

Since 2000, the Confederation's budget for education (contributions to vocational education and training, the cantonal universities and the universities of applied sciences, and for the FIT domain, which is 100% financed by the Confederation), research (in particular SNSF and CTI) and innovation (smaller individual budgets) have been combined and submitted to parliament for approval jointly. This amalgamation has the effect of providing a certain amount of budget coordination and a minimal coordination of strategy. However, the ERI strategy operates at a highly abstract level and provides, in the best case, a general framework for the budget claims. Although strategic controlling has been introduced, it is in a difficult position as, due to the nature of the projects, effective checks are only possible in the medium to long term. An initial step towards a systematic innovation policy has therefore been taken, but it is still a long way from being integrated in a coherent policy.

6.4 Swiss Innovation Policy: Characteristics and Challenges

6.4.1 *Principles of Switzerland's Innovation Policy Concept*

From what has been described, the political debates, and the policy in practice, it is apparent that Switzerland's innovation policy operates according to certain principles, whether implicit or explicit in terms of normative rules. Of course these are not recognised unreservedly by all participants – some are very controversial and some are not always respected. There is nevertheless a Swiss "innovation policy paradigm", which can be described as follows:

First, innovation is the result of entrepreneurial business activity and is thus the fundamental task of companies par excellence. The private sector carries the main responsibility for innovation processes. It needs to have sufficient leeway and not be constrained unnecessarily by regulatory policy rules. The state is only active in a subsidiary way. Its role is to create a favourable framework and attractive conditions, whether by means of innovation-friendly regulations or in the form of material inputs such as infrastructure in education and research. It gives no indirect fiscal incentives for R&D activities and none are requested.

Second, competition is a major incentive for innovation. Companies are not the only ones to compete. Increasingly, universities are competing against one another. Innovation policy has to recognise this and allow for existing or newly evolving competition between public and private actors in the innovation system. It should ensure that there are as few competitive distortions as possible arising as a result of government involvement in education and research.

Third, innovation policy has to be aware of the connections and interactions between many policy areas and their actors. Innovation policy affects many policy areas (cross-sectional policy), particularly, education and research policy, competition and labour market policy, location and regional policy, and financial policy. Major impulses for innovation may emanate from sectoral policies such as health, environment, energy and transport policy. This requires coordination, or at least the absence of any mutual negative influence.

Fourth, the aim of innovation policy is to strengthen companies' innovative performance in Switzerland so that added value, income and jobs can be created and the economy may grow. Particular attention has to be paid to efficiency, quality and synergies, in conjunction with actors such as companies, university institutions and the state. Nowadays, innovations very often occur in networks of large and small firms, suppliers and buyers, public and private research and development facilities, and education institutions, associations and authorities. Innovation policy should contribute to the simplification and improvement of cooperation in such networks. Companies' international competitiveness is therefore, to a large extent, based on the quality of this cooperation and networking. The attractiveness of a location for innovative companies, researchers and professional people depends entirely on favourable conditions to innovate successfully.

Fifth, state innovation promotion is geared towards excellence. It aims to offer people opportunities in their specific areas of activity in order to develop themselves and their talents fully. One goal is to achieve the top internationally competitive performance in selected areas.

Sixth, innovation policy measures need to be oriented towards effectiveness and performance. State financial support should also always have a performance component and be awarded in competition. Innovation promotion should strengthen the flexibility and adaptability of actors in business and university institutions, and support structural change. This entails promoting fast implementation and dissemination of new knowledge: dissemination-oriented innovation policy. Due to their economic importance, export-oriented SMEs and start-up companies are one of the policy's particular target groups.

Finally, innovation promotion works mainly according to the bottom-up principle. Individual teams or companies should seize the initiative for R&D activities and assume responsibility. Individual projects are promoted directly on request. There is no promotion programme in business-related applied R&D. This does not a priori exclude a political focus on strategically important themes (priorities), especially, in the area of pure research. However, this kind of focus would be achieved, if at all, by means of a kind of follow-up policy in terms of taking up and consolidating positive trends that can be identified on the basis of bottom-up developments and trends.[17]

6.4.2 Challenges to the Uniqueness of Swiss Innovation Policy

This section discusses whether the outlined policy model will be sufficient to achieve a high innovation performance level in the future. In order to answer these questions, four major issues for Swiss innovation policy will be explored.

6.4.2.1 Transfer of Competence for Innovation Policy Between Different Institutional Levels and Competition Between Them

Changes can be identified with respect to the political players and institutions which are in charge of promoting innovation. Policy decisions are not only made at a national level, but also at cantonal and, to an increasing degree, also at supranational level through, for example, the European Commission. The consequences of parallel processes of the regionalisation and internationalisation of innovation policy are complex: activities are undertaken sometimes in cooperation, but quite often in competition and with little consistency. Moreover, the structures and decision making procedures of these institutions are often highly dependent on each other, forming a "multi-level system" with associated problems due to a lack of overview and transparency (e.g., Grande 2000).

Therefore the role of innovation policy at a national level and its associated policy principles are challenged from two sides (Hotz-Hart et al. 2003):

[17] See the development of the Raster tunnel microscope in Zurich at IBM and the subsequent promotional activities of the SNSF in nanotechnology or analogue research results and the increased promotion introduced later in connection with the superconductors and semiconductors.

- Regional (or canton) initiatives in the promotion of innovation play a significant role which might even be increased in the future.[18] These initiatives are intended to raise the attractiveness of the region. In doing so, regional initiatives show little hesitation to use direct state intervention or subsidies, which violate the principles established at a national policy level.
- EU innovation policy initiatives and a transfer of authorities to a supra – or multinational body raises similar questions. Although Switzerland is not a member of the EU there is a direct impact of EU-innovation policy. Switzerland is a full member of the seventh Framework Programme. In several respects, principles of promotion activities within this framework are in contradiction to established principles of Swiss innovation promotion. A major controversy is over direct payments to companies in order to promote R&D-projects which are common for EU-policy and all European countries. For companies which operate at an international level, innovation promotion policies of different nations offer competing incentives for the allocation of their activities via tax relief, subsidies for salaries of R&D personal, or direct payments for R&D projects as mentioned above. Therefore, companies located in Switzerland must evaluate and decide where they should apply for support with R&D-proposals in the country, or whether they should change locations and go abroad. Within Switzerland, pressure is increasing that promotion policy should be amended in such a way that Swiss companies have "equal opportunities" compared to their competitors. Certain interest groups claim that Switzerland should align with international habits and practices of promoting innovation and therefore should introduce instruments which up to now were considered inappropriate.[19]

In both cases, the question is whether there will be an adjustment of policy principles or a coexistence of different policy concepts. Is coexistence and long term stability possible? Switzerland might try to have access to and cooperate with EU-programmes and instruments. However, in the context outlined above, basic principles of Swiss innovation policy on a federal level, such as no direct payments for companies or bottom-up orientation, would be challenged and the uniqueness of the Swiss approach threatened.

[18] See regional innovation promotion strategies of central Switzerland, Western Switzerland, Greater Zurich Area, or multinational activities in the region of Basel.

[19] 2005 Johann N. Schneider-Ammann, president of Swissmem, The Swiss Mechanical and Electrical Engineering Industries, and Member of the National Council suggested in the national parliament (05.3539) equal treatment in the promotion of R&D at universities and in private business. He claimed that private partners in R&D-projects shall be eligible to get state subsidies. Up to now, the Innovation Promotion Agency – CTI has no right to do so which is – in the eyes of Schneider-Ammann – a disadvantage from the point of view of Swiss mechanical and electrical engineering industries and for innovation activities in Switzerland in general.

6.4.2.2 Setting Focal Points and Priorities

In relative terms, taking, for example, R&D expenditures per head or as a share of GDP, the Swiss economy achieves an excellent performance; but in absolute terms (e.g., overall expenditure on R&D) there are clear limits. The existing high degree of diversification of innovative activities is one of the strengths of the Swiss economy. It reduces risks and allows fast adjustment to changes in the market place. Nonetheless, certain developments and innovations require a critical mass, such as a minimal amount of resources which can be spent for one R&D project. Globalization, with its increase in transparency and mobility, open and widely penetrated markets and – as a result – the structural changes in the economy, along with the small size of the country, will have major consequences for the Swiss economy. The high degree of diversification can hardly be maintained in the long run. The international dynamics and pressures mentioned are likely to cause a reduction in diversification through a specialisation of Swiss businesses.[20]

In principle, structural change and setting priorities should be conducted via competition in the free marketplace. However, this principle does not clarify the interplay between innovation promotion by public authorities and private initiatives sufficiently. In the process of priority setting, several contributions of policy authorities are possible and may be even necessary. In Switzerland the following tendencies can be identified or are expected:

- A typical response of companies and/or universities to these challenges is to switch to other, mainly neighbouring fields where they look for a new, adequate niche in order to exploit their advantage of small size. In order to do so successfully, a high degree of creativity to find such niches, readiness to adapt, flexibility, and risk taking are needed. Up to now, the Swiss NIS could repeatedly implement such a strategy successfully. Nonetheless, the noted risk-aversion could make it difficult to pursue this strategy in a fast changing world.
- The promotion of innovation can focus on knowledge management instead of promoting particular technologies. In offering intelligent services, policy authorities can facilitate adoption of "state of the art" and "good practice" in science and technology. Diffusion oriented innovation policy was the dominant policy concept in the 1990s. Its instrumental focus was on strengthening training and education in selected fields.[21]
- Public promotion agencies do not intend to set priorities top-down. However, they work with positive trends they can identify according to bottom-up initiatives of universities and companies. They support and strengthen these

[20] See, for example, the development of the company Zellweger which was well established in textile machinery. After substantial restructuring and downsizing a new company arose, Uster Technologies, which is focused on testing and measuring textiles and spinning products. Therefore, the result of the restructuring was a highly specialised function within the textile business.

[21] For the typology see Ergas (1987).

efforts in the manner of a "follow-up" policy. Public means increase and accelerate the efforts to progress in this field.

- Public authorities offer services or financial support in order to bring innovative players together. They function as moderators in coalition building with respect to innovative networks, and ease the setting up of focal points.[22] Only through joint efforts of public and private agents – partnership where necessary, and competition where suitable – can the NIS achieve high innovation performance.

Switzerland as a location for innovation is under pressure from international forces to converge with and adjust to widely common preferences and products. There are pressures to abolish long standing advantages of the location Switzerland and to equalise what was up to now specialisation and a distinct, widely distinguished profile. In the future, however, the economy in Switzerland can only be successful if it shows a strong profile with idiosyncrasies and, therefore, specialisation. The core principles necessary to achieve this are decentralisation, being close to the market place and to the customers, and bottom-up. In Switzerland, the major, successfully globalised companies in the field of pharmaceuticals, life science or in the financial sector need no more than a liberal framework and ample room to manoeuvre for their business activities.

6.4.2.3 Institutional Differentiation of R&D Activities

According to the concept of "open innovation", companies decide consciously which innovation they want to develop independently in-house, which they want to elaborate in cooperation with external third parties (such as other companies and/or universities), and which they might buy completely. As a result of this trend, differentiation between institutions specialising in R&D and the development of technologies is taking place. A new market for these particular services has arisen with the international participation of institutions from the public and private sector and new institutions with a mixed public-private base (Arora et al. 2001). As a consequence, universities and their laboratories find themselves in a new, increasingly competitive environment. Rival private organisations can also offer R&D-services, or education and training; they can be profit or non-profit oriented, an example being the rising sector of Knowledge Intensive Business Services (KIBS). Because of this, it might be increasingly difficult for universities as a whole, or even for a single department, to fund R&D-activities via third party contributions. The business model of "entrepreneurial universities" (Etzkowitz 2004) will develop where R&D-services and science and technology transfer can gain commercial benefits. In any case, many universities will position themselves anew with respect

[22] See the project on systems biology and is implementation through SystemX.ch in the region of Basel with the participation of ETHZ, Universities of Basel and Zurich, the pharmaceutical industry and the Swiss federation.

to the market place. Therefore, there will be consequences for the public promotion of innovation. Universities should be granted sufficient autonomy. Principles of public funding education, R&D and innovation, regulations of intellectual property rights, and licensing will have to be adjusted or newly designed.

6.4.3 Pattern of Innovation Policy and Policy Coordination

In the following section, the pattern of technology and innovation policy in Switzerland will be assessed and positioned in view of the international policy debate. In the classification of Dolata (2004) mentioned in the introductory section of this article, the fourth paradigm is the most appropriate to describe the situation in Switzerland. Policy intends to restructure the infrastructure and the institutions of its own NIS according to the needs of the world market to make it more attractive to foreign players. This intention will be implemented through initiatives in the areas of competition, networking and priority setting. The target groups of such a policy are not only single, important agents such as large companies, but also new entrants and networks. Since the report on technology policy by the Federal Government in 1992 (Bundesrat 1992), the Federal Administration has repeatedly elaborated upon concepts of state-of-the-art innovation policy. In the 1990s, under pressure after a long period of slow growth for the Swiss economy, the promotion and diffusion of new technologies was seen as a key element to strengthen Switzerland's international competitiveness. It was therefore a cornerstone of a growth-oriented economic policy. Policy attention and points of departure moved from R&D, via technology, to innovation policy.[23] The promotion of knowledge and its diffusion and implementation in products and services which are successful in the marketplace, were at the heart of these policy concepts. The detailed reports and recommendations have had, however, very little, if any, impact.

Innovation policy has been regarded as a cross-sectional task. Confederation, cantons and relevant associations, as well as ministries and offices within the public administration, are expected to work together efficiently in order to implement a consistent and effective innovation policy. This needs substantial coordination among the actors involved. Such a policy concept cannot be clear cut, but must be open to interpretation. Quite often it will be obstructed by legitimate, but partisan and often divergent, interests of the institutions involved. Federal authorities have tried repeatedly to develop innovation policy bottom-up and jointly with the relevant players within the economy and academia. In practice, every such effort has remained "under construction", permanently in search of a pragmatic solution which supports the dynamics of the economy and science, taking the most recent context into account.

[23] Lundvall and Borras (2006) identify a similar development in other OECD countries.

In fact, innovation promotion at the federal level has taken place in specific areas such as environment or energy policy, where each ministry or office can assert and defend its self-determination. Therefore, policy is ruled by the particular interests of the institutions involved. Coordination or adjustment with respect to an overall objective, such as improving the innovation performance of the economy as a whole, has always remained rudimentary. If it took place at all, it was done so informally between the institutions in charge.[24] For example, decision-makers in the institutions involved may have tacitly shared a common policy view or agreed among themselves on particular measures they intended to implement, such as the promotion of R&D projects in the fields of environment or energy. The system of R&D in Switzerland enjoys a large degree of autonomy and self-determination. The scientific community has a major impact on setting priorities, if any are set. The promotion of R&D has a subsidiary role through financing single projects and small sized programmes limited in time. The findings from the case studies of Freiburghaus et al. (1991) are still valid to a large extent today. The content of R&D is the result of the bottom-up principle and can therefore only be determined retrospectively. Mainly these subjects are promoted as proposed by individual scientists or academic research teams. The capabilities of large institutions to direct R&D activities, principally the Federal Institute of Technology (ETH/FIT), are substantially better than those of the public promotion agencies such as the Swiss Science Foundation.[25] Therefore, these institutions, and via them, the established R&D structures, decide upon the allocation of finance and personnel in R&D. Hardly any fresh impulses for new directions in R&D are given by public promotion agencies. The most substantial initiative from a federal agency in this field was the build-up of a long term research structure in Switzerland through the implementation and support of National Centres of Competence in Research.[26] This overall policy concept allows a high degree of decentralised initiatives. However, it runs the risk of reproducing the intentions and interests of the established structures.

[24] Often such a process is not directed or moderated. It is a kind of self-coordination between the units involved. An example could be the growth report 2008 of the State Secretary of Economic Affairs (SECO). The programme to stimulate economic growth is based on the list of measures which where decided earlier within the programme for the legislation. Some adequate measures where selected and put together to the growth programme afterwards.

[25] 2006 the Swiss Federal Institute for Technology as a whole counted for 40% of the resources for R&D of the overall federal budget for R&D (of a total of CHF 2.2 billion). The Swiss Science Foundation had a share of 20% and the Innovation Promotion Agency – CTI 5%. Therefore, ETH is a major force in the federal R&D-policy (ETH Board 2010).

[26] National Centers of Competence in Research (NCCR) promote long-term research projects (up to 10 years) in areas of vital strategic importance for the development of science in Switzerland, for the economy of the country, and for Swiss society.

6.5 Conclusions

There is some indication that the reserved tactics and the limitation to a few basic principles and small scale measures and programmes of innovation promotion has contributed to the success of the Swiss NIS. Policy activities have not always been consistent, nor have they always implemented their vision according to the rules, a fact which can be seen in the slow liberalisation of the markets for telecommunication, energy and postal services Nonetheless, there is great freedom for decentralised units such as universities and companies to take the initiative and make their own decisions. They can and must take responsibility and risks within a competitive framework; this encourages their motivation but also limits damage if wrong decisions are taken. At the same time, public authorities focus on strengthening R&D infrastructure, education, and training, keeping the labour market flexible and allowing open access to major international activities. This represents an implicit innovation policy. Such a policy refrains from setting national goals and top-down initiatives of innovation promotion. Altogether, public authorities provide a sound base for the high performance of the NIS.

Swiss economic policy explicitly mentions the promotion of growth (Staatssekretariat für Wirtschaft SECO 2008, 2003). Support for innovation plays a minor role, subordinated to the growth objective. This remains in contrast to most countries in Europe, and is part of the uniqueness of Switzerland. But reflection is needed to see how relevant growth is for a highly developed economy such as the Swiss one. The arguments of the intense discussion about "quality of growth" in the 1980s (Expertenkommission "Qualitatives Wachstum" des Eidgenössischen Volkswirtschaftsdepartements 1985) lost appeal in the slow growth phase of the 1990s. Although more recently growth has been critically commented on under the heading of "sustainability", its political impact has been minor.

During many difficult phases of economic development, the NIS of Switzerland has proven to be highly flexible and adaptable to changes in the international context. However, there are many substantial challenges and projects which have yet to be solved and implemented in several subsystems of the NIS. Some of the major issues are in the educational system: nationwide harmonisation of the compulsory school system and the intended integrated regulation of the tertiary sector (in discussion since 1997 and planned for implementation in 2013), both of which will most likely come under scrutiny of a public referendum.[27] Other issues concern the system of R&D, particularly amendments to the laws on R&D. With respect to public finances for education, R&D and innovation, there is pressure for cuts at all levels and in all subsystems so that annual budgets will be frozen. Furthermore, free movement on the labour market between Switzerland and EU member states will be challenged through a public referendum. As different

[27] Up to September 2010, the nationwide harmonisation of the compulsory school system was accepted in 15 counties and rejected in seven. The political debate is still in process.

subjects are connected through EU-treaties with Switzerland, the participation in the framework programme will also be at stake in such a referendum. With respect to these challenges, the unique excellence of Switzerland is threatened and leadership in politics is essential. In the future, special efforts from all participants will be necessary in order to keep the uniqueness and high performance of the NIS of Switzerland.

References

Arora, A., Fosburi, A., & Gambardella, A. (2001). *Markets for technology*. Cambridge, MA: MIT Press.

Arvanitis, S., Hollenstein, H., & Marmet, D. (2005). *Internationale Wettbewerfäshigkeit: Wo steht der Standort Schweiz? Eine Analyse auf sektoraler Ebene*. Zurich: vdf-Hochschulverlag.

Arvanitis, S., Kubli, U., Sydow, N., & Wörter, M. (2006). *Knowledge and technology transfer between universities and private enterprises in Switzerland – An analysis based on firm and institute data, study on behalf of the eth-board. Synthesis report*. Zurich: ETH Zurich, Institute for Business Cycle Research (KOF).

Arvanitis, S., Kubli, U., Sydow, N., & Wörter, M. (2007). *Innovationsaktivitäten in der Schweizer Wirtschaft – Eine Analyse der Ergebnisse der Innovationsserhebung 2005*. Bern: Staatssekretariat für Wirtschaft.

Bieri, O., Balthasar, A., Feller-Länzlinger, R., Ebersberger, B., Edler, J., & Ruhland, S. (2005). *Evaluation der schweizerischen Beteiligung am 5. und 6. Forschungsrahmenprogramm der Europäischen Union sowie des Informationsnetzwerkes Euresearch. Interface Politstudien CH-Luzern in Zusammenarbeit mit Fraunhofer-Institut für System- und Innovationsforschung (ISI), D-Karlsruhe*. Bern: Staatssekretariat für Bildung und Forschung SBF.

ETH Board (2010). *Intermediate evaluation of the ETH Domain 2010: Self-assessment report compiled for the intermediate evaluation*. Zurich: Staff ETH Board.

Bundesamt für Berufsbildung und Technologie. (2010). *Fakten und Zahlen*. Bern: Berufsbildung in der Schweiz.

Bundesamt für Statistik. (2006). *Innovationsbericht*. Switzerland: Neuenburg.

Bundesrat. (1992). *Technologiepolitik des Bundes*. Bern: Bundesamt für Konjunkturfragen.

Bundesrat (2006). *Bericht über die Förderung von Frauen als Unternehmerinnen in Erfüllung des Postulats 03.3153 Fetz*. Bern: Staatssekretariat für Wirtschaft.

Chesbrough, H. W. (2003). *Open innovation: The new imperative for creating and profiting from technology*. Cambridge, MA: Harvard Business Press.

Dohse, D. (2001). Technologiepolitik auf neuen Pfaden. Einige Anmerkungen zur regionenorientierten Innovationspolitik der Bundesregierung. *Raumforschung und Raumordnung, 59*, 446–455.

Dolata, U. (2004). *Unfassbare Technologien, internationale Innovationsverläufe und ausdifferenzierte Politikregime, Perspektiven nationaler Technologie- und Innovationspolitiken. artec-paper Nr. 110*. Bremen: Universität Bremen.

Edquist, C. (Ed.). (1997). *Systems of innovation: Technologies, institutions, and organisations*. London: Pinter.

Eidgenössisches Volkswirtschaftsdepartement. (2003). *InnoNation Schweiz: Aktionsplan des Eidgenössischen Volkswirtschaftsdepartements zur Förderung von Innovation und Unternehmertum*. Bern: Bundesamt für Berufsbildung und Technologie.

Ergas, H. (1987). The importance of technology policy. In P. Dasgupta & P. Stoneman (Eds.), *Economic policy and technological performance* (pp. 56–96). Cambridge/New York: Cambridge University Press.

Etzkowitz, H. (2004). The evolution of the entrepreneurial university. *International Journal of Technology and Globalization, 1*(1), 64–77.

European Union (2010). European innovation scoreboard. http://www.proinno-europe.eu/innometrics/page/innovation-union-scoreboard-2010. Accessed 10 Mar 2011.

Eurostat (2010). Community innovation statistics (CIS). http://epp.eurostat.ec.europa.eu/portal/page/portal/microdata/cis. Accessed 10 Oct 2010.

Expertenkommission "Qualitatives Wachstum" des Eidgenössischen Volkswirtschaftsdepartements. (1985). *Bericht qualitatives wachstum.* Bern: Bundesamt für Konjunkturfragen.

Federal Office for Statistics (2010). Portal statistik Schweiz. www.statistik.admin.ch.

Freiburghaus, D., Balthasar, A., Zimmermann, W., & Knöpfel, C. (1991). *Technik-Standort Schweiz: Von der Forschungs- zur Technologiepolitik.* Bern, Stuttgart, Wien: Haupt.

Grande, E. (2000). *The erosion of state capacity and the European innovation policy dilemma: A comparison of German and EU information technology policies* (Political Science Series 70). Vienna: Institute for Advanced Studies.

Hotz-Hart, B., Good, B., Küchler, C., & Reuter-Hofer, A. (2003). *Innovation Schweiz: Herausforderungen für Wirtschaft und Politik.* Zurich: Rüegger Verlag.

Hotz-Hart, B., Dümmler, P., Good, B., Grunt, M., Reuter-Hofer, A., & Schmuki, D. (2006). *Exzellent anders! Die Schweiz als Innovationshost.* Zürich: Rüegger Verlag.

ITM. (2006). *Effect of CTI start-up label measures from 1998–2005, phase I, final report.* St. Gallen: Institute of Technology Management.

Lundvall, B.-Å. (1988). Innovation as an interactive process: from user-producer interaction to the National Innovation Systems. In G. Dosi, C. Freeman, R. R. Nelson, G. Silverberg, & L. Soete (Eds.), *Technical change and economic theory* (pp. 349–369). London: Pinter.

Lundvall, B.-Å. (2007). *National innovation systems: Analytical focusing device and policy learning tool.* Working paper, R2007:004. Östersund: ITPS Swedish Institute for Growth Policy Studies.

Nelson, R. R. (Ed.). (1993). *National systems of innovation: A comparative analysis.* Oxford: Oxford University Press.

OECD. (2009). Communications outlook 2009. Paris: Organisation for Economic Co-operation and Development.

OECD. (2010). *Education at a glance 2010.* Paris: Organisation for Economic Co-operation and Development.

Oelkers, J. (2005). *Der Handlungsspielraum der Bildungspolitik: Eine wissenschaftliche Perspektive.* Paper presented at the Nationale PISA-Tagung,Neuchâtel, Bundesamt für Statistik, 25 November 2005

Staatssekretariat für Bildung und Forschung. (2007). *Bibliometrische Untersuchung zur Forschung in der Schweiz.* Bern: Staatssekretariat für Bildung und Forschung SBF.

Staatssekretariat für Wirtschaft SECO. (2003). *Grundlagen der Wirtschaftspolitik, Nr. 3. Der Wachstumsbericht.* Bern: Eidgenössisches Volkwirstchaftsdepartment.

Staatssekretariat für Wirtschaft SECO. (2008). *Wachstumspolitik 2008–2011, Maßnahmen zur weiteren Stärkung des Schweizer Wirtschaftswachstums. Grundlagen der Wirtschaftspolitik Nr. 15D, Bericht des Bundesrates vom 2. April 2008 in Erfüllung der Motion 01.3089 "Wachstumspolitik. Sieben Massnahmen".* Bern: Staatssekretariat für Wirtschaft SECO.

Swiss National Science Foundation (2010). Webportal. http://www.snf.ch/E/ABOUT-US/Pages/default.aspx. Accessed 15 Oct 2010.

Volery, T., Bergmann, H., Haour, G., & Leleux, B. (2007). *Bericht 2007 zum Unternehmertum in der Schweiz und weltweit.* St. Gallen, Lausanne: Global Entrepreneurship Monitor.

Werle, R. (2001). Liberalisierung und politische Techniksteuerung. In G. Simonis & R. Martinsen (Eds.), *Politik und Technik. Analysen zum Verhältnis von technologischem, politischem und staatlichem Wandel am Anfang des 21* (Jahrhunderts (Vol. Politische Vierteljahresschrift, Sonderheft 31), pp. 407–422). Wiesbaden: Westdeutscher Verlag.

Weyer, J., Kirchner, U., & Riedl, L. (1997). *Technik, die Gesellschaft schafft*. Edition Sigma: Berlin.
World Economic Forum. (2008). *The global competitiveness report 2007–2008*. Geneva: World Economic Forum.
World Economic Forum. (2010). *The global competitiveness report 2009–2010*. Geneva: World Economic Forum.

Chapter 7
The Politics of Innovation: Analysing Inter-organisational Networks Around German Innovation Policy Advisory Bodies

Matthias Orlowski

7.1 The High-Tech Strategy for Germany

In August 2006, the German Federal Government adopted the *High-Tech Strategy for Germany* (HTS). This overarching national strategy for innovation policies presents a joint effort of all governmental departments to improve Germany's innovation capacity in order to enhance economic growth and employment in high-tech industries. In order to accomplish the ambitious goal to make Germany the "most innovative economy in the world," the HTS defines several key technologies where companies and research facilities will be granted financial support and better access to venture capital. The financial efforts are accompanied by regulative measures that are supposed to improve the framework conditions of companies and research facilities that work in innovative branches. These regulative efforts include a wide range of measures like corporate tax reform, reducing administrative burdens, and measures to better secure intellectual property rights (BMBF 2006). The Excellence Initiative, the Higher Education Pact 2020, and the Joint Initiative for Research and Innovation are all important parts of the HTS and have brought major changes to the education and research system (BMBF 2009). The major goal of these reforms was to increase competition among German universities and research institutions for public funding and to attract more researchers and scientists from abroad.

Besides these financial and regulative means, one key goal of the HTS is to advance connectivity between the societal subsystems of industry, science, and politics in order to speed up innovation processes. At the national level, the government established the *Council for Innovation and Growth* (CIG) and the *Industry-Science Research Alliance* (ISA) in order to bring together important representatives of the three societal subsystems. Besides advising and supporting

M. Orlowski (✉)
Berlin Graduate School of Social Sciences, Berlin, Germany
e-mail: matthias.orlowski@hu-berlin.de

J.M. Bauer et al. (eds.), *Innovation Policy and Governance in High-Tech Industries*,
DOI 10.1007/978-3-642-12563-8_7, © Springer-Verlag Berlin Heidelberg 2012

the government in the implementation of various High-Tech Strategy programmes, the 17 members of the *Council* as well as the 18 members of the *Research Alliance* are meant to establish an exemplary network of connections between the realms of industry, science, and politics.

Members of the bodies serve in an advisory function. They are also instrumental in developing the intended linkages between the three societal subsystems. As they have an impact on the execution of these duties, the role of the bodies' members in their own organisation and their existing affiliations with other organisations are important. Interests stemming from their primary occupation will likely influence their positions on innovation policy. Furthermore, these scientific, industrial, and political organisations are supposed to be linked via the CIG and the ISA. In order to assess what kind of interests have access to German innovation policy via the two advisory bodies, and whether or not they do indeed improve connectivity between industry, science, and politics, this article provides an analysis of the inter-organisational network that surrounded the CIG and the ISA in 2006.

A systemic approach to innovation suggests that the bodies should preferably be staffed with representatives of various economic sectors and different societal subsystems. Such a composition would enhance the integration of the innovation system and assure a balance of interests within the consultation process. Applying interlocking directorate analysis to operationalise relations between organisations, this article explores whether or not the constitution of the two bodies meets these criteria.

As the following analysis shows, the CIG and the ISA fail to do so. The results indicate a preponderance of certain economic interests within the network. Furthermore, it turns out that there exist numerous other links between organisations within the three subsystems outside the advisory bodies. This weakens the importance of CIG's and ISA's integrating function, since they barely constitute unique, unprecedented links between industry, science, and politics.

7.2 Systems Theory, Innovation Systems and Interlocking Directorates

Modern societies have differentiated into various subsystems due to division of labour. According to functionalist system theory (Parsons 1951; Münch 1982; Luhmann 1984) each subsystem performs specific functions for other subsystems, as well as for society as a whole. Assuming that the actors of each subsystem act on a specific maxim while fulfilling their particular functions, one can refine the systemic premise of functional differentiation to the level of individual and collective actors (Schimank 2005; Lange and Schimank 2004). The scientific and educational system is engaged in the differentiation between true and false knowledge. Its actors therefore seek to maximise approved knowledge. Authoritative decisions about values are the major task of the political system (Easton 1957). Since a certain degree of power is required to enforce these values, maximising power is the principal concern of actors in the political system. Profit maximising is the leading principle for actors belonging to the economic subsystem. According to these kinds of maxims, which lead actors in their activities, one can identify numerous societal subsystems. However, since the scientific, the political, and the economic system constitute the core elements that shape national innovation systems, I will confine myself to these three for now.

As noted above, the societal subsystems do not operate in complete isolation from each other. In order to be able to fulfil its function, each subsystem relies on the others' functional compliance. This presupposes the integration of the distinct subsystems. Any particular subsystem's needs have to be recognised by the others and the latter's actors need incentives to meet these requirements (Schimank 1996). Coping with these interdependencies can only take place at the level of collective or individual actors. Principal mechanisms to communicate needs are mutual observation, purposeful interaction, or negotiations (Lange and Schimank 2004, p. 20). The basis of all these coordination mechanisms is an information flow between actors of at least two different subsystems.

7.2.1 The Innovation System

The integration of different societal subsystems becomes particularly important with regard to the innovation system. Successful innovation policy often requires the coordination of knowledge and actors from multiple subsystems. One has to note that the innovation system does not constitute a subsystem in the way discussed above. Actors in the innovation system seek to fulfil a common function. They create, store, and transfer knowledge, skills, and artefacts, which define new technologies (Metcalfe 1995). However, the actors follow different maxims while pursuing these tasks. The innovation system is the intersection of different societal subsystems (Lundvall 1992; Freeman and Soete 1997). It can be seen as a heuristic device, an arrangement to cope with the complex causal interrelations that characterise innovation processes rather than as a distinct societal subsystem.

Innovation processes often require the application of fundamental research in order to develop new technologies, which enable the creation of new, innovative products. However, this process is not linear but shaped by complex, repeated interactions between various actors from different societal subsystems (Lundvall 1992). The scientific, political, and economic subsystems constitute the core elements of the innovation system. In Germany, universities and public research institutions belong to the scientific subsystem. They mainly conduct basic research. These institutions do not seek profits but have an interest in the production of knowledge as such. The political system fundamentally shapes the overall framework within which the innovation process takes place. Particular projects are publicly funded, the development of certain technologies can be prohibited at the outset, and actors are constricted by legally binding norms. Finally, in market economies, the application of new technologies in order to develop new products and the subsequent commercialisation of these products are typically – though not exclusively – carried out by profit-seeking private enterprises.

However, actors from other societal subsystems have a significant influence on innovation processes too. This is particularly reflected in the demand for products that result from these processes. Anticipated negative impacts on the environment and on society as a whole are as important for the demand for innovative products as the benefits attributed to the products. Consumers, as well as groups and institutions concerned with technological impact assessment, therefore, also have to be considered as being part of the innovation system, even though they only have an indirect effect on innovation processes which is mediated by the three core subsystems (Dolata 2007). Since the Council for Innovation and Growth as well as the Industry-Science Research Alliance are supposed to enhance the integration of all subsystems constituting the innovation system, the inclusion of important actors from the societal areas identified here, is one criterion to assess the prospective effectiveness of the two bodies in attaining their goals.

7.2.2 Pluralism of Knowledge, Values and Interests

Both, the CIG and the ISA are practical advisory bodies, based on the corporate model, rather than a scientific one. This means that the bodies' members' knowledge is based on experience rather than on theory-led research. Various factors in a person's environment shape its experiences. On the one hand, values and interests of individuals are partly derived from the maxims that are dominant in the societal subsystems in which the particular individual occupies a position (Heinrichs 2002; Mayntz 2006). On the other hand, the concrete subject matter of a person's occupation fundamentally shapes her knowledge. Occupational images, again, are not congruent with societal subsystems. They rather constitute a common denominator between positions in different subsystems. The knowledge of a research assistant in the computer sciences, for instance, is largely congruent with that of a software engineer who works for a private company. Thus, occupational images are subsystem spanning. At the same time, however, they vary within one subsystem.

One can assume that the members of the *Council* and the *Research Alliance* do not make their recommendations based on "objective" knowledge that is free of interests. Their values, interests, and knowledge are likely influenced by their primary occupational position. In order to avoid a preponderance of particular interests within the two advisory bodies, they should therefore be staffed with representatives of various sectors and subsystems. Such a pluralism of knowledge, values, and interests within the advisory bodies is the only way to assure a balance of interests within the consultation process. This pluralism is the second criterion, on the basis of which the prospective effectiveness of the CIG and the ISA shall be assessed.

7.2.3 Interlocking Directorates

The interlocking directorates approach focuses on the membership of one person in various organisations and the resulting linkages between the corresponding organisations. "An interlocking directorate occurs, when a person affiliated with one organisation sits on the board of directors of another organization" (Mizruchi 1996).

In Germany, as well as in the United States, the first interlocking directorate studies examined the linkages between banks and large corporations. These studies were of high political interest since they provided indicators for potential cartelisation processes (Fennema and Schijf 1978–1979). Until the 1950s, most analysis of interlocking directorates dealt with the linkages among financial institutions, those among corporations, and those between financial institutions and corporations (Wallich 1905; Sweezy 1953). The fact that they were used as an indicator for cartelisation reveals that in these studies, interlocks were interpreted as a means for cooperation and collusion among enterprises. Particularly in the case of interlocks between corporations and banks, the practice of sharing board members enhances an organisation's control over some sources of uncertainty in its environment (Pennings 1980; Mizruchi and Stearns 1988). Many of these studies are based upon organisational theory and resource dependency models. These approaches are still going strong today in the context of various research questions (Geletkanycz et al. 2001; Carpenter and Westphal 2001; Ruigrok et al. 2006). They typically focus on the impact of interlocking directorates on the performance of organisations. Consequently, the underlying assumption is that information flow between organisations is enhanced through interlocks, which then enables them to adapt quickly to changing circumstances (Dalton et al. 1999).

Since the mid-1950s, sociologists have become increasingly interested in researching interlocking directorates too. Within Marxist and elite theoretical frameworks, they have investigated the socio-economic backgrounds of major interlockers (Domhoff 1967; Silva et al. 2006), social integration through joint board membership and the corresponding social cohesion (Mills 1956; Scott 1991; Carroll and Shaw 2001; Carroll and Colin 2003). In these studies, the focus is less on the organisations linked by interlocking directorates but rather on the interlockers themselves.

All in all, interlocking directorates studies address a wide range of substantive questions. Rather than being a matter of common knowledge, each study derives the causes and consequences of interlocking directorates from an underlying theoretical approach. In principle, the emergence of interlocks has an ambivalent character. Their origin has to be examined for every particular case, for example, by exploring whose motivation and what kind of underlying motives have led to the interlock (Mizruchi 1996). However, in terms of their consequences, all theoretical approaches to interlocking directorate analysis share one common assumption. They all presuppose an information flow between the involved persons and their respective organisations that results from interlocking directorates. The theoretical differences then stem from varying propositions about the consequences of this information flow. The exchange of information is seen as independent of the causes that led to interlocking directorates and has been verified empirically as well (Davis 1991; Davis and Greve 1997). Thus, I reduce my assumption to the lowest common denominator of interlocking directorate theories, assuming that interlocks provide a valid indicator for relations between organisations. They do at least provide an information channel between the involved organisations.

Summarising these theoretical considerations, one can state that the integration of societal subsystems is crucial for a society's innovative capacity. This integration requires the exchange of information between actors from different subsystems. Interlocking directorates provide an information channel between organisations and collective actors respectively. If two organisations from different societal subsystems are interlocked, the likelihood of information exchange between subsystems is increased. Thus, interlocking directorates can be regarded as a means for societal integration. The actual communication, however, always involves real persons. The knowledge, values, and interests of these people influence what kind of information they exchange, how they exchange it, and, finally, how they perceive and interpret it. Knowledge, values, and interests, in turn, are fundamentally shaped by the subsystem in which a particular person's main professional affiliation can be located, and the concrete occupational subject matter of that position. In regard to a balanced consultation of the German Federal Government and the hoped for improvement of the connectivity between industry, science, and politics, two aspects regarding the composition of the two advisory bodies appear to be crucial: the representation of various organisations stemming from different societal subsystems and the incorporation of various occupational images. This is not only due to pragmatic considerations regarding innovation systems, but can also be derived from normative considerations in regard to democratic theory. Whether or not these requirements have been met in the constitution of the *Council for Innovation and Growth* and the *Industry-Science Research Alliance* shall be examined in the remainder of this article.

7.3 Operationalisation and Data

The interlocks created by the advisory bodies' members constitute the network that is analysed below. Therefore, a list of all organisations the *Council*'s and the *Research Alliance*'s members are also affiliated with was the point of departure when gathering the network data. For every single member, these organisations were identified via publicly accessible curriculum vitas and several databases containing information about the board members of corporations, foundations, and other associations. The results were then crosschecked against the annual reports of the identified organisations. These organisations' additional board members were gathered from the annual reports as well. In order to examine the questions stated above, the definition of interlocking directorates had to be modified and broadened. For corporations with separate advisory and management boards,[1] research institutions, foundations, and state institutions, all members of the advisory bodies, as well as those of the management bodies, were considered as potential interlockers. In the cases of the German Parliament, the Federal Government, and the policy fora included in the study, all members were incorporated in the dataset. Thus, an affiliation matrix that maps the membership of every person to various organisations could be constructed. By multiplying the transposed affiliation matrix with the original one, a square adjacency matrix was calculated which contains information about the number of shared members among all organisations.[2] Since neither intensity nor direction of the relationships will be addressed in the following analysis, the adjacency matrix was dichotomised and linkages were considered to be undirected.[3]

Organisations were attributed to societal subsystems according to the main purpose stated in their charter. Profit-seeking corporations belong to the economic subsystem, non-profit research institutions to the scientific subsystem, and state institutions to the political system. Foundations, interest groups, and fora that provide places to discuss policy proposals have a unique character. Since they do not commonly pursue one unambiguous purpose, they cannot be ascribed to a particular subsystem. Rather they are considered to be intermediaries between the subsystems, each one striving for a particular set of goals. They are therefore regarded as distinct types of organisations. Consequently, the nominal variable "purpose" was classified into six categories: corporation, research institution, state institution, interest group, non-profit organisation, and policy forum. Based on the industrial classification of the *German Federal Statistical Office* (Destatis 2002), the nominal variable "branch" was defined in order to proxy the occupational subject matter associated with each organisation. Each organisation was ascribed to one of the 19 branch categories.[4]

---

[1] According to German law, corporations exceeding a certain number of employees are obliged to install an advisory board which monitors the board of directors.

[2] The affiliation matrix X has all persons of the dataset in the first column and all organisations in the first row. Cells indicate the affiliation of a person with an organisation by 1 and contain 0 otherwise. The multiplication of the transposed matrix X' (that is the same matrix but with organisations in rows and persons in columns) with X results in a so-called adjacency matrix that has all organisations in rows as well as in columns. The cells of this affiliation matrix indicate the number of shared board members (the intensity of their connection). See Wasserman and Faust (Wasserman and Faust 1994) and Faust (Faust 2005) for details on this procedure.

[3] It is not plausible that the communication between two organisations increases proportional to the number of shared board members. Therefore, the intensity of connections (the number of shared members) is not interpreted here. Furthermore, since an interlock is only considered as a channel of communication without stating whether organisation A communicates with B or vice versa, links are considered undirected. All that matters here is whether a connection exists (indicated by 1 in the adjacency matrix) or not (indicated by 0).

[4] Two independent coders were asked to assign all organisations to the 19 categories of the variable branch. Intercoder reliability according to Perreault and Leigh (Perreault and Leigh 1989) was high. It was 0.99 for the coding of "purpose" and 0.98 for "branch".

7.4 The Politics of Innovation

The first section of the analysis deals with the question whether or not pluralism of knowledge, values, and interests is ensured in the *Council* and the *Research Alliance*. This is done by examining the network's composition with respect to the identified variables. In the second section, the integrating function of the two advisory bodies will be assessed based upon the linkages among the participating organisations.

7.4.1 The Composition of the Network

All in all, 180 organisations are linked through interlocking directorates to the *Council of Innovation and Growth*, the *Industry-Science Research Alliance*, or both bodies. 25 research institutions are part of the network including major players like the *German Research Foundation* and the *Max-Plank-Society*. The fact that these institutions are incorporated in the innovation policy network certainly is an important aspect, as they broaden the range of interests and players represented.

Besides the German Parliament and the Federal Government, the Bund-Länder Commission for Educational Planning and Research Promotion and the KfW Banking Group represent state institutions in the network.

More than half of all organisations are corporations. Among them, there are major holdings like *Volkswagen*, *Daimler*, or *Henkel* as well as small and medium-sized enterprises from various sectors. The large proportion of financial institutions within the group of corporations is noticeable. This reflects the fact that financial institutions are generally heavily interlocked with other corporations from all kinds of branches, as noted in previous interlocking directorate studies (Mizruchi and Stearns 1988; Pennings 1980). The lion's share of corporations is engaged in machinery and vehicle construction. Since this branch is a medium-rather than high-tech[5] sector, such a strong representation within the network could be seen as counteracting the intended purpose. The high proportion of corporations from this sector indicates that this traditionally strong branch within the German economy also managed to assert itself within the high-tech political advisory bodies. The same holds for the German power-supply industry. All major energy providers except *Eon* are included in the inter-organisational network.

The representation of the machinery and vehicle construction industries' interests is further strengthened by the presence of 6 (out of 16) interest groups from this sector, including the labour union *IG Metall*. Thus, there is a risk that

[5] The technology intensity of particular branches was derived from the ratio of research and development expenses to production output and value added within the particular industries, according to the OECD's classification of industries based on technology (OECD 2005).

these interests dominate in the consultation process. Besides five inter-sectoral interest groups, the *IG Metall*, and the five employer organisations from the machinery and vehicle construction industry, there are two interest groups each representing one of the branches "energy" and "automation". These branches also are mainly medium-tech, but traditionally exert high influence within German politics and in the economy. The only interest group that belongs to high-tech industries is *BITCOM*, an employers association representing the concerns of the information technology and telecommunications industry. Furthermore, it is noteworthy that there is only one labour union, *IG Metall*, among all interest groups in the network. All others are employer associations, including the two major ones, the *Federation of German Industries* and the *Confederation of German Employers' Associations*. This uneven representation raises concerns about the balance between capital and labour interests within the CIG and the ISA.

Consumer associations, ecological groups, as well as groups and institutions concerned with technological impact assessment, are not included in the network. Given their importance for the innovation system as outlined above, this is a serious deficit in the constitution of the two advisory bodies. Although there are commissions and bodies advising the German government in these regards, their incorporation into the innovation policy network certainly would have enriched the quality of the consultation process.

Examining the composition of the networks around the *Council* and the *Research Alliance* separately allows us to assess whether they complement each other or whether they form redundant linkages. One hundred organisations are directly linked to the CIG, 32 of which also belong to the network surrounding the ISA. The major research institutions and five national, inter-sectoral policy fora are linked to both advisory bodies. Only 7 of the 104 corporations in the overall network are related to both the *Council* and to the *Research Alliance*. However, among those seven, some belong to the largest holdings of the German economy, namely the *Allianz Group*, *Siemens*, *ThyssenKrupp*, and the *Lufthansa Systems Group*. Except for the *Allianz Group*, these corporations spend a high share of their revenue on research and development. Therefore, the fact that they are linked to the CIG as well as the ISA might be due to their overall economic and political power as much as their high-tech profile.

Sixty-four corporations are included in the network surrounding the *Council*. Thus, in this sub-network the share of private enterprises is higher than in the overall network. Furthermore, corporations from medium- to high-tech sectors are dominating. Almost all corporations from high-tech sectors deal with information technology. In contrast, other technologies with great potential, like optics or nanotechnology, are only sparsely represented.

With the establishment of the *Research Alliance*, 84 organisations have been incorporated into the new innovation policy network. The 12 newly integrated research institutions are particularly striking. The *German Research Center for Artificial Intelligence* and the *Laser Center Hannover* could contribute important knowledge about the development of high-end technologies. With regard to the newly integrated corporations, the diversification of high-tech sectors ought to be mentioned. Enterprises that deal with optics and measurement engineering are solely embedded via the ISA. Since these branches were identified as cutting-edge fields within the High-Tech Strategy and therefore chosen for special promotion, companies that belong to these branches certainly should be able to address their concerns within the advisory bodies. The *Research Alliance* therefore binds important additional branches into the network. There are 12 interest groups added to the network by the ISA as well. Most of them represent the machinery and vehicle construction sector, the interests of which predominate within the overall network. The only labour union is also incorporated via the *Research Alliance*.

In sum, both advisory bodies connect important actors from industry, science, and politics. Both the major research institutions and various corporations stemming from trendsetting industries are involved in the consultation process. Thereby, major holdings, as well as small and medium-sized enterprises, are represented. However, in order to incorporate all important groups of the innovation system, representatives of powerful German holdings could have been substituted by spokespersons of consumer groups and agencies concerned with technology impact assessment. By establishing the *Industry-Science Research Alliance* as a second advisory body, corporations from other important branches and research institutions were added to the innovation policy network. Even though the establishment of this second body did partly correct the existing disequilibrium in the interests represented, in the overall network the interests of capital predominate, in particular those of the machinery and vehicle construction sector.

7.4.2 The Integration of Industry, Science, and Politics

This section examines the integrating function of the *Council* and the *Research Alliance*. It will analyse the overall network of interlocking directorates through the application of various network analytical concepts. The focus here is on the second research question, asking whether the advisory bodies do indeed improve the connectivity between industry, science, and politics. In other words: Did they actually constitute unique, unprecedented links between the three societal subsystems?

The degree of integration in the overall network can be assessed using the concept network's density.[6] 12% of all possible interlocks are realised among all organisations included in the network. Since the interpretation of this figure as such is problematic, it shall just serve as a benchmark for now. Examining the path

[6] The network's density d is the share of realised links L (here: observed interlocks) in all possible links between all nodes (here: organisations) in a network. Since the connections are considered unweighted and undirected, the maximum number of links is given as $g(g-1)/2$ where g is the number of nodes in the network. Therefore $d = 2L/g(g-1)$ (Wasserman and Faust 1994).

Table 7.1 Average densities in blocks according to "purpose"

	Non-profit organizations	Corporations	Interest groups	Research institutions	Policy fora	State institutions
Non-profit organizations	0.229	0.100	0.113	0.221	0.414	0.298
Corporations		0.082	0.108	0.080	0.263	0.038
Interest groups			0.158	0.070	0.325	0.047
Research institutions				0.240	0.344	0.130
Policy fora					0.644	0.500
State institutions						0.833

length[7] between all organisations provides another means to assess the cohesion of the network. Since all organisations in the network are either connected to the CIG, the ISA, or both, the potential maximum path length in the network is three. The observed average path distance in the overall network is 2.05. This suggests that the number of organisations that has to pass the maximum path length to communicate with other organisations via interlocks is higher than the number of directly connected organisations. Indeed, only 10% of all organisations are directly linked. In nearly 20% of all dyads the path length is three. The remaining dyads are connected via one intermediary organisation.

The theoretical considerations suggest that the observed medium degree of cohesion is due to the relative isolation of the societal subsystems. Accordingly, the connectivity among organisations stemming from the same subsystem should be higher than that between organisations from different systems. One would expect that linkages of the latter kind are conveyed by intermediary organisations. In order to test this hypothesis, a block model form of the original adjacency matrix was calculated.[8] Blocks were built according to the variable "purpose" and the densities of relations within each block, as well as for all links between the different blocks were computed (see Table 7.1). At first glance, one notes that connectivity among state institutions is almost perfect. Whereas the density of linkages between state institutions and research institutions is on an average level, connectivity between state institutions and corporations remains below it. According to the theoretical considerations, the figures of the diagonal array in Table 7.1 should be higher than the remaining ones, and they should be higher than the average 12% as well.

However, this is not true in all cases. The fact that the density of linkages between the particular blocks and the policy fora is higher than that within each block is due to the method of collecting data. Even if one neglects this, one has to note that the measured values still do not conform to the theoretical assumptions. The density of

[7] Path length is the number of links on the shortest connection between two nodes in a network.

[8] The block model simply sorts the nodes of the adjacency matrix according to a variable that provides information about nodes' characteristics. One can thus analyse the network components that only consist of nodes with the particular characteristic used to "block" the model or those parts of the network that link different blocks.

linkages among corporations, in particular, is remarkably low. First of all, this result derives from the huge differences in the number of actors that were ascribed to each block. The number of possible links among all actors within one block grows exponentially with an increasing number of actors attributed to it. Therefore, the concept of network density is inadequate to compare the connectivity of blocks with high differences in regard to the number of actors ascribed to each block.

The E-I-Index according to Krackhardt and Stern (1988) provides another, more suited concept to assess whether or not the systemic structure is reflected in the interlocking directorates network surrounding the *Council* and the *Research Alliance*. This index represents the ratio of the number of links within a particular block to the number of links that block members have with actors outside that same block. E-I-Index values in Table 7.2 show that for most types of organisations, linkages to organisations belonging to other blocks outnumber those within their own block. In the case of corporations, the number of internal and external relations is roughly equal with an E-I-Index value of 0.008. Among the external links, those to policy fora, including CIG and ISA, predominate. Only 2% of all external links are connections to the political system. This is the smallest share in external relations of the economic system. The remaining external ties are more or less equally distributed between interest groups, non-profit organisations, and research institutions.

Although there are direct links between corporations and organisations ascribed to the other societal subsystems, most external relations connect corporations with intermediary organisations. Within the scientific subsystem, connectivity is relatively high too. Nevertheless, external links to other subsystems and intermediary organisations predominate for research institutions as well. The high connectivity between industry and science attracts particular attention. After all, 46% of all external links connect research institutions with corporations. The marginal connection of the scientific subsystem to interest groups is not surprising, given that their primary purpose is to represent economic interests rather than scientific ones. The actors of the scientific subsystem are also engaged in non-profit organisations.

Table 7.2 E-I-Index and the distribution of external links

	Non-profit organizations	Corporations	Interest groups	Research institutions	Policy fora	State institutions
Non-profit organizations	96	291 (45%)	38 (8%)	116 (24%)	87 (18%)	25 (5%)
Corporations	219 (24%)	882	180 (20%)	208 (23%)	274 (31%)	16 (2%)
Interest groups	38 (13%)	180 (60%)	38	28 (9%)	52 (17%)	3 (1%)
Research institutions	116 (26%)	208 (46%)	28 (6%)	144	86 (19%)	13 (3%)
Policy fora	87 (17%)	274 (53%)	52 (10%)	86 (17%)	58	20 (4%)
State institutions	25 (32%)	16 (21%)	3 (4%)	13 (17%)	20 (26%)	10
Total	581	1,779	339	595	577	87
External links	485	897	301	451	519	77
E-I-Index	0.670	0.008	0.776	0.516	0.799	0.770

The figures in brackets indicate the share of external links to the block in the column, in all external links for each particular row. The figures in the diagonal present the number of links within each block

The share of all external links to these organisations roughly equals those of the economic system. This indicates the intermediary character of non-profit organisations. The linkages to state institutions account for the smallest share in all external links of the scientific system, which is again due to the small number of state institutions in the overall network.

This fact also explains the high external orientation of the political system. As indicated by Table 7.1, state institutions are relatively highly connected among each other. Table 7.2 shows that the external links of the political system are more or less equally distributed across the other blocks. However, there are relatively few links to interest groups, whereas the number of links to non-profit organisations is above the average. Like the economic and the scientific subsystems, the political system is directly linked to other societal subsystems and to intermediary organisations. The intermediaries all show a similar pattern of linkages. In the three blocks, external linkages prevail at a high level, once again indicating their integrating function. About half of all external links connect intermediaries to the economic system. This applies to all kinds of intermediary organisations. The proportion of any one intermediary block to other intermediary blocks ranges between 8% and 18%. The small share of linkages connecting intermediaries with state institutions reflects the already familiar fact that only few state players are represented in the network.

In sum, there are links within each block as well as among the five blocks. The E-I-Index does not indicate that the structure of the overall network is differentiated into isolated subsystems. Thus, the medium degree of cohesion is not due to the expected structural holes that would mirror the societal subsystems.

Building a block model using the matrix of the path distances between all organisations, allows us to calculate average path lengths for each block and for the connections among the blocks. The average path distance between research institutions and corporations is 2.11. This indicates that most research institutions and corporations communicate via third organisations. The average path length for the linkages between research institutions and state institutions, as well as for those between state institutions and corporations, is just under two. In these cases, then, the proportion of direct links is higher. Yet there are many organisations in each block that are only linked to others via third ones. Furthermore, the block modelling of path distances points out that, after removing the CRG and the ISA from the network, policy fora still have an average path distance below two to all other blocks. The other fora are directly linked to multiple other organisations, thus having a major integrating effect in the overall network.

Both, the *Council* and the *Research Alliance* were ascribed to the block of policy fora. In order to assess the actual integrating function of these two separately, one has to compare the cohesion of the original network with the network after the removal of the two advisory bodies. Only nine organisations become isolated after removing both bodies. This indicates that communication via interlocking directorates is possible beyond the CIG and the ISA for the lion's share of all organisations. The maximum path length, however, increases to five. Yet, with only 0.1% of all dyads realising the maximum path length, this only affects a few organisations. Average path distance increases by 11% from 2.05 to 2.28, which means that communication channels lengthen. In particular, links between corporations are weakened by the removal of the two advisory bodies from the overall network. The resulting increase in the path distance between corporations and all other blocks is above average. In contrast, the average path length of non-profit organisations remains relatively constant. This again indicates their role as intermediaries. However, the average path distance between interest groups and all other kinds of organisations decrease most strongly after removing the *Council* and the *Alliance*. This is true, even though interest groups are also regarded as intermediaries. The comparison of the original network with a network lacking the two advisory bodies finally shows that the CIG and the ISA shorten the information channels between the observed organisations. However, for almost all organisations the information exchange via interlocking directorates remains possible. This applies to both links within subsystems, as well as to those across boundaries.

Applying betweenness centrality according to Freeman allows us to assess what kind of organisations mediate indirect information flows (Freeman 1978–1979). One possible interpretation of this measure is that it provides information about the extent to which a particular organisation controls the communication between two others. It measures all minimum paths between two nodes that pass through a third. Since the information flow through interlocking directorates is considered as a means of societal integration, one can assume that organisations with a high betweenness centrality perform a major integrating function. This is because they provide the shortest information channel between many other organisations. Calculating the average betweenness centrality for the block model shows that policy fora by far have the highest degree of centrality in that sense. This is not surprising of course, since the two advisory bodies that constituted the starting point of the network survey were ascribed to this block. Either the *Council* or the *Research Alliance* has direct links to all other organisations, which explains the high centrality of this group at a first glance. For this reason, the average betweenness centrality of policy fora decreases sharply from 1051 to 595 after removing the CIG and the ISA. However, this block remains the most central one, followed by the non-profit organisations (Table 7.3).

After removing the two advisory bodies, the average centrality of non-profit organisations increases by 164% from 87 to 231. This sharp increase indicates that they function as intermediaries between other organisations as soon as the shortest

Table 7.3 Average betweenness centrality in blocks

	Original network	Network without CIG and ISA	Change	Percentage change (%)
Non-profit organizations	87.40	230.59	143.19	164
Corporations	13.88	23.74	9.86	71
Interest groups	77.72	178.73	101.01	130
Research institutions	74.72	129.11	54.39	73
Policy fora	1,050.82	594.93	−455.90	−43
State institutions	23.25	29.90	6.66	29

path between those organisations does not pass through the *Council* or the *Research Alliance* anymore. The same is true for interest groups, whose average betweenness centrality rises from 78 to 179 after removing the CIG and the ISA. All in all, those groups regarded as intermediaries between societal subsystems indeed have a higher degree of betweenness centrality than organisations belonging to a particular subsystem. After hypothetically removing the *Council for Innovation and Growth* and the *Industry-Science Research Alliance*, their mediating role further increases. Thus, one can infer that intermediaries contribute remarkably to the integration of societal subsystems.

A hypothetical removal of all intermediary organisations from the network additionally reveals their overall integrative function. The resulting network consists of five distinct and isolated components. In 34% of all dyads, the involved organisations cannot reach each other at all. Yet the largest component still consists of 107 organisations linked by interlocking directorates. This analysis illustrates that a very large proportion of the investigated organisations stemming from industry, science, and politics are directly connected via interlocking directorates. All others are integrated through intermediaries. Furthermore, the intermediary organisations significantly shorten the lengths of communication channels. The average path distance in the network without any intermediary organisations increases to 2.24. This is an increase of 18% as compared to the original network. Consequently, policy fora, non-profit organisations, and interest groups play an important integrating role.

Examining the subjects addressed in policy fora other than the CIG and the ISA reveals that they all deal with innovation policy issues. For example, the *Forum für Zukunftsenergien* (Forum for Future Energy Sources) and the *German Energy Agency* focus on energy efficiency and promote the development of renewable energy sources. Other fora were established to discuss policies that should improve Germany's innovative capacity in general. All these fora proclaim one of their major concerns to be the improvement of connectivity between industry, science, and politics. Apart from the implementation of the High-Tech Strategy, they thus serve essentially the same purposes as the two advisory bodies examined in this article. Members of the German Federal Government are represented in the *Rationalisierungs -und Innovationszentrum der Deutschen Wirtschaft* (German Industry's Rationalisation and Innovation Centre) and the *German Energy Agency*. Since all policy fora included in the network are strongly connected by interlocking directorates beyond the CIG and the ISA, the government could gain information on innovation policy issues from all fora with a maximum path distance of two.

The *Council* and the *Research Alliance* enhance the connectivity of industry, science, and politics. Yet, as shown in Fig. 7.1, numerous actors from the three different societal subsystems are directly linked as well. Furthermore, there are various intermediary organisations that also contribute to the integration of the three subsystems. Since only organisations with direct links to either the CIG or the ISA were included in the analysis presented in this article, the question rises whether a broader analysis of connectivity among important actors would come to a different conclusion. In particular, one might ask whether a more inclusive

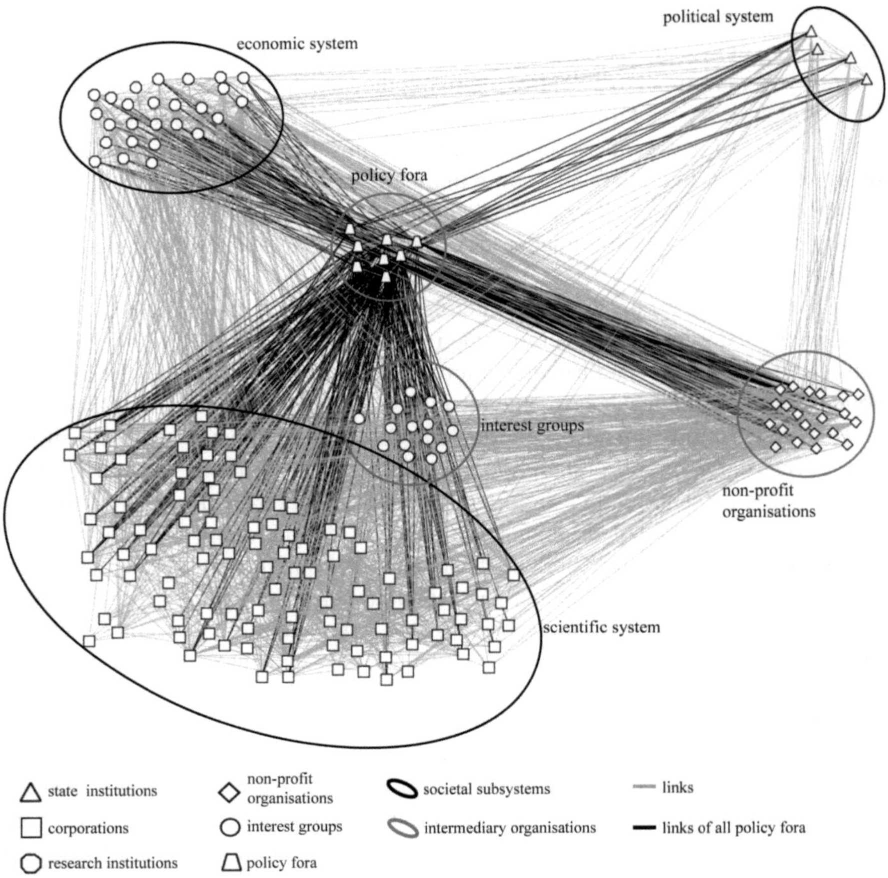

Fig. 7.1 Connectivity between societal subsystems beyond the CIG and the ISA, and the role of intermediaries

analysis of the connectivity between actors in the innovation system would reveal that the integrating function of existing policy fora at least resembles the integrating effect of the recently established ones. Since the pre-existing fora essentially deal with the same issues as the newer ones, intensifying cooperation with them would have been an alternative to the establishment of new advisory bodies. In doing so, the government would have foreclosed an additional gateway for particular interests to the political system.

7.5 Conclusions

The functionally differentiated society relies on the performance of various societal subsystems. They, in turn, mutually depend on each other's performance in order to fulfil their function. In order to cope with these interdependencies, the integration of the different subsystems is necessary. This integration is accomplished by information exchange and communication between individual and collective actors. Since interlocking directorates can be interpreted as a channel of communication between organisations, they provide a means of societal integration, given that they link two organisations stemming from different societal subsystems. With regard to the innovation system, the integration of three societal subsystems, namely the political, the scientific, and the economic system, gains major importance.

The analysis of the interlocking directorate network surrounding the *Council for Innovation and Growth* and the *Industry-Science Research Alliance* shows that several direct links between actors from industry, science, and politics in Germany indeed exist. Besides the two advisory bodies, there are several other policy fora that integrate those organisations that are not directly linked to each other. Thus, intensifying the cooperation with these fora would have been an alternative to the newly establish additional advisory bodies for the German Federal Government. Considering the prior existence of other fora with similar purposes, one certainly cannot speak of "a new culture of strategic cooperation" as stated by Annette Schavan, the German Federal Minister for Education and Research (Schavan 2006, own translation). Even though the *Council* and the *Research Alliance* integrate some important actors that can be ascribed to the innovation system, they fail to integrate important groups like consumer associations, ecological groups, and groups concerned with technological impact assessment. Furthermore, capital interests and the interests of traditionally powerful industries, regardless of their research and development efforts, are over-represented in the two advisory bodies. Since there are no societal groups to outbalance this preponderance included in the network, the responsibility to level out these imbalances during the political process is ultimately up to the decision makers in the political system.

References

BMBF. (2006). *Die Hightech-Strategie für Deutschland*. Bonn: Federal Ministry of Education and Research.

BMBF. (2009). *Research and innovation for Germany: Results and outlook*. Berlin: Federal Ministry of Education and Research.

Carpenter, M. A., & Westphal, J. D. (2001). The strategic context of external network ties: Examining the impact of director appointments on board involvement in strategic decision making. *The Academy of Management Journal, 44*(4), 639–660.

Carroll, W. K., & Colin, C. (2003). The network of global corporations and elite policy groups: A structure for transnational capitalist class formation? *Global Networks, 3*(1), 29–57.

Carroll, W. K., & Shaw, M. (2001). Consolidating a neoliberal policy bloc in Canada, 1976 to 1996. *Canadian Public Policy – Analyse de Politiques, 27*(2), 195–217.

Dalton, D. R., Daily, C. M., Johnson, J. L., & Ellstrand, A. E. (1999). Number of directors and financial performance: A meta-analysis. *The Academy of Management Journal, 42*(6), 674–686.

Davis, G. F. (1991). Agents without principles? The spread of the poison pill through the intercorporate network. *Administrative Science Quarterly, 36*(4), 583–613.

Davis, G. F., & Greve, H. R. (1997). Corporate elite networks and governance changes in the 1980s. *The American Journal of Sociology, 103*(1), 1–37.
Destatis. (2002). *Klassifikation der Wirtschaftszweige, Ausgabe 2003 (WZ 2003)*. Wiesbaden: Statistisches Bundesamt (German Federal Statistical Office).
Dolata, U. (2007). *Technik und sektoraler Wandel: Technologische Eingriffstiefe, sektorale Adaptionsfähigkeit und soziotechnische Transformationsmuster*. MPIfG Discussion Paper 07/3. Köln: Max-Planck-Institut für Gesellschaftsforschung.
Domhoff, G. W. (1967). *Who rules America?* Englewood Cliffs: Prentice-Hall.
Easton, D. (1957). An approach to the analysis of political systems. *World Politics, 9*(3), 383–400.
Faust, K. (2005). Using correspondence analysis for joint displays of affiliation networks. In P. J. Carrington, J. Scott, & S. Wasserman (Eds.), *Models and methods in social network analysis* (pp. 117–147). Cambridge, MA/New York: Cambridge University Press.
Fennema, M., & Schijf, H. (1978–1979). Analysing interlocking directorates: Theory and methods. *Social Networks, 1*(4), 297–332.
Freeman, L. C. (1978–1979). Centrality in social networks conceptual clarification. *Social Networks, 1*(3), 215–239.
Freeman, C., & Soete, L. (1997). *The economics of industrial innovation*. London: Pinter.
Geletkanycz, M. A., Boyd, B. K., & Finkelstein, S. (2001). The strategic value of CEO external directorate networks: Implications for CEO compensation. *Strategic Management Journal, 22*(9), 889–898.
Heinrichs, H. (2002). *Politikberatung in der Wissensgesellschaft. Eine Analyse umweltpolitischer Beratungssysteme*. Wiesbaden: Deutscher Universitäts-Verlag.
Krackhardt, D., & Stern, R. N. (1988). Informal networks and organizational crises: An experimental simulation. *Social Psychology Quarterly, 51*(2), 123–140.
Lange, S., & Schimank, U. (2004). *Governance und gesellschaftliche Integration*. Wiesbaden: VS-Verlag.
Luhmann, N. (1984). *Soziale Systeme. Grundriß einer allgemeinen Theorie*. Frankfurt am Main: Suhrkamp.
Lundvall, B.-Å. (1992). *National systems of innovation. Towards a theory of innovation and interactive learning*. London: Pinter.
Mayntz, R. (2006). Die Organisation wissenschaftlicher Politikberatung in Deutschland. In H. A. der Wissenschaften (Ed.), *Polikberatung in Deutschland* (pp. 115–122). Wiesbaden: VS-Verlag.
Metcalfe, S. (1995). The economic foundations of technology policy: equilibrium and evolutionary perspectives. In P. Stoneman (Ed.), *Handbook of the economics of innovation and technical change* (pp. 409–512). London: Blackwell.
Mills, C. W. (1956). *The power elite*. Oxford: Oxford University Press.
Mizruchi, M. S. (1996). What do interlocks do? An analysis, critique, and assessment of research on interlocking directorates. *Annual Review of Sociology, 22*, 271–298.
Mizruchi, M. S., & Stearns, L. B. (1988). A longitudinal study of the formation of interlocking directorates. *Administrative Science Quarterly, 33*(2), 194–210.
Münch, R. (1982). *Theorie des Handelns: Zur Rekonstruktion der Beiträge von Talcott Parsons, Emile Durkheim und Max Weber*. Frankfurt am Main: Suhrkamp.
OECD. (2005). Annex A: classification of manufacturing industries based on technology. In OECD (Ed.), *OECD science, technology and industry scoreboard 2005* (pp. 181–184). Paris: Organisation for Economic Co-Operation and Development.
Parsons, T. (1951). *The social system*. London: Routledge.
Pennings, J. M. (1980). *Interlocking directorates*. San Francisco: Jossey-Bass.
Perreault, W. D., & Leigh, E. L. (1989). Reliability of nominal data based on qualitative judgements. *Journal of Marketing Research, 26*(2), 135–148.
Ruigrok, W., Peck, S. I., & Keller, H. (2006). Board characteristics and involvement in strategic decision making: Evidence from Swiss companies. *Journal of Management Studies, 43*(5), 1201–1226.

Schavan, A. (2006). *Forschung schnell in gute Produkte umsetzen. Press release 23 June 2006*. Bonn: Federal Ministry of Education and Research.
Schimank, U. (1996). *Theorien gesellschaftlicher Differenzierung*. Opladen: Leske & Budrich.
Schimank, U. (2005). *Differenzierung und Integration der modernen Gesellschaft. Beiträge zur akteurszentrierten Differenzierungstheorie 1*. Wiesbaden: VS-Verlag.
Scott, J. (1991). Networks of corporate power: A comparative assessment. *Annual Review of Sociology, 17*, 181–203.
Silva, F., Majluf, N., & Paredes, R. M. (2006). Family ties, interlocking directors and performance of business groups in emerging countries: The case of Chile. *Journal of Business Research, 59*(3), 316–321.
Sweezy, P. M. (1953). *The present as history, essays and reviews on capitalism and socialism*. New York/London: Monthly Review Press.
Wallich, P. (1905). *Die Konzentration im deutschen Bankwesen*. Stuttgart: Cotta.
Wasserman, S., & Faust, K. (Eds.). (1994). *Social network analysis: Methods and applications*. Cambridge/New York: Cambridge University Press.

Part III
Sectoral Perspectives on Innovation Policy

Chapter 8
Power Games in Space: The German High-Tech Strategy and European Space Policy

Johannes Weyer and Volker Schneider

8.1 Introduction

In 2006, the German government published a national High-Tech Strategy (HTS) aiming at a better coordination of its research and innovation policy, and to increase its financial support to research and development in high-tech sectors. The new government, from 2005 to 2009 a "large coalition" between the Christian Democratic Parties (CDU/CSU) and the Social Democrats (SPD), announced an increase in public research spending to 3% of GDP until 2010 and planned to also create 1.5 million new jobs. Its most spectacular component was to concentrate research spending on 17 particularly defined economic sectors which, in the long run, would strengthen Germany's international competitiveness (BMBF 2006).

Among these strategic areas, the space sector plays a prominent role, since the largest share of the HTS budget (25% of the total programme budget during the government's 4-year legislative period, equivalent to Euro 3.9 billion) was devoted to space technologies (see Fig. 8.1). This is almost twice the amount of resources devoted to energy, and triple the money spent on information and communication technology. This concentration on large scale technology indicates that one of the major traditional facets of Germany innovation policy, in which governmental decisions are essential in the selection and the support of specific technological fields, still plays an important role, despite the strategy paper's rhetoric emphasizing clusters, networks, and entrepreneurship.

The policy formation of the German High-Tech Strategy appears thus as an interesting riddle. Does it really represent a significant change in German innovation policy, or is it more a case of selling old wine in new bottles through

J. Weyer (✉)
Dortmund University of Technology, Dortmund, Germany
e-mail: johannes.weyer@tu-dortmund.de

V. Schneider
University of Konstanz, Konstanz, Germany

J.M. Bauer et al. (eds.), *Innovation Policy and Governance in High-Tech Industries*,
DOI 10.1007/978-3-642-12563-8_8,

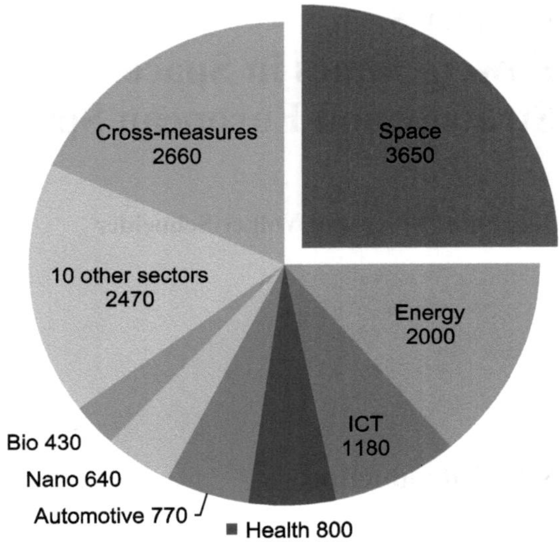

Fig. 8.1 Funding of space technology and other high-tech sectors within the new German innovation strategy (million Euros)

a clever marketing campaign? How far has this new strategy been shaped by the new Government and the policy orientations of its constituent parties? Or is it rather an example of bureaucratic politics and turf wars inside the Governmental administration? Further influential factors may emanate from the international level: What role did the EU play in this innovation policy strategy? Which effects had technological competition at the global and European level on the strategy and how was it influenced by European industrial policy-making trying to cope with the challenges of intensified competition?

The goal of this article is to untangle the complex process of formulating this policy strategy. This is accomplished by analyzing it as a multi-level development that is simultaneously shaped by national, European and global processes. To borrow Norton Long's metaphor, the German High-Tech Strategy is embedded in an "ecology of games" that are simultaneously played at national, European and international policy arenas (Long 1958; see also Dutton et al., Chap. 3). Each game has its specific logic and constellation of players, and, in addition, the various games are nested and overlapping. In this article we try to unveil the political logic of the various game constellations. Our main argument is that in this policy programme, big technology is used as an instrument to establish and to expand policy domains in the interest of particular administrative and industrial actors.

In the following section we first give a short outline of the German HTS and the position of space policy as a component of this strategy. In later subsections we interpret major policy motives from a bureaucratic politics and industrial policy perspective and also point to European developments as supranational determinants of this policy process. We end with an assessment of German research and innovation policy, finding that concentration on large technological systems is still its major orientation.

8.2 The German "High-Tech Strategy": A New Role of the State?

On August 29, 2006, the cabinet of the new government of Conservatives and Social Democrats approved a High-Tech Strategy (HTS) to implement an integrated approach to innovation policy in which a dozen of ministries were involved. The collective strategy would be coordinated by the Ministry of Research and Education. Although the HTS could be seen as a policy adjustment of the new government, the idea first emerged in 2004 when the two major players, the research ministry and the ministry of economics, announced they were going to coordinate their activities via a "high-tech master plan" (Dohse 2005).

Based on an enlarged funding budget, the HTS involved a detailed plan for a broad spectrum of events and policy measures. The major components and sequences are depicted in a milestone that was published in the strategy paper (see Fig. 8.2).

If one compares the strategy paper with the various research and development reports of the Federal Ministry of Research and Technology (BMFT), it becomes evident that the HTS was not a completely new approach to innovation, in which older policy instruments were replaced by newer ones, but rather an "integrated

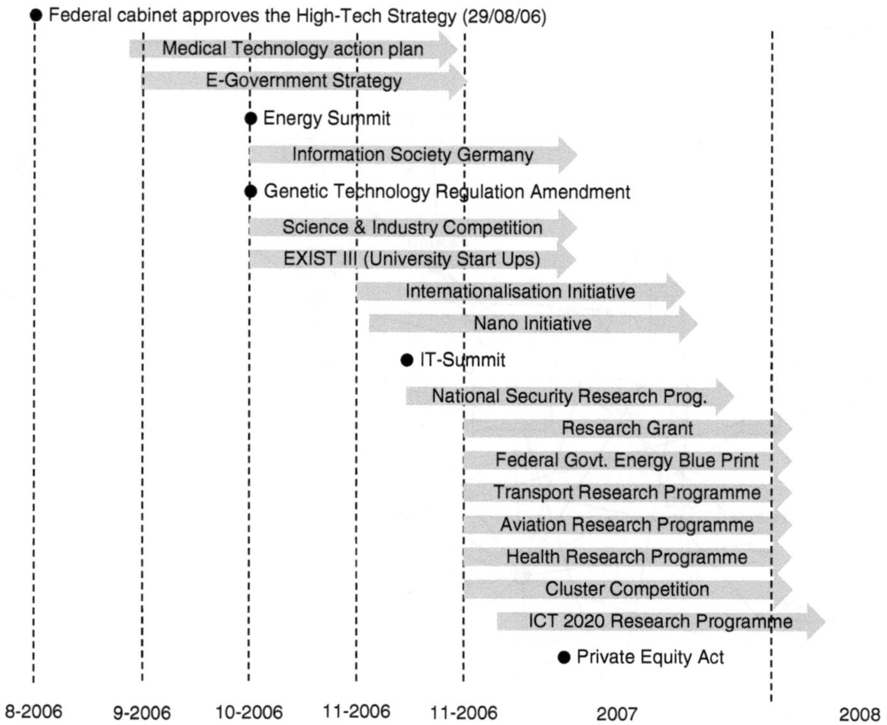

Fig. 8.2 Milestones for the high-tech strategy

programme marketing" in which a complex combination of existing and planned activities was presented as a new and coherent programme.

An innovative facet was undoubtedly the extension of purely distributive measures (research funding) by a series of regulatory and institutional components. The strategy thus combined increased R&D funding with reforms of tax law and the introduction of a law facilitating private equity to attract venture capital. Also an overhaul of Germany's Law Pertaining to Companies with Limited Liability would reduce the regulatory and financial burden on small and medium companies. The strategy thus integrated a diverse spectrum of policy measures related to the general conditions of innovation, but also to the private and public demand for innovative products. The ultimate goal was to accelerate development, market access and the diffusion of innovation.

From a relational perspective, the German HTS consists of a complex network of actors and policy measures related to a variety of high-tech sectors and to some cross-cutting activities. Figure 8.3 visualises the multiple actors and technological fields that are involved in the HTS with methods borrowed from political network analysis (Brandes et al. 1999; Schneider et al. 2009). Actors are depicted with points, and squared grey areas denote the 17 high-tech sectors. Links between the points and areas indicate the joint involvement of actors in these fields of R&D. The geometric position of actors indicates the closeness of actors to the various R&D fields in which they are involved. The size of the areas is relative to their financial support in the 4-years high-tech budget.

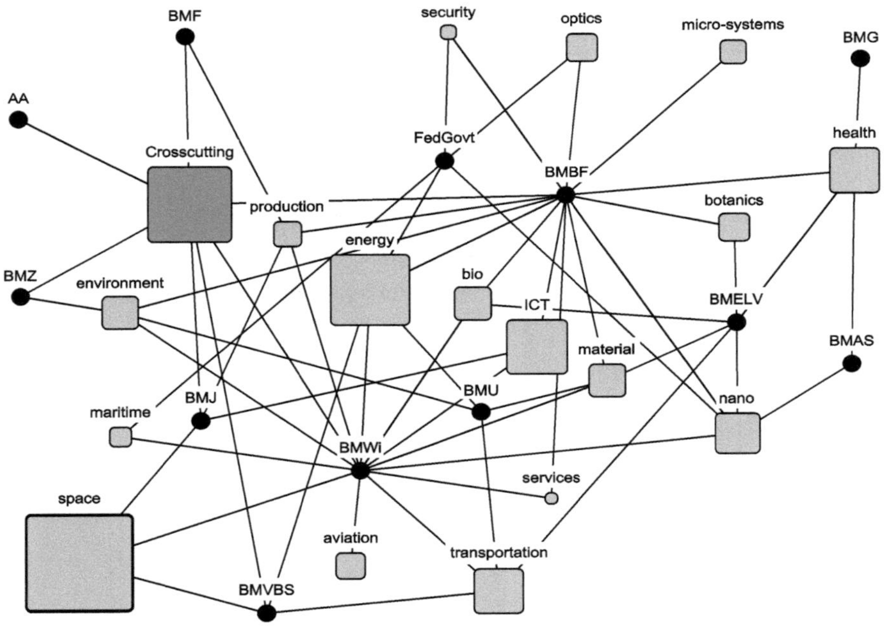

Fig. 8.3 Participation of governmental actors in sectoral high-tech measures

The most prominent actors in this respect are the Federal Ministry of Education and Research (BMBF, "research ministry") and the Federal Ministry of Economics and Technology (BMWi, "economics ministry"). In the overall programme, space technology covers the largest amount of resources, but in relational terms it is rather peripheral. Only three actors – the ministries of economics, of justice (BMJ) and of transportation (BMVBS) – are involved in this subfield.

A major challenge in this complex network of actors and activities is coherence and integration. To improve the coordination of this heterogeneous complex, the German government created two institutional platforms for information exchange and policy advice: On the one hand, there is the Industry-Science Research Alliance, composed of representatives from the industrial and science sectors, while on the other hand, there is the Council for Innovation and Growth as an advisory body to the chancellor. It is composed of prominent scientists (for an in-depth analysis of these bodies see Orlowski, Chap. 7).

Although the programme looks eclectic and is garnished with modern rhetoric, it is nevertheless shaped by a more or less coherent background theory. Its basic philosophy was influenced by Frieder Meyer-Krahmer, the permanent secretary of both the former social-democratic minister, Edelgard Bulmahn, and the new Christian democratic minister, Annette Schavan. Similar to Mayer-Krahmer's contribution to a special issue on "National policies in the age of globalization," (Meyer-Krahmer 2005) published *before* the election in late 2005, the programmatic government paper called for a paradigm change in research and innovation policy. The strategy paper argued that focus should be shifted to emerging lead-markets within the context of complex technologies gaining prominence in future. Similarly, the high-tech manifest stressed this strategy as the only means to maintain and advance Germany's competitive position in the global innovation race.

The new strategy also implied a redefinition of the role of the state. Since the 1960s, a basic feature of German innovation policy has been to support large technological projects and systems. Main examples include nuclear and space technology, highspeed trains, information technology, and others. Often, such large-scale projects were carried out by a small number of big companies (Weyer 1993b; Weyer et al. 1997). In the mid-1990s the former (conservative) minister for research, Jürgen Rüttgers, introduced a new pattern of innovation policy by the BMBF competitions "Mobility in conurbations" (1996), "BioRegio" (1996) and "InnoRegio" (1999) (Conrad 2007; Dohse 2005). For the first time German innovation policy refrained from selecting particular technologies but primarily promoted regional networks and clusters to better stimulate self-organised learning processes and to spur innovation at the regional level.

The HTS of 2006 can be seen as the partial perpetuation of this new policy orientation, which was then combined with the traditional emphasis on large technology. This approach was extended to *all* fields of innovation policy in *all* federal ministries. The major governmental task was now to provide institutional infrastructures for coordination and information exchange in order to facilitate the emergence of lead-markets. This was no longer a direct governmental intervention into these processes. Most of the measures proposed in the strategy paper thus concentrated on indirect incentives and facilitators.

In addition, the High-Tech Strategy failed to define particular objectives. Apart from general goals, such as the protection of nature, the fight against poverty, and the struggle for peace in the world, neither technological nor political targets were explicitly defined. The paper did not even explicate the concept of "high-tech," as the long list of activities shows. It ranges from space, nano- and biotechnology to the promotion of SMEs, e-government, and even gender issues.

8.3 Bureaucratic Politics and the Industrial Policy Game

In structural terms, the German HTS indicated a major shift in research and innovation policy in three respects: (1) A change from direct subsidies for big technology to indirect incentives for regional clusters (at least at the programmatic level); (2) a further diffusion of governmental activities into different sectors of society (social policy, domestic construction policy, transportation policy and others) as a new objective of innovation policy; and (3) a change in actor positions: The research ministry was no longer the single key player in the field, but shared responsibilities with other core actors such as the ministry of economics (BMWi). The latter changes especially undermined the strategic position of the research ministry.

After the new Federal Government took office in autumn 2005, two major decisions were made that weakened the position of the BMBF. Since the 1970s, the BMBF had been the key actor in the field of research and innovation. Other ministries such as the ministry of defence or the ministry of economics only had peripheral positions, for instance in the field of military and aviation technology. Although there had been many attempts following each change in government during the 1980s and 1990s to split up the BMBF and to create a space ministry or a ministry for industrial policy, every chancellor since then had been resistant to these pressures.

However, in late 2005, the BMBF had to abandon many of its former competences in higher education as a result of the federalism reform, which aimed at a fortification of the role of the Federal States. In addition, the jurisdiction of the ministry of economics was broadened to a ministry of economics *and technology* by transferring the responsibility for high-tech fields such as aerospace, information technology, energy and transportation from the research ministry to the ministry of economics. This competence shift was based on the organizational directive of chancellor Angela Merkel of 5 November 2005 (BMBF 2006, 2007). The financial impact of this reorganization was that the BMBF lost a significant part of its budget in 2005. Figure 8.4 shows the evolution of the German research budget with respect to the major sponsoring institutions. Data are based on various research reports (BMBF 2007, 2010).

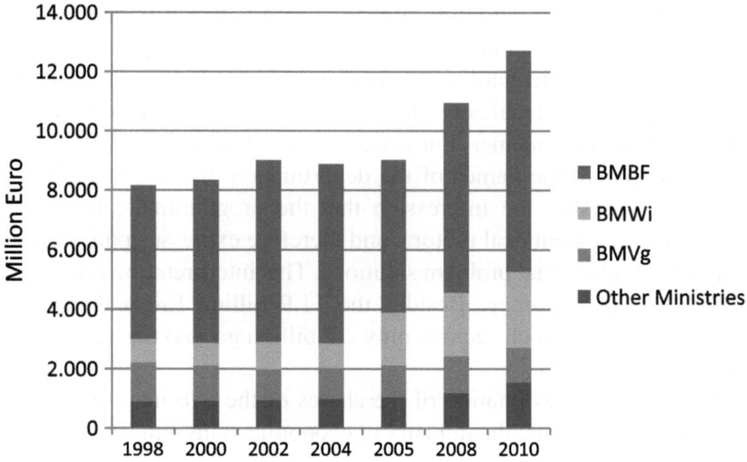

Fig. 8.4 Public R&D funding by major German institutions (in million Euro)

Above all, the BMBF lost its policy leadership in research and innovation. This had frequently been disputed during the last few decades, but was never really endangered. This reorganisation thus can be regarded as a redistribution of policy responsibilities which might result in a complete dissolution of this particular research policy domain. *Research policy,* in this perspective, is extended to an integrated *industrial policy* in which research and innovation are only subfields and partial sequences of an encompassing industrial policy geared towards competitiveness and innovation. This notion of industrial policy was frequently used by Michael Glos, minister for economics 2005–2009, arguing in the tradition of Franz-Josef Strauß and Edmund Stoiber. Both were former Premier Ministers of Bavaria and strong advocates of *economic statism* or *neo-mercantilist economic policy*.

From a "bureaucratic politics" (Allison and Halperin 1972) perspective, the High-Tech Strategy can be regarded as a reaction to BMBF's loss of competences. It served as a strategic move to regain media attention and opinion leadership in the field, and to claim the role of a programmatic think-tank for the entire federal innovation policy.

However, the new policy strategy also implied some risk. Since it was aimed towards a variety of fields of application – a long-standing demand of sociological innovation research (Meyer-Krahmer and Kuntze 1992; Weyer et al. 1997) – this strategy may have unintentionally contributed to the dissipation of this particular research policy domain. This is due to the fact that other ministries, such as the ministry of economics, the ministry of environment and the transportation ministry also came into play. Furthermore, a policy without precise objectives (such as, for instance, building a nuclear plant) and a policy which predominantly has to rely on indirect incentives is difficult to evaluate. The demand for additional funds (six billion Euros for the period 2006–2009) can thus be considered as a strategic move by the BMBF to regain influence in a field where the risk to become insignificant had increased.

If we compare existing activities and the strategic plan depicted in the government high-tech paper, some interesting patterns appear. The paper itemises 17 high-tech sectors, from space technology to services, which can be seen as an inventory of existing programmes. Ironically, these programmes are in most cases presented in a rather conventional manner. For instance, the labelling of the technology field mostly corresponds to the names of the departmental division in their respective ministry. This nourishes the impression that the programmatic foreword in the strategy paper is to a great deal rhetoric and therefore expresses political marketing goals rather than substantial problem solutions. This interpretation is also supported by the distribution of money: Besides the 11.9 billion Euros dedicated to the aforementioned 17 high-tech sectors, only 2.7 billion go into the new cross-section measures (BMBF 2006).

In addition, a rough estimation of the shares of the two main ministries in the high-tech budget shows that the ministry of economic gains almost two-thirds of the funds for strategic sectors (61%), while the research ministry only receives about 22% of that amount. In contrast, one of the main measures based on the new paradigm of innovation policy (clusters and regions) is only supported with 600 million Euros (4%). Given the fact that most of the regional clusters will probably be situated within the specific high-tech fields such as biotechnology or information technology, the new components of innovation policy are almost insignificant.

Table 8.1 also shows that space and energy take the lion's share and cover about half of the high-tech sector's budget. We take this as a sign that the traditional big technology orientation is persisting. Ironically, it was those two sectors which gave birth to the research ministry in 1962. At that time, the ministry of nuclear energy was upgraded to work as a general ministry for research and a second division for space policy was added (Weyer 1993a, 2006). Since the identity of the BMBF had been strongly shaped by the two large divisions for space and atomic energy, the loss of these two main pillars must indicate a fundamental change in the identity of the ministry.

The current transfer of resources and competences in both fields towards the ministry of economics reveals that, besides the modern rhetoric, traditional patterns of innovation policy are persisting and probably will be reinforced in future. A paradigmatic example for this trend is the national and European space policy, which is one of the major fields in which the new industrial policy comes into play.

8.4 The Space Sector Between Science and Industrial Policy

Since the 1960s, German space policy has been shaped by three major interest conflicts: (1) The clash between a national and an international orientation of space projects; (2) the contradiction between the construction of rockets and manned spacecraft as demanded by the space industry versus the participation in satellite missions as demanded by the community of space scientists; and (3) the conflict between a European and a transatlantic cooperation in space. The conflict was also

Table 8.1 Participation of the ministries of research (BMBF) and economy (BMWi) in the German high tech strategy (2006–2009, million Euros)

	Total	BMBF	BMWi
High-tech sectors			
Space	3,650		3,650
Energy	2,000		1,600
ICT	1,180		1,180
Health	800	600	
Transportation	770		385
Nanotechnology	640	480	
Biotechnology	430	215	
Material	420	210	
Environment	420	210	
Optics	310	310	
Botanics	300	150	
Aviation	270		270
Production	250	125	
Micro-systems	220	220	
Maritime technology	150		150
Security	80	80	
Services	50	25	
Sub-total	11,940	2,625	7,235
Cross-sectoral measures			
SME	1,840		1,840
Clusters and regions	600	600	
Startups	220		220
Sub-total	2,660	600	2,060
Total high-tech and cross-sectoral	14,600	3,225	9,295

Sources: BMBF (2006, p. 104); organisation charts of the ministries. The distribution of funds between the departments is estimated on the basis of the information on the ministries related to the 17 sectors

over peaceful missions with purely scientific objectives versus commercial aims that were mostly envisioned by big technology projects (Weyer 2006).

All important decisions – e.g., contribution to the Post-Apollo-Programme in the 1970s or the plans for the European space plane Hermes in the 1980s – were accompanied by fierce debates. The BMBF always had to find a proper balance between the *Transatlantics* and the *Europeans* in order to maintain the identity of German space policy.

Despite numerous conflicts, the space policy carried out by the BMBF can be evaluated as partly successful, since the ministry succeeded in the consolidation of a supportive institutional sector in order to gain autonomy vis-à-vis the BMWi and the ministry of defence (BMVg). This even applies if some of the space projects – from Azur to Symphonie, and from Spacelab to Columbus – have to be seen as failures, at least if one compares their achievements to the initial targets and promises (Weyer 2006).

However, since the 1990s we can observe a steady transformation in German space policy in favour of extended objectives and accelerating Europeanization. Space policy has turned increasingly into a policy game that has become nested in European and global power games, particularly in the fields of industrial policy and security politics. During the last decade, the European Union (EU) gradually superseded the European Space Agency (ESA) as the major player in the field, claiming new resources and competences. A lighthouse project for this new policy is Galileo, the European system of satellite navigation. Galileo has served the interests in security and industrial policy alike, since it is expected to support new lead-markets at the global level, and at the same time has provided for technologies which the European armies need for global crisis intervention (Härpfer 2003; Geiger 2005; Weyer 2005).

The German High-Tech Strategy mirrored this new situation by claiming a "leading role" for Germany in the field of satellite navigation and earth observation (Galileo and GMES satellites).[1] However, the international space station (ISS) and the European Ariane 5 require large amounts of funds. Strong financial support is a necessity as international competition in the launcher market is stiff, and the over-designed European high-tech rocket is difficult to market. It is most notably in the field of rockets that the German strategy paper claims to "guarantee a European access to space of its own" and to develop new launcher systems – again with German "leadership". Apart from the activities of the ESA and the EU, a "national space programme" shall serve to "promote the enforcement of German objectives" within European programmes (BMBF 2006). This is the strategy of re-nationalizing international activities which has become a common practice since the 1960s. The German space policy thus reinforces the traditional pattern of public support of big technology in order to expand political power and prestige (Weyer 1993a).

While there are similar projects that are executed in public-private partnership, the space policy in its plurality mirrors the old paradigm, and recent developments even display a switch back to old patterns. After the failure to assure a substantial industrial participation in the Galileo project, the EU returned to the traditional procedure of state-driven construction of big technology in the summer of 2007. This "rollback" becomes manifest if we look at the planning in this policy area that is depicted in Table 8.2.

In late 2005, when the new course in space policy was set, the Germany space industry promoted two ideas: First, the construction of the manned and winged space plane *Kliper* in German-Russian cooperation, thus guaranteeing an independent access to the space station after the retirement of the US shuttle fleet in 2010 – as well as a resumption of German activities in the area of hypersonic space planes, which had been cancelled in 1996 (Weyer 1992, 2006).[2] Second, a European mission "back to the moon and finally further" (Frankfurter Allgemeine Zeitung,

[1] GMES – Global Monitoring for Environment and Security.

[2] The ESA council rejected this project at its meeting in December 2005 (Frankfurter Allgemeine Zeitung, 8 December 2005, p. 36).

Table 8.2 German space manifesto 2005–2008

	Million €	Share	Old paradigm	New paradigm	Comments
Earth observation, incl. meteorology	744	22.8	X	(X)	GMES etc.
Space station	614	18.8	X	–	Manned spaceflight
Space exploration	550	16.8	X	–	Unmanned satellites, e.g. missions to planets
Space transportation	433	13.2	X	–	Ariane 5 and successional systems
Communication, navigation	296	9.1	X	(X)	Galileo etc.
Management, general budget, etc.	274	8.4	–	–	–
Space research	200	6.1	X	–	Predominantly manned missions (ISS)
Technology for space systems	157	4.8	X	–	Basic technology for satellites and carrier systems
Total sum	3,268	100			

Source: BMBF (2006, 2007)

1 December 2005, p. 41) conducted "with German leadership" (Gemsa 2007), which was mainly justified through reference to the activities of space nations, as well as with arguments like "it is dangerous not to participate" (Manfred Fuchs/ OHB, quoted in Frankfurter Allgemeine Zeitung, 1 December 2005, p. 42).

These two proposals can be seen as a strategy to re-nationalise space policy and to expand the budget for space research, which had to face cutbacks, particularly in manned projects, during the red-green Federal Government (1998–2005). However, it is remarkable that the German space lobby, in its attempt to demand "additional funds" (Gemsa 2007) put specific emphasis on large scale projects – mostly manned – with almost no commercial impact. The discourse in this context is strongly based on political arguments such as prestige and international competition rather than economic or business considerations.

The space industry's new advocacy displays a remarkable new self-esteem, even if most major projects in manned spaceflight during the 1980s failed. The current state of the ISS is far behind the flamboyant promises of the 1990s, where the goal was to station a crew of about 30 astronauts permanently in space. The German space industry, despite all the extra public funding, has been unable to warrant independent European access to space via its own launcher. In light of this incapacity, it seems unrealistic that the German space industry could shoulder even larger projects such as the *moonshot* (Weyer 2004, 2005).

The new self-esteem of the German space industry can be related to the expectation that, after the change in Government in 2005, the minister of technology, Michael Glos, would revitalise the neo-mercantilist industrial policy approach mentioned above. This approach actually was applied by Franz-Josef Strauß during the 1950s and 1960s. Strauß was the long-time head of the CSU, the Bavarian part of the German Christian Democrats, and the former Premier Minister of Bavaria. In his early political career, he was Federal Minister of Nuclear Energy (1955) and also the German defence minister between 1956 and 1962. His economic orientation differed sharply from Ludwig Erhard's liberal economic policy, the founder of the German social market economy. Strauß' economic philosophy was closer to French neo-mercantilism or "Colbertisme high-tech" (Cohen 1992).

Franz Josef Strauß had the political standing to create the German aerospace industry, of which a major part is located in Bavaria. Other regional concentrations are in Hamburg, Bremen and Berlin-Brandenburg. Although this industry is still very small compared to other economic sectors (i.e. covering only about 1 per cent of employment in manufacturing), it acquired European dimensions during the 1980 and 1990s (Hornschild and Wieland 1997).

A first, large step in industrial restructuring was a merger of the traditional companies MBB, Dornier and others into the new corporation, Daimler Aerospace (DASA), which was established in 1989. Initially, the aim was to create a national champion for keeping up with American and French space companies. But later it was realised that only an integrated, multinational approach could face American competition. The German and French Governments thus supported a merger of DASA, the French Aérospatiale-Matra and the Spanish Construcciones Aeronáuticas S.A. (CASA) in the new European Aeronautic Defence and Space Company (EADS). The German head office of EADS is located in the Munich area. Responsibilities for space technology within EADS are concentrated in EADS Astrium. Through this politically sponsored restructuration, EADS turned into Europe's largest aerospace corporation and the second largest aerospace corporation in the world.

Based on the new structure, the European space industries increased their world market shares during the last decade slightly, vis-à-vis the U.S. Figure 8.5 shows the evolution of export markets in the world for the U.S. and the major European space

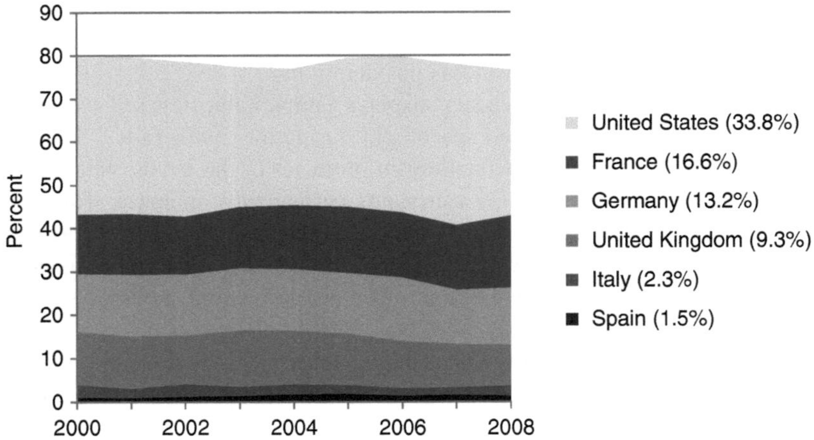

Fig. 8.5 World market shares of major national aerospace industries (in %, 2008 values in parentheses). Source: OECD STAN indicators 2009

nations. The American export market share declined from 36.2% in 2000 to 33.8% in 2008 and the British share dropped from 12.2% to 9.3%. While France and Germany had similar world market shares at the beginning (13%), eight years later the French clearly performed better. The European approach to industrial restructuring is therefore not without conflicts about the distribution of gains and achievements.

8.5 European Space Policy and the Galileo Project

The shift from pure research funding in the space sector to state-directed industrial policy has its clearest expression at the European level. Space policy is one of the latest policy domains that has been "Europeanised" during the last decade (an almost complete list of Europeanised policy areas is given by Graziano and Vink 2007). While European cooperation in the space sector has existed for half a century, only the Treaty of Lisbon (decided in 2007, in force since 2009) gave the EU formal supra-national policy making competencies in this policy area (Wouters 2009). In the following subsection we will outline this Europeanisation process in more detail.

8.5.1 The New European Space Policy

European cooperation in outer space began in 1962 when the European Space Research Organization (ESRO) and the European Launcher Development Organization (ELDO) were founded. In 1975, the organizations were merged into the European Space Agency (ESA), created at that time by the 10 member states of the European Community and three other European countries. The European space programme received a new boost in the same year when the Single Market Programme was prepared in the context of the European Single Act. At the ESA's ministerial conference in Rome in early 1985, a new and ambitious programme for the decade 1985–1995 was decided upon – to "expand Europe's autonomous capability and competitiveness in all sectors of space activities" (Langerfeux 1986). In budgetary terms this implied a doubling of the amount spent in the decade before.

Reimar Luest, former president of the Max-Planck-Society in Germany and ESA Director General 1984–1990, stated in summer 1986: "On the political level, I hope that Europe's space activities will be backed more strongly than ever since European politicians now have recognised we must develop into a space power and must obtain autonomy in space" (Aviation Week & Space Technology, 9 June 1986).

Quite similar to other newly Europeanised areas such as telecommunications, major driving forces in this policy change during the 1980s were the European Commission and European industry representatives (Hayward 1994; Jones 1996). In 1986, Eurospace, the corporatist representative of the European space industry, published an optimistic market study on the European space sector, and the European Commission published a document entitled "The Community and Space: A Coherent Approach" in July 1988 (European Commission 1988). Both reports emphasised the commercial value of space. The Commission's paper expressed it bluntly: "The era of the conquest of space has given way to an era of space exploitation" (p. 1). It emphasised telecommunication and earth observation as major areas for space applications. The Commission requested that the Council accept the need for the Community to play a more active role in space matters, but at that time ESA remained the single platform of European space cooperation. Unlike the aforementioned telecommunications policy in which the Commission succeeded in its strive for extended policy competencies (Schneider and Werle 2007), space policy remained primarily a national undertaking.

The situation only changed a decade later when the European Commission tried another push. In a communication to the Council and the European Parliament it formulated a proposal entitled "The European Union and Space: fostering applications, markets and industrial competitiveness" with emphasis on the strategic importance of this high-tech sector (European Commission 1996). In the following year the Council issued a resolution on the "reinforcement of the synergy between the European Space Agency and the European Community" and the new push gained momentum in February 1999. The Commission proposed a European public-private partnership (PPP) to finance and build a new satellite navigation constellation called Galileo. The new system would give the EU industry a more competitive position in the increasingly lucrative satellite navigation market.

In the year 2000, the Council presented a resolution "on a European space strategy" (Council of the European Union 2000) in which the Commission was asked to set up a joint task force with the ESA to develop plans for this new policy, in which, besides industrial policy, security and defence policy perspectives should also be integrated. Starting with various recommendations and expert reports, in the following years the institutional machinery of the EU entered the new European space policy domain step by step (Mazurelle et al. 2009). In January 2003, the Commission presented a Green Paper on "European Space Policy" which, after intense consultation with industry and other important socio-political actors, was succeeded in November the same year by a White Paper on "Space: a new European frontier for an expanding Union – An action plan for implementing the European Space policy" (European Commission 2003).

In the White Paper, the commission "put space in Europe's policy toolbox" (European Commission 2003) to better achieve economical, ecological and security targets. Besides satellite communication and earth observation (for civil and military use), the paper intended to "strengthen industrial performance" (p. 9). The paper also called Galileo, "the first major space project launched under the aegis of the EU" (European Commission 2003).

This approach crossed new borders since European space projects had been, hitherto, executed only with national management or within the framework of the European Space Agency (Cavallo 2000). By its charter, the ESA was bound to peacefully explore space and was therefore constrained in the pursuit of commercial or political goals – even if such considerations inevitably played a part in the planning of space missions. To implement the new policy architecture, the White Paper suggested a re-arrangement of competences. While the EU would then support space-based solutions with technological standardization, legal harmonization of procurement, and the distribution of R&D money, the ESA would be responsible for technological and management know-how. The ESA thus turned into a service provider for an ambitious industrial policy within the European Union.

The White Paper was followed by a framework agreement between the ESA and the EU, which was passed in November 2003 and went into force in the beginning of 2004. The contractual partners agreed on an "efficient and mutually beneficial cooperation (. . .) to link demand for services and applications" in economy, politics and society (ESA–EU 2003).

The commitment to this kind of cooperation was a radical policy change within the ESA, since the agency had never been officially involved in industrial policy up to that point. In addition, the new EU leadership meant a loss of autonomy for the ESA. A strategy paper from 2005 outlined the new division of labour, in which the EU would "define the priorities and requirements for space based systems", whereas the ESA was now responsible to "support the technical specification of the space segment" (European Commission 2007a).

Three points in the Framework Agreement accentuate this readjustment: (1) The participation by the EU in "optional programmes" of the ESA[3] now allowed an active shaping of major projects. (2) In the case of EU financed projects this meant a turn away from the traditional "fair return" principle. ESA rules guaranteed that industrial contracts were in proportion to the budget contributions of their members. (3) The setup of a European Space Council to coordinate ESA and EU actions displaced the ESA Minister's Council as the ultimate political authority in European space policy.

The "Resolution on the European Space Policy," adopted by the European Space Council on its fourth meeting in May 2007 (Council of the European Union 2007), outlined major directions and corner stones of future space policy in the EU. European leadership as a space power was an important claim, and space was considered as a "strategic choice" in the quest for "independence" and "readiness to assume global responsibilities" (European Commission 2007b; see also ESA Industry Portal 2007). Such goals were embedded in an institutional framework to improve coordination (European Commission 2007b). Terms like "single, coherent framework policy" (ibid.) signaled the intention to bundle and centralise competencies within the European Commission. Its industrial policy orientation stressed the synergy between defence and civil space programmes and technologies. The Commission argued in favour of a technology-push model in

[3] In contrast to a must-agenda, which is financed pro-rata by *all* ESA member states, a facultative agenda (like Ariane or Columbus) offers to voluntarily participate; this makes agenda setting more flexible and helps to better meet the interest of the members.

which space is seen as a lead market where public authorities can create conditions for industry-led innovation (European Commission 2007b). In such a perspective, governments at the national and supranational level have to actively push technology development in the long-term interests of its industries.

As mentioned above, a key element in the European policy change is Galileo, which in 2007 was seen as a "strategic infrastructure" (Logsdon 2008; European Commission 2007b). In the following subsection we will outline the history of Galileo in order to understand why this project – despite its failure as a PPP – still serves as a lighthouse or "flagship" project for European space policy.

8.5.2 The Contentious History of the Satellite System Galileo

In this section we deal with the question of how a large technical project or system can be used as an instrument for the build-up and consolidation of a national and supranational policy domain (Froehlich 2010). Galileo is a satellite-based system for location and navigation, similar to the US global positioning system, GPS. The European system, which consists of 30 satellites, was planned to be launched in 2008. However, because of a number of frictions, the deadline has been postponed to the year 2014.

From its very beginning, Galileo was presented as an alternative and independent solution to the American GPS. Its signals would be of a superior quality and precision in order to be used for purposes where reliability and safety were crucial, e.g. guiding airplanes precisely during the landing (Taverna 2003).[4] Originally the cost of development was estimated at about 3.4 billion Euros, a third of which should be provided by the EU and the ESA (Taverna and Wall 2006, Frankfurter Allgemeine Zeitung 29 December 2005, p. 11). This was a technical and institutional innovation regarding both the expected participation of the industry and the new division of labour between the two European bodies, which, for the first time, should jointly finance a large space project (European Commission 2003).

From the outset Galileo was considered as a major step towards European "sovereignty and independence" (Hein 2000), since it promised to reduce Europe's dependence on the American GPS, which can be shut down for civilian applications in case of war. Furthermore, the EU expected to expand markets in telematics and navigation with a number of new jobs. Industrial and social impacts were used to legitimise a mega project which never had big supporters in industry. This reluctance is understandable since GPS signals are disseminated free of charge and the

[4] The main difference between the two systems is the availability of the signal, which is 90 percent (GPS) or 99 percent (Galileo) respectively (Geiger 2005).

US has refrained from jamming since 2000.[5] From a rational perspective, neither users nor manufacturers of navigation systems could identify an urgent need for a second system. Most of the above mentioned services are also provided by the current GPS technology. Bernd Gottschalk, chairman of the German automobile manufacturers' association, argued that he could not see any additional benefit of a European system, and that his association would not spend "a penny for Galileo" (quoted by Schiffhauer 2003).

This critique applies when GPS and Galileo services are transmitted on the same frequencies. In this case, there would absolutely no need for a second system. But if the signals are aired on different wave lengths, more powerful and expensive receivers are needed. A closer examination of this power game on frequencies is interesting, as it casts a light on the political logic of this kind of technology competition. Originally, Galileo was conceived as a "civilian system under civilian control" (Council of the European Union 2001), only complementary to the military system GPS. Such a project gained public support, especially with the German public and the former research minister Edelgard Bulmahn, who criticised the militarization of space fiercely and instead demanded commercial applications. Thus, the original marketing strategy for Galileo was a successful approach to the mobilization of political support, although some experts, even in the early years of the project, had emphasised Galileo's importance for security policy (Härpfer 2003). In this line of argumentation, Galileo and the earth monitoring system GMES appeared as ingredients to an emerging world power that were necessary for global crisis intervention (e.g., logistic support for operations in remote regions) when the localisation of people and the deployment of precision weapons and other smart devices would be required. The Kosovo war in the year 1999 was the final turning point for Europe, since it explicitly demonstrated the gaps of military reconnaissance and guidance and gave support to the idea of a European system.

The plans for a second independent system for satellite navigation was not only criticised by users and producers, but also triggered a sharp reaction from the U.S. Government. The Americans argued that precise navigation signals broadcasted by the European agencies could be misused by terrorists or rogue states. Although Galileo provided five classes of services with different signal accuracy, even the signal of the lowest, freely available class was highly precise. The German think-tank *Stiftung Wissenschaft und Politik* warned hauntingly not to release this dual-use technique at global scale, since its misuse could "hardly be controlled" (Geiger 2005).

After a long dispute, an agreement on "interoperability" of the two systems was achieved in 2004, including the mutual recognition of both systems and the separation of the frequencies. However, this agreement also provided that the U.S. would have access to Galileo's control centre (Härpfer 2003, Frankfurter Allgemeine Zeitung 27 December 2004, p. 15). The compatibility of the two systems and the

---

[5] The stop of the artificial distortion of GPS in 2000 was, among other game plans, motivated by the strategy to hamper Europe's plans to establish an independent satellite navigation system. Even in the past, the U.S. had adopted this strategy to thwart European attempts of independence, generously offering cooperation, e.g. in the case of Azur (1960s), Ariane (1970s) or ISS (1980s).

cooperation on military issues thus compromised the European strategy to gain independency from the U.S. – be it in the commercial or in the military arena.

The history of the Galileo project is thus characterised by power games at three levels: between Europe and the U.S.; within Europe (especially between the EU commission and the national governments); and finally between government and industry. Often, the project nearly failed, since European governments could not agree on the distribution of costs and the allocation of rights and duties. The project could finally be launched in 2003 when Germany was awarded the role of "system leader," which also included responsibility for the major mission control centre in Oberpfaffenhofen (Frankfurter Allgemeine Zeitung 29 December 2005, p. 11, Winkelhage 2006). During 2005, Germany blocked the negotiations for some time. The major reason was a dispute about the inclusion of "an all-German entity in the industry consortium" – in order to "guarantee Germany (...) a strong say in how the system is run" (Taverna and Wall 2006).

The original plans for Galileo as a PPP finally failed (Nardon 2009; Plattard 2008). In May 2007, the European Commissioner for Transport, Jacques Barrot, had to confess that his plan to establish Galileo in cooperation with the European industry could not be realised. He withdrew the call for tenders for the billion Euro project. The industry's reluctance can be explained by various factors: the financial risk of constructing a service liable to pay costs, when the same service would be offered by the dominant competitor for free; the risk of liability resulting from the guaranteed availability (e.g., in case of the crash of a plane guided by Galileo); the political pressure – motivated by the idea of a fair participation of all countries – to merge the two competing industry syndicates was not effective (cf. interview with EADS-chairman Thomas Enders, in Frankfurter Allgemeine Zeitung 11 May 2007, p. 15). The major conflict of interest was economic. Why should an alliance of private companies set up a large technical infrastructure that would be partly offered as a public good – to the state for military purposes, and to the public who would use the basic services for free?

To break up the stalemate, two options seemed to be equally plausible: the return to the traditional pattern of a state-driven development of infrastructure systems (based on public procurement) and its pure legitimisation through Galileo's international security potential, or the legitimation of large public funding through positive externalities as infrastructures to economic and industrial development (Zervos and Siegel 2008). Although the Commission stated, in 2005, that there were no plans for the military use of Galileo (Frankfurter Allgemeine Zeitung 14 November 2005), key European actors followed both paths. The fact that the EU Commission supported the development of the project at any price is a strong sign of the consolidation of this new policy domain. In the meantime, European space policy had been successfully established as a new institutional niche at the expense of ESA and the national governments.

8.6 Political Logics in the Promotion of Big Technology

Although some state-driven large scale technological projects have failed (e.g., the U.S. space shuttle, the Germany Transrapid, and the French Concord), the promotion of big technology still seems to be an attractive option for policy makers, as the case of Galileo shows. In this section we will try to explain this tendency towards large scale technology through an "explanation sketch" that goes beyond the perspectives of "rational industrial rent-seekers" or "irrational politicians." We will point to a complex combination of political, economic, and technical logics within the ecology of games in this type of high technology development.

In the literature on large technical projects there are a number of references to irrational governmental action and policy-making processes. For instance, John Logsdon called the U.S.-Shuttle a "policy failure" (1986) and Henning Klodt (1987) stressed the counterproductive effects of state intervention in fields such as science and technology in general. Our puzzle is to explain why authorities decide on such projects that are expensive, economically useless and even risky, as in the case of nuclear energy. To explain such policies simply through powerful space lobbyists or clever rent-seekers would be easy. A more complex explanation is provided by Otto Keck (1988), who assumes that there are rational decision makers, but also emphasises information asymmetries between major players, i.e. government and business. A more differentiated perspective on development problems of mega-projects, provided by Nils Bruzelius et al. (2002), specifies typical social and technical flaws, such as the insufficient reviews of project proposals and especially inter-role conflicts among governmental actors involved in the funding and the regulation of large technical projects.

In the following paragraphs we will concentrate on an explanation sketch which includes two approaches: First, a version of systems theory focusing on mechanisms of inter-systemic communication and subsystem-specific action orientations (Weyer 1993b). Second, we will apply an institution and actor-centred perspective of modernisation theory to analyse the development of large technological infrastructure systems (Mayntz 2001; Weyer 2005; Schneider and Mayntz 1995).

1. Taking the perspective of systems theory, space policy is part of the political system of modern societies. Within this particular subsystem the actors involved primarily follow the logics of politics, i.e. maintaining and expanding their particular power position in competition with other actors (mostly organizations) in the same field. Big technology projects in this perspective are useful instruments and "stakes" in power games to attract media attention and to gain public visibility. There is an undisputable relationship between the big symbolic meaning of such projects and their enormous financial needs. If an actor discovers a niche – such as space policy on the European level – a large-scale space project may be a useful symbol for the enforcement of his strategy.
   In this perspective, big technology is a stake in political power games. It can be used as an instrument for the preservation of power (e.g., maintaining the American monopoly in satellite navigation) but also for the expansion of one's own domain (e.g., establishing a European space policy motivated by security issues). This dynamic can culminate in the apparent paradox that large scale technological projects do not necessarily need technological or economic success. In contrast, more crucial are the specific side effects a project has within the political system, i.e. strengthening of actor positions and domain defence.

A prime example for such an effect is the first German satellite Azur. Although the satellite was already technologically defective when launched in 1969, it still helped to create a German space industry and expand the domain of the research ministry (in detail Weyer 1993a). Technological decision-making shaped by politics in this field is thus linked to short-term feedbacks in the political system rather than to long-term success in other social subsystems such as science and economy.

2. The second line of argumentation uses modernisation theory in the analysis of large-scale technical systems based on Renate Mayntz's actor-centred and institutionalist approach. Mayntz assumes a "complex interdependence" of technology development and evolution of the modern state, in which predominantly "military technology and modern technological infrastructure played a determining part" (2001). According to Mayntz there is a certain structural morphotropism between "the modern central state and the large technical systems." Both have not only "stimulated their respective growth reciprocally" but also "mutually promoted the trend towards centralization" (ibid.). The state did not only sponsor the development and extension of industrial monopolies but – in most cases – has also been responsible for the provision and operation of infrastructural systems. This is an indicator for the tight link between big technology and the modern intervention state.
   In this perspective it is very plausible that the EU relies on the efficacy of the above-mentioned mechanisms in order to strengthen its position in the science and technology policy arena. The EU commission challenges its member states by creating supranational legitimacy for this policy domain, and it is supported by national ministries and industrial lobbyists in the pursuit of this strategy. National interests work hand in hand with actors at the supranational level in the expansion of EU competences in this policy domain, and to centralise programmes and functions in Brussels (Edler and Kuhlmann 2005; Kaiser and Prange 2002). In such a context, the EU was able to decide for an exclusive sponsorship of Galileo without much criticism from its member states. In a long-term perspective, the satellite system thus could have a similar impact on Europe's sovereignty as the development of the railway had on the genesis of the nation state.

8.7 Conclusions

In this article we have shown that German innovation policy in the space sector is shaped by multiple processes at the national, European and international level, and by multiple action orientations related to politics and the economy, but also to technological contingencies. Innovation policy in the space sector can be seen as a "stake" in a variety of overlapping games played at national and international levels. We have shown that a major determinant at the German level is the intra-bureaucratic conflict between the research and economics ministries. Each organization has its specific goals and histories, and each is connected to different supporters and stakeholders. This national game is "over-determined" by a European game, in which core institutions of the EU strive for the extension of their competences and resources. The European push in space policy itself is embedded in a game of global competition and strategic positioning with regard to emerging technologies and lead-markets. This global context is then used to bet on large-scale technical and risky projects.

The fact that the German HTS has put such a great emphasis on the space sector indicates that the orientation towards large technical projects is still prominent within German innovation policy. Big technology is always risky and in many cases inefficient in economic terms. The main message of our analysis therefore is that political calculus dominates these processes. In order to understand the implications of these findings, one has to consider the various planning intervals of politics. The potential failure of long term projects only has an impact on the following generations of politicians. Politicians can make their mark in contemporary politics while taking measures to cope with challenges at their time. A striking example of this is the reaction of the Bush administration to the Shuttle crisis in 2004. On the one hand, President Bush publicly announced the termination of the Shuttle-programme in 2010, while, on the other hand, he promised to plan for manned space-flights to the Mars in the year 2024. Through such a long-term project he was able to evade a performance test of his current policies.

Large technological projects, regardless of technological, ecological and economical risks, are appealing options for politicians, as they conserve "the illusion to control and govern suchlike processes" (Weyer 2005, p. 23). The small number of actors involved, under the patronage of politicians, predominantly creates this illusion. Furthermore these projects can be protected against external disruption. Economic efficiency is often not required and, in many cases, new technologies also open up new territories and spaces. Thus politicians can create the impression of "being able to activate and to govern" (Weyer 2005, p. 24). In traditional markets and industries, where actors, networks and evaluation standards are already established, governing is much more complicated. In addition, the diversity and heterogeneity of actor constellations in innovation settings makes it very difficult to evaluate the impact of public governance.

References

Allison, G., & Halperin, M. (1972). Bureaucratic politics: A paradigm and some policy implications. *World Politics: A Quarterly Journal of International Relations, 24*, 40–79.

BMBF. (2006). *The hightech strategy for Germany*. Bonn/Berlin: Federal Ministry of Education and Research.

BMBF. (2007). *Report of the federal government on research 2006*. Bonn/Berlin: Federal Ministry of Education and Research.

BMBF. (2010). *Federal report on research and innovation 2010*. Bonn/Berlin: Federal Ministry of Education and Research.

Brandes, U., Kenis, P., Raab, J., Schneider, V., & Wagner, D. (1999). Explorations in the visualization of policy networks. *Journal of Theoretical Politics, 11*(1), 75–106.

Bruzelius, N., Flyvbjerg, B., & Rothengatter, W. (2002). Big decisions, big risks: Improving accountability in mega projects. *Transport Policy, 9*(2), 143–154.

Cavallo, G. (2000). The European space agency and its successful science programme. *European Review, 8*(2), 243–262.

Cohen, E. (1992). *Le colbertisme high-tech*. Paris: Hachette.

Conrad, J. (2007). The role of public policy in promoting technical innovations: The case of the regional innovation network InnoPlanta. *STI-Studies, 3*, 67–85.

Council of the European Union. (2000). Council resolution of 16 November on a European space strategy (2000/C 371/02). *Official Journal of the European Communities. Information and Notices, 43*(Part 371), 2–3.

Council of the European Union. (2001). Council resolution of 5 April 2001 on Galileo (2991/C 157/01). *Official Journal of the European Communities. Information and Notices, 44*(Part 157), 1–3.

Council of the European Union. (2007). Council resolution on the European space policy, 25 may 2007 (10037/07). *Official Journal – European Union Information and Notices C, 50*(136), 1–5.

Dohse, D. (2005). Clusterorientierte Technologiepolitik in Deutschland: Konzepte und Erfahrungen. *Technikfolgenabschätzung – Theorie und Praxis, 14*(Nr. 1/März 2005), 33–41.

Edler, J., & Kuhlmann, S. (2005). Towards one system? The European research area initiative of research systems and the changing leeway of national policies. *Technikfolgenabschätzung – Theorie und Praxis, 14*(Nr. 1/März 2005), 59–68.

ESA–EU. (2003). Framework agreement between the European community and the European space agency (25.11.2003). *Zeitschrift Fur Luft und Weltraumrecht, 53*(Part 1), 89–93.

European Commission. (1988). *The Community and space: a coherent approach, COM(88) 417final*. Luxembourg: Office for Official Publications of the European Communities.

European Commission (1996). *The European union and space: Fostering applications, markets and industrial competitiveness* COM(96) 617 final (04.12.1996), Brussels.

European Commission. (2003). *Space: A new European frontier for an expanding Union. An action plan for implementing the European space policy (white paper), COM(2003) 673 final*. Luxembourg: Publications Office.

European Commission. (2007a). *European space policy: Communication from the commission to the council and the European parliament, COM(2007) 212, 26.04.2007*. Luxembourg: Office for Official Publications of the European Communities.

European Commission (2007b). *Space council welcomes historic European space policy*. (20 June 2007).

Froehlich, A. (2010). Space and the complexity of European rules and policies: The common projects Galileo and GMES–precedence for a new European legal approach? *Acta Astronautica, 66*(7–8), 1262–1265.

Geiger, G. (2005). *Europas weltraumgestützte sicherheit. Aufgaben und probleme der satellitensysteme Galileo und GMES*. Berlin: Stiftung Wissenschaft und Politik. SWP-Studie 2005, S 27, September 2005.

Gemsa, T. (2007). Deutschland will zum Mond (Teil 1). *Luft-und Raumfahrt*, 2/2007, 36–38.

Graziano, P., & Vink, M. (2007). *Europeanization: New research agendas*. London: Palgrave Macmillan.

Härpfer, S. (2003). Mogelpackung Galileo. *Blätter für deutsche und internationale Politik*, 10/2003, 1172–1175.

Hayward, K. (1994). European Union policy and the European aerospace industry. *Journal of European Public Policy, 1*(3), 347–365.

Hein, G. W. (2000). *Wirtschaftsstrategische und sicherheitspolitische Bedeutung des europäischen Satellitennavigationssystems GALILEO und seine Auswirkungen auf die zivile Infrastruktur. Zusammenfassung und Ergebnis einer interdisziplinären Studie* No place/publisher.

Hornschild, K., & Wieland, B. (1997). Is the German and European space industry fit to face the future? *Economic Bulletin, 34*(12), 3–12.

ESA Industry Portal (2007). Europe's space policy becomes a reality today. (22 May 2007)

Jones, C. (1996). Aerospace. In H. Kassim & A. Menon (Eds.), *The European Union and national industrial policy* (pp. 88–105). London: Routledge.

Kaiser, R., & Prange, H. (2002). A new concept of deepening European integration?–The European research area and the emerging role of policy coordination in a multi-level governance system. *European Integration online Papers (EIoP), 6*, 18.

Keck, O. (1988). A theory of white elephants: Asymmetric information in government support for technology. *Research Policy, 17*, 187–201.

Klodt, H. (1987). Mehr Sternschnuppen als Sternstunden. Eine kritische Bilanz der staatlichen Forschungsförderung, Frankfurter Allgemeine Zeitung, 31 October 1987, pp. 15.

Langerfeux, P. (1986). Europe's commitment to space – Europe aims for space independence. *Aerospace America*, 49.

Logsdon, J. M. (1986). The space shuttle program: A policy failure. *Science, 232*(4754), 1099–1105.

Logsdon, J. M. (2008). The new European space policy as seen from across the Atlantic. *Yearbook on Space Policy, 2006*(2007), 167–181.

Long, N. (1958). The local community as an ecology of games. *The American Journal of Sociology, 64*(3), 251–261.

Mayntz, R. (2001). Triebkräfte der Technikentwicklung und die Rolle des Staates. In G. Simonis, R. Martinsen, & T. Saretzki (Eds.), *Politik und Technik. Analysen zum Verhältnis von technologischem, politischem und staatlichem Wandel am Anfang des 21. Jahrhunderts* (Vol. PVS Sonderheft, pp. 3–18, Vol. 31). Wiesbaden: Westdeutscher Verlag.

Mazurelle, F., Wouters, J., & Thiebaut, W. (2009). The evolution of European space governance: Policy, legal and institutional implications. *International Organizations Law Review, 6*(1), 155–189.

Meyer-Krahmer, F. (2005). Handlungsspielräume und Modernisierungserfordernisse nationaler Technologie-und Innovationspolitik. *Technikfolgenabschätzung – Theorie und Praxis, 14*(Nr. 1/März 2005), 12–17.

Meyer-Krahmer, F., & Kuntze, U. (1992). Bestandsaufnahme der Forschungs- und Technologiepolitik. In K. Grimmer, J. Häusler, S. Kuhlmann, & G. Simonis (Eds.), *Politische Techniksteuerung* (pp. 95–118). Opladen: Leske + Budrich.

Nardon, L. (2009). Galileo and the issue of public funding. *Yearbook on Space Policy, 2007*(2008), 125–137.

Plattard, S. (2008). What's the problem with Europe's flagships Galileo and GMES? *Yearbook on Space Policy, 2006*(2007), 153–166.

Schiffhauer, N. (2003). Völker, hört die Signale. *Technology Review* (Nov. 2003), 126–129.

Schneider, V., & Mayntz, R. (1995). Akteurzentrierter Institutionalismus in der Technikforschung. *Fragestellungen und Erklärungsansätze. Jahrbuch Technik und Gesellschaft, 8*, 107–130.

Schneider, V., & Werle, R. (2007). Telecommunications policy. In P. Graziano & M. Vink (Eds.), *Europeanization: New research agendas* (pp. 266–280). London: Palgrave Macmillan.

Schneider, V., Janning, F., Leifeld, P., & Malang, T. (2009). *Politiknetzwerke. Modelle, anwendungen und visualisierungen.* Wiesbaden: VS-Verlag.

Taverna, M. A. (2003). European challenger. Go-ahead for Galileo may force U.S. leaders to stake a new position on satnav systems. *Aviation Week & Space Technology, 08.09.2003*, 61–62.

Taverna, M. A., & Wall, R. (2006). Almost a go for Galileo. *Aviation Week & Space Technology* (25.09.2006), 56–57.

Weyer, J. (1992). Der Raumtransporter SÄNGER als Instrument deutscher Großmachtpolitik? Gutachten, erstellt im Auftrag des Büros für Technikfolgenabschätzung des Deutschen Bundestages. *Materialien zum TAB-Arbeitsbericht* (p. 49). Bonn.

Weyer, J. (1993a). *Akteurstrategien und strukturelle Eigendynamiken* (*Raumfahrt in Westdeutschland 1945–1965* (13.10.2003 00:00:0)). Göttingen: Otto Schwartz.

Weyer, J. (1993b). System und Akteur. *Zum Nutzen zweier soziologischer Paradigmen bei der Erklärung erfolgreichen Scheiterns. Kölner Zeitschrift für Soziologie und Sozialpsychologie, 45*, 1–22.

Weyer, J. (2004). Innovationen fördern – aber wie? Zur Rolle des Staates in der Innovationspolitik. In M. Rasch & D. Bleidick (Eds.), *Technikgeschichte im Ruhrgebiet – Technikgeschichte für das Ruhrgebiet* (pp. 278–294). Essen: Klartext Verlag.

Weyer, J. (2005). Staatliche Förderung von Großtechnikprojekten. Ein dysfunktionaler Anachronismus im Zeitalter der Globalisierung? *Technikfolgenabschätzung – Theorie und Praxis, 14*(Nr. 1, März 2005), 18–25.

Weyer, J. (2006). Die Raumfahrtpolitik des Bundesforschungsministeriums. In P. Weingart & N. C. Taubert (Eds.), *Das Wissensministerium – Ein halbes Jahrhundert Forschungs-und Bildungspolitik in Deutschland* (pp. 64–91). Weilerswist: Velbrück.

Weyer, J., Kirchner, U., Riedl, L., & Schmidt, J. F. K. (1997). *Technik, die Gesellschaft schafft. Soziale Netzwerke als Ort der Technikgenese*. Berlin: Edition Sigma.

Winkelhage, J. (2006). Galileo soll zur Jobmaschine werden. *Frankfurter Allgemeine FAZ.Net*, (19.01.2006).

Wouters, J. (2009). Space in the treaty of Lisbon. *Yearbook on Space Policy, 2007*(2008), 116–124.

Zervos, V., & Siegel, D. (2008). Technology, security, and policy implications of future transatlantic partnerships in space: Lessons from Galileo. *Research Policy, 37*(9), 1630–1642.

Chapter 9
Global Strategies and Policy Arrangements: Institutional Drivers for Innovation in the Wind Turbine Industry

Karsten Ronit

9.1 Introduction

Wind power is a very old source of energy. As Braudel (1992, p. 3) notes, "The West experienced its first mechanical revolution in the eleventh, twelfth and thirteenth centuries. Not so much a revolution, perhaps, as a whole series of slow changes brought about by the increased number of wind- and watermills". Much later, windmills were replaced by other technologies, especially in the wake of the industrial revolution, when fossil fuels became the primary source of energy. Wind power, however, has been rediscovered as one of the many sustainable forms of energy. It has been increasingly recognised as a valid alternative energy source that may contribute to the solving of a range of environmental problems and to reversing the dangers of global climate change.

Over the last few decades, wind energy has experienced quite an astonishing revival, from being viewed as a somewhat redundant and archaic form of energy to achieving a more prominent role in post-industrial societies. The wind turbine industry has undergone interesting scientific developments over the last few decades that qualify it as a high-tech-industry. Significant research input has been directed to, for instance, the technology of wind turbine and intelligent wind power systems and their placement. Parallel to this progress, the industry's commercial and organizational development has linked different actors along the production chain and created new relations to scientific communities and public institutions within and across nations.

Given the prevalence of non-renewable resources, however, it is evident that the potential of wind energy has not yet been fully exhausted. Conventional forms of energy remain extremely important and will continue to be the chief components in the overall supply of energy in the decades to come, and some of them can even appear

K. Ronit (✉)
University of Copenhagen, Denmark
e-mail: kronit@youmail.dk

J.M. Bauer et al. (eds.), *Innovation Policy and Governance in High-Tech Industries*,
DOI 10.1007/978-3-642-12563-8_9, © Springer-Verlag Berlin Heidelberg 2012

in a cleaner form (Jaccard 2005). Several factors account for both the opportunities and the barriers found in the wind energy sector. To gain a better grasp of these institutional factors it is necessary to analyse how different actors create such drivers.

From the outset, it must be recognised that there is no single driving actor, because all actors are embedded in a larger system of innovation. In this system a variety of actors and their interactions play a key role. Thus, this chapter focuses on the agenda-setting powers and on their capacities for building innovation systems in the area of wind technology. It also analyses three groups of actors whose efforts and relations are crucial for the advancement of this industry.

First, while major initiatives are launched by states, the activities of relevant intergovernmental organizations need to be discussed as well, given the global availability of wind and the strong current interest in global environmental problems. A number of these organizations belong to the UN system, but some forums that create various incentives relevant for wind power can also be found outside this framework.

Second, the emergence of such a new policy field also hinges on business, especially the wind turbine industry, which seeks to influence public policy in ways that stimulate further innovation and create new demand for wind energy. At the same time, opposition must be expected from traditional producers of energy who have established links with decision-makers and may be in a better position to leverage public policy.

Third, civil society organizations with an interest in sustainable energy resources play a key role. Usually, environmental groups are very critical of specific industries (a fact that must also be taken into account), but environmental groups can become allies with specific corporations and industries in the renewables sector, and thus contribute to the further advancement of business. In sum, an assessment of the wind turbine industry – both its prospects and impediments – requires an analysis of multiple players and their interactions.

This chapter is organised into four major sections. After the introduction, a theoretical sketch of relevant literatures is given. It discusses the contribution of various intergovernmental organizations, central business players, as well as civil society actors to the development of energy and climate policy, with a special emphasis on wind energy. In the following section, a closer look is taken at how intergovernmental organizations, business, and civil society today address wind energy as a relevant energy source and what remedies are suggested to formulate a coherent innovation strategy. The chapter finishes with a conclusion.

9.2 Innovation Systems in Wind Energy – International Agencies, Business and Civil Society

The development of national innovation systems in the field of wind energy is important to the development of the industry. Several countries have experimented with economic, regulatory and institutional models, and have gained important experiences that have been translated into relevant strategies (Toke 2007). The successful transfer of such models can save costs, but the introduction of new models takes time, new institutions must be created, and old ones must be dismantled. Occasionally, such national models may even be transferred, take root at the international level, and be promoted by international organizations due to the fact that some systems are considered particularly successful and more easily transferable (Szarka 2007). The use of wind energy on a global scale is important to the wind turbine industry because it makes the industry less dependent on domestic markets; in addition, new markets can be uncovered, providing a stronger financial basis for further innovations in this particular technology. In fact, the globalization of this industry is very promising, since wind is a source of energy that is globally available.

Without sidestepping the relevance of a comparative focus in which variation across countries is duly recognised (Lewis and Viser 2007; Bechberger et al. 2008), a coordinated take is required in cases where global problem-solving is encouraged. Indeed, there seems to be a particular demand for the coordination of innovation systems, and even a coherent approach when global solutions are deemed necessary; nascent global innovation systems may here develop and gain prominence. Therefore, the study of innovation systems must be closely linked to the development of global public policy (Reinicke 1998).

In recent years the wind turbine industry has been closely linked to climate change problems and climate policy, and it is very difficult to understand the role of the industry without understanding this general framework. National and regional efforts have been particularly helpful in this context because they further support global initiatives. The adoption of global strategies is needed, however, to identify global aspects of climate change, devise relevant programmes, and build new and stable institutions to implement policy. This seems to be particularly relevant for the development of innovation systems relating to wind energy.

Climate policy is a complex field entailing a horizontal and vertical dimension. Horizontally, climate policy is a structure that spans and interacts with a variety of other smaller and larger policy fields. The general principles relating to climate are debated and implemented across these fields. Vertically, climate policy can be broken down into different sectors, each having its bearing on particular industries and innovation systems; in our context, renewables is a broad kind of sector, which also includes wind energy policy.

National governments remain key players in a period of economic, societal and political globalization but new global institutions and new global innovation systems also emerge. Intergovernmental organizations arise to coordinate the policies of states and mitigate conflicts; they are highly interesting in the context of climate change, but they also develop some special capacities in relation to states (Finnemore and Barnett 2004; Hawkins et al. 2004; Biermann and Siebenhüner 2009). Three theoretical issues are particularly relevant in relation to public drivers in the development of certain policy fields and industries.

First, states are principals that delegate tasks to agents in the form of international organizations, but uncertainty arises as to what extent states are able to control these secretariats or whether they develop some degree of autonomy. There is a strong tradition of emphasizing the directing hand of states, but the argument that the bureaucracies and expert bodies involved develop their own perspectives has gained increasing recognition. Because of their autonomous expertise and value systems, these organizations are in many cases capable of formulating new initiatives and making special contributions to problem-solving. This is relevant in our case because some special agencies today concentrate on climate issues.

Second, and closely related to the first aspect, there is a large range of success among the states which act as principals of intergovernmental organizations. Some states may be more successful in advancing their particular regulatory model and innovation strategy when dealing with other states and, consequently, may leave a significant imprint on the work of intergovernmental organizations. This imprint, of course, also includes climate, energy, and wind issues where global coordination is required.

Third, coordination between intergovernmental organizations is essential in the formation of global policies. With several international organizations involved in some aspect of energy and climate policy, there is a risk that agencies with smaller or larger stakes in climate and energy issues will formulate their own innovation strategies and employ different tools, and that experiences will not be sufficiently shared, making priorities harder to set. To avoid such fragmentation, stronger coordination and new forums are often demanded. Indeed, climate policy has seen the emergence of such new bodies.

Business is a decisive factor in relation to global policies (Braithwaite and Drahos 2000), including climate policy, wind energy and innovation systems (Levy and Newell 2005). Policy is not simply the set of issues agreed upon by states and intergovernmental actors; their policies are typically formulated and implemented in a process in which business delivers key input. However, the wind turbine industry is not alone on the scene, and the way production is organised both creates and solves serious climate problems. In the context of this chapter, however, we can briefly note that business creates many negative externalities, making it very difficult to form a coherent business opinion, and we will go on to analyse the ways in which the wind turbine industry is actually involved in problem solving. Here, of course, a major focus is on the role of the wind turbine industry in the context of the broader energy sector, the alternative wind technologies, and the emergence of a special innovation system around this industry.

Three issues deserve special attention. First, it will be interesting to examine what kind of political actions characterise this particular industry. Is it a fragmented industry with each corporation caring about its own performance in the market, or do we see collective actions in which corporations unite to represent the industry more generally? Second, it is of great importance to establish whether the industry – corporations and associations – actively leverages public policy in the context of states and, in particular, intergovernmental organizations, or whether it is instead concentrating on technological and commercial issues and the dissemination of knowledge (also an important factor in the innovation system). Third, we will study whether the industry prefers to develop its own strategy, formulate its own codes of conduct, and exchange with public agencies, or whether alliances are forged with various civil society organizations as important stakeholders.

Finally, we will direct our attention toward a group of actors that have entered global politics across a rich variety of fields, namely civil society organizations of various sorts, including consumers, environmentalists, and other actors, occasionally influencing business (Keck and Sikkink 1997; Bandy and Smith 2005; della Porta et al. 2006). These organizations have neither the political authority held by states in the traditional political realm, nor the authority of corporations in the market place; however, they have moral power, often possess expert knowledge, and can articulate certain interests in relation to particular forums (Betsill and Correll 2007), including renewables and wind energy.

Therefore, we examine three major issues with regard to civil society actors. First, we discuss whether specialised and relatively resourceful civil society organizations have emerged in climate politics, with special regard to issues of wind energy, taking into account that collective action by civil society is complicated and not necessarily trimmed to address such specific questions as wind power.

Second, there is also a good chance of civil society organizations taking part in relevant policy processes if they are organised in a coherent format. If there is no clear organization of a civil society group with a serious interest in energy problems, or if several organizations seek to represent more or less identical, or even conflicting, ideas, then it can be difficult to accommodate civil society organizations in the decision-making structures of intergovernmental organizations. It matters, for instance, whether there is a distinct preference among civil society organizations to actively contribute to the development of innovation strategies, instead of merely delivering general statements.

Third, civil society organizations not only influence the small field of wind energy through the participation in official forums, but may also engage in cooperation with business. In fact, organizations and activist groups that are strong advocates of alternative and sustainable forms of energy should, in principle, be able to forge alliances with the renewables sector and the wind turbine industry.

In sum, a systemic perspective on innovation must include the complex set-up of public and private actors to get a better grasp of the institutional drivers for wind power. A focus on regulatory frameworks and incentives created by governments and intergovernmental organizations – a "state perspective" – is essential, but is far from exhaustive. Nor is a focus on the development and application of new technologies and the economic performance of corporations – a "market perspective" – sufficient to analyse the institutional drivers in wind energy.

Important aspects are squeezed out in such "state" and "market" perspectives. To mention two brief examples: First, this "state-market" dichotomy, often found in innovation studies, sidelines the fact that corporations are often organised collectively, in forms that seek to represent consolidated industry interests, and in some cases have their own intelligence systems that are active in providing information to their members. Second, another aspect missing in this dichotomy is the role of civil society groups that become involved in policy processes, and whose interests must be accounted for by international agencies and business players alike. In other words, a variety of private organizations also contribute to political strategy in a global setting (Ronit and Schneider 2000). Thus, in a systemic perspective, we recognise the complexity of differing actor qualities, as well as the general institutional pattern in which actors are embedded; as we shall see, these factors are also important in relation to the wind turbine industry.

Furthermore, intergovernmental organizations, business and civil society are not unitary actors but composite entities, each of them often characterised by various priorities, dilemmas and a wide range of multilateral arrangements (Ronit 2007). It is essential to include these interfaces in the study of innovation strategies (Steiner et al. 2006). As far as the public side is concerned, the climate change issue has stirred many conflicts between states which have set different priorities in relation to fossil fuels, renewables, and wind energy. Business is split in regard to the appropriateness of sustainable energy. Moreover, the wind turbine industry is but one of several renewables. Civil society has many priorities and is not always able to act in concert.

This systemic approach builds upon a combination of different and disparate literatures. Unfortunately, and ironically, these strands of research are often very isolated from each other. The role of international agencies is studied independently of the various private actors surrounding them; studies on business concentrate on corporate behavior in the market and analysis of civil society tends to highlight the cooperation between different parts of civil society. Whereas scholarly studies of these actors bear witness to a strong division of labour, real life brings these actors together in efforts to create global innovation systems with regard to wind energy.

9.3 Wind Energy and International Agencies

From the outset, it must be emphasised that wind energy policy, as a small and emergent policy field, is located within the more encompassing and complex realms of energy and climate policy that create various institutional underpinnings for wind energy. The conditions of the wind turbine industry cannot be understood without this context. Political strategies have come a long way since the 1970s, when the wind turbine industry started to experience a renaissance after centuries of virtual non-existence. In the past, the wind turbine industry was primarily an affair for pioneers, some driven by a combination of commercial interests and idealistic motives. The energy crises, however, produced a political and public awareness of using traditional energy sparingly, allowing wind energy to become a relevant alternative, although in the past, wind turbine industry was primarily an affair for pioneers, some driven by a combination of commercial interests and idealistic motives.

However, renewables and wind energy first became embedded in the larger field of energy policy. The most solid structure in relation to energy was the International Energy Agency (IEA), established in 1974 by the Organisation for Economic Cooperation and Development (OECD), but it encompasses only OECD member countries. As a special entity, the Renewable Energy Unit is today concerned with wind energy plus other renewables, but the IEA's work covers all forms of energy, so wind energy must compete for attention and organizational resources with other renewables, as well as with conventional energy, such as oil, coal and nuclear power. Consequently, a number of opportunities and barriers characterise the development of wind energy policy within IEA. It is worth noting that other organizations and forums are also important for the development of an innovation system in relation to wind energy. The IEA coordinates with some of them.

In countless ways, energy policy and wind energy are linked and have become increasingly related to climate policy. Climate change from the emission of carbon dioxide and other greenhouse gases has emphasised the need to develop renewables in a systematic fashion. Wind energy is one of the renewables that attracts attention, and it is evident that wind energy has been significantly nourished by this agenda. Squarely put, the more urgency with which climate change is considered, and the more the emission of carbon dioxide is seen as an overriding problem of climate change, the greater the need for effective measures of sustainable energy production.

As a policy field, climate policy gained momentum through the 1970s and led to the first World Climate Conference hosted by the World Meteorological Organization (WMO) in 1979. Major steps forward in policy development were taken in the late 1980s and early 1990s in a number of UN policy processes. At the Earth Summit in Rio de Janeiro in 1992, the important United Nations Framework Convention on Climate Change (UNFCCC), to which a secretariat is attached, became involved in the implementation of this convention and the Kyoto Protocol from 1997. The UNFCCC has, in a way, developed a life of its own and has organised a long series of conferences since the early 1990s – in Berlin, Geneva, Kyoto, Buenos Aires, Bonn, The Hague, Bonn, Marrakech, New Delhi, Milan, Buenos Aires, Montreal, Nairobi and, most recently, Bali, Poznan, Copenhagen, Cancun, and Durban. The Earth Summit also led to other steps: It adopted the Agenda 21 action plan in a process which later created the United Nations Commission on Sustainable Development and the World Summit on Sustainable Development in 2002. Energy issues were here given attention in a much larger social and economic framework, a framework less concrete but certainly important in its own right.

In relation to the implementation of the UNFCCC, the Intergovernmental Panel on Climate Change (IPCC) has also been an important body (IPCC 2004). It publishes assessment reports at regular intervals and gains considerable public attention. The IPCC was established in 1988 by two permanent UN organizations, namely the World Meteorological Organization and the United Nations Environment Program (UNEP) – the two primary organizations in the UN family currently involved in climate issues.

It is interesting to note, though, that the IEA as a lead organization in energy was not engaged in the creation of the IPCC and, therefore, did not leave a noteworthy imprint on its organization and strategy. This does not suggest that energy has been squeezed out, but it indicates that it is only one of several issue areas filtered into the work of the IPPC. It goes without saying that wind energy is still linked to these agendas as one of several questions.

The IPCC bases its activities on scientific knowledge, providing relevant information to governments and other interested parties on the many aspects of climate policy through its influential assessment reports. Its first report played a key part in bringing climate issues to the agenda of the Earth Summit. Given the current controversies around climate change, it is not astonishing that views on the IPCC's status differ – some stakeholders see IPCC as too neutral and some see it as too partisan. In any case, it is quite clear that its reports attract much attention and can be considered essential input to discussions and negotiations around climate issues.

Although the key efforts of the IPCC and its working groups are centered on assessing climate change, there is also room for identifying possible steps towards mitigating it. In this context, the introduction of renewables, especially wind energy, is encouraged: "Non-hydro renewable energy-supply technologies – particularly solar, wind, geothermal, and biomass – are currently small overall contributors to global heat and electricity supply, but are increasing most rapidly. Costs, as well as social and environmental barriers, are restricting this growth. Therefore, increased rates of deployment may need supportive government policies and measures" (Metz et al. 2007, p. 253).

Most intergovernmental organizations approach renewables on a rather broad basis and, therefore, one cannot reasonably claim that efforts are targeted and directed at wind energy as such. Although major initiatives are not hammered out in these organizations, the broader agenda-setting is of huge importance, as it spills over into national domains and has a bearing on national innovation systems. The special division of labour between global and domestic institutions is important. It is difficult to imagine that national agencies could perform the same kind of scientific investigations and coordinating work as international agencies. But national institutions need this global policy transfer for launching their own initiatives and for giving priority to the renewables sector. At the national level we often find disagreement on the degree and kind of support offered to the wind turbine industry. Yet public policy spans many measures involving technology, education, research, taxation, and public investment, and is backed by arguments about reliable energy supply, environmental benefits, halting climate change, improving competitiveness, etc. The specific design and mix of all these measures and arguments varies across nations.

At the intergovernmental level, a range of bodies are concerned with climate and energy issues. Although there is no single unit coordinating these policies, advances have been made in coordinating public policy. Some intergovernmental organizations have become involved in coordinating energy and climate issues – new policies have been adopted, in particular, those manifested through the climate convention. New entities have also emerged, in particular, the IPCC, with the purpose of seeking further coordination in an otherwise fragmented area.

The most recent institutional innovation in the intergovernmental realm is manifested through the creation of the International Renewable Energy Agency

(IRENA) in January 2009, the formation of which has been strongly supported by the World Council for Renewable Energy (WCRE). Whereas some of the already mentioned agencies have a rather broad field to cover, the IRENA is fully committed to the renewables sector, and, headquartered in Abu Dhabi, it became fully operational in 2011. The agency was under preparation for the last decades and the creation of new international agencies is not a straightforward process: Some spaces are already occupied by existing agencies that either have similar tasks to manage or have competing priorities to defend. It is not a UN body but it liaises with the UN and other intergovernmental organizations, and it goes without saying that the body is welcomed by a variety of stakeholders.

These developments offer new opportunities for the creation of an innovation system linked to wind energy as a small and slowly emerging policy field. In fact, wind energy policy is beginning to develop its own agendas, strategies and institutional frameworks, albeit with a strong attachment to the more encompassing fields of energy and climate policy. In this context, the participation of private parties is helpful.

The organizations and forums mentioned here have a number of mechanisms for cooperating with industry and civil society. For example, as an intergovernmental forum, the UNFCCC is highly focused on the participation of states. Public attention has noted strong conflict between them and the vital role of national programmes, but there is also room for involving stakeholders. Various stakeholders are sometimes nominated as part of the national delegations to the conferences of the parties (COP). We also find observer organizations, such as Greenpeace and World Wide Fund for Nature (WWF). In addition, many organizations have together created the Climate Action Network (CAN), which is an active part of these conferences. In the context of the IPCC, less significance was attributed to the formal participation of stakeholders in the early days of the IPCC, but in preparing the 2002 session of the IPCC, a need for "a deeper engagement with industry and NGOs – possible formation of informal IPCC-industry and IPCC-NGO task groups" (IPCC 2002, p. 3) was stressed as an explicit goal.

In other words, an innovation system emerging around states and intergovernmental organizations has been extended to include a larger group of affected parties in business as well as in civil society. At the same time, the role of these private parties in the innovation system is to deliver relevant inputs into traditional public policy. Important institution-building has also evolved in the private sector, where we find more specialised actors in the area of wind energy.

9.4 Wind Energy and Business

The return of wind energy in modern times has been closely linked to energy and climate agendas and received impulses from public policy, but the industry has also made its own significant contributions to the development of this particular field.

The fact that the wind turbine industry follows its own paths has also led to a specific organization of business interests. Both broad and narrow organization of interests emerged. The oldest organization in the broad domain of energy is undoubtedly the World Energy Council (WEC), established 1924 as the World Power Conference. It covers both the conventional forms of energy and renewables. The Council is an important organization in putting energy on the political agenda and is also helpful in highlighting the role of new and alternative forms of energy, including wind energy, which, however, must seriously compete for attention with other forms of renewable energy. In fact, some of the forces opposing the wind turbine industry are found within the WEC.

The WEC is not purely a business association, however, although it embodies a strong industry element. It is based on having national committees as members, and the members are supposed to rally behind a broad range of interests, including industry, various regulatory agencies (irrespective of whether industry is in public or private hands), groups from the scientific community, and civil society organizations, including consumers and environmentalists. In sum, WEC operates with a very open membership policy. This openness does not suggest that all interests are equally represented, but it does show that there is an interest in building alliances with non-industry groups to represent the energy sector.

The lead organization in representing business in the area of wind energy, only recently created in 2005, is the Global Wind Energy Council (GWEC). Confronted with a set of global challenges, the industry has been organised in a global format, with the following national and regional founding members: American Wind Energy Association, Australian Wind Energy Association, Canadian Wind Energy Association, Chinese Renewable Energy Industries Association, European Wind Energy Association, Indian Wind Turbine Manufacturers Association, Japan Wind Association, and Japan Power Association – a constituency that documents the outreach of the association.

However, membership has been extended to other parts of the globe, including Africa and South America, at the same time as the membership base has been broadened in the stronghold regions through corporate membership. Thus, the key firms in the industry are organised via direct membership and are at the same time represented through national and regional associations. This organizational format allows for the combination of different corporate strategies. Even though entities other than industry can become members of the GWEC, few have done so, and strong alliances are not forged at this level.

Prior to the formation of the GWEC, many national and regional associations had already been established and were typically linked to domestic policy-making and innovation systems. The formation of the new global association stimulated industry action and has given national and regional associations an important global forum. Climate change and the demand for renewables are of course issues debated in these domestic contexts, but agenda-setting and policy development have increasingly shifted to the global level. It is at this level that strong influence needs to be exerted for setting general priorities in public policy through the commitments of states to work seriously with climate change and mitigating problems. Whether it is possible to follow-up on these global developments hinges on the wind turbine industry, in particular the ability to pose as an appropriate alternative that can offer stable and cheap energy supplies.

To influence global policy developments, the GWEC is engaged in various activities that form part of the emerging global innovation system. First, the global development of the wind turbine industry is monitored closely, and the current capacity of the industry and its considerable future potentials are illustrated with statistics and reports. It shows that firms are no longer novices and that wind power is not merely an idealistic enterprise, but a serious and prospering industry which has much to offer and could be integrated more thoroughly into global energy and climate strategies.

Second, policy proposals are presented through annual reports and policy recommendations, and through commenting on studies provided by intergovernmental organizations, as well as other public and private actors in the field of climate policy, energy policy, and wind energy policy. In this context, the GWEC seeks to have a stronger dialogue with the IEA.

Third, in addition to these forms of leverage, information that facilitates the analysis of the market environment and improves competitiveness is provided to association members. Timely and relevant information of this kind is not always obtainable outside of the GWEC, and such activities help create basic logistics in the wind turbine industry.

Although GWEC activities fan out in different directions, we must observe broader organizational formats with a bearing on wind energy. While energy and wind energy associations are primarily concerned with energy and climate change, the industry also is represented in more encompassing associations. Of the many groups engaged in climate issues, the World Business Council of Sustainable Development (WBCSD) (dating back to 1995, when the Business Council for Sustainable Development and the World Industry Council for the Environment merged) is important. Its mission is to encourage firms to adopt new corporate practices to reduce emissions and to take part in other sustainable strategies. Through sharing experiences, disseminating information, and benchmarking, the WBCSD assists in the creation of key elements in the innovation system.

At the same time, the WBCSD seeks to influence and engage with public and private players in the global realm to influence public policy. It is not a spokesperson for the energy industry per se, or even for the wind turbine industry. Rather, most of its efforts seek to create framework conditions which have implications for the wind turbine industry. Membership in the WBCSD is by invitation only – thus, the association chooses its own members, not the other way round. Interestingly the WBCSD has not taken steps to articulate the interests of the wind turbine industry as it is more focused on using conventional energies with care, a development which can, in the long run, facilitate the strengthening of renewables.

The WBCSD draws members from a broad spectrum of businesses, particularly very large corporations whose successful handling of energy problems can be an important contribution to problem-solving. Indeed, "Energy and Climate" is one of its four focus areas, which include a plethora of concrete projects that help corporations save energy and introduce sustainable energy. The WBCSD also maintains a continuous dialogue with the IPCC and the UNFCCC, in whose conference activities and project implementation the WBCSD is engaged.

Furthermore, the WBCSD is building relations with various stakeholders, and its strategy "acknowledges that the world is shifting towards partnerships between government, business and civil society to address the major challenges" (WBCSD 2010). The International Union for Conservation of Nature and Natural Resources (ICUN) and the World Wide Fund for Nature (WWF) are listed among some of its exchange partners. Because corporate social responsibility initiatives are encouraged, relations can also be forged at the level of single firms. Following this model, innovation is more likely to take place through pioneering firms rather than through industry-wide incentives and rules. However, some broader initiatives relevant to sustainable business in general, but also with a view to including civic groups, have emerged, with the WBCSD as one of the key initiators of the 2009 World Business Summit on Climate Change.

Additional specialised organizations in the field of wind energy exist, illustrating the bewildering diversity in the organizational landscape. For instance, the World Wind Energy Association (WWEA), founded in 2001, has brought attention to the role of wind technology and is open not only for corporate membership, but also for scientists and other interested parties. Indeed, it is a forum that brings together different actors with an interest in promoting wind power in general. It has also taken an active part in the formation of the World Council for Renewable Energy (WCRE), founded in 2001, and the International Renewable Energy Alliance (IREA), established in 2004. These groups show the fairly recent emergence of the global structures in business, but they are not all designed as proper business associations. As we shall see in the next section, a major feature of the field of wind energy is the fact that a global innovation system is not built around governmental institutions and the industry alone, but embraces a broader set of actors.

9.5 Wind Energy and Civil Society

A vast number of civil society organizations pay attention to climate policy and wind energy. The Climate Action Network includes a significant group of small and large organizations but the single key player is Greenpeace International, which for many years has been involved in a range of environmental issues. The adoption of the UNFCCC, referred to above, has further stimulated Greenpeace initiatives and contributed to the development of relevant strategies. In addition, energy policy has been embraced, as has the more specific issue of wind energy. Greenpeace is a staunch opponent of nuclear energy, coal and other conventional forms of energy, and an equally strong advocate of renewable energy, including wind energy.

Strategies are elaborated to enhance the role of sustainable energy, and a special climate and energy unit fulfils this purpose. In cases where Greenpeace does not have sufficient expertise, collaboration is sought with the expert community and other stakeholders (Greenpeace 2000). It is interesting to note that such studies are often far more concrete than those produced by the business community, and that they are in no way merely an idealistic statement about the general desirability of sustainable energy (Greenpeace 2004). Such studies detail the economic, environmental and social benefits; they not only reflect the concerns of civil society but also envisage relevant business strategies and practical technological steps. These analyses do not always have a global perspective, but are sometimes limited to regional issues, predominantly European, where the biggest advances have been made (EREC and Greenpeace 2007).

Greenpeace is very much involved in campaigning, and also participates in forums organised around intergovernmental conferences, often linking the two. Several of these joint studies bring Greenpeace into collaboration with the wind turbine industry. The Global Wind Energy Outlook (Greenpeace and GWEC 2006) is a voluminous report that covers a very broad issue area in relation to wind energy.

As outlined above, the GWEC is the primary organization in the representation of the wind turbine industry, and the collaboration between an important environmental organization and a central business association stresses the willingness of both entities to enter into a mutually beneficial cooperation. In other words, strategies are not built around a single commercial or environmental perspective but expressed in a combined effort that is more easily amenable to public policy that tries to align different concerns in its innovation strategy. Because business associations and civil society organizations represent different constituencies, but also speak to and reverberate different opinions, there is a greater likelihood that such initiatives will also be welcomed by politicians with different backgrounds at domestic and international levels. Traditional conflicts between "left" and "right" and between "economy" and "nature" can more easily be mitigated once business and civil society succeed in developing a dialogue and building alliances.

Another major civil society organization with a strong engagement in climate policy, energy policy and wind policy is the WWF. Its focus on climate issues goes back a long way, but the Earth Summit in 1992 provided a new impetus to its strategy. For many other private and public organizations, however, the energy policy and wind energy debates are of more recent origin: Policies have been increasingly elaborated and substantiated through studies in which the WWF has been involved, but policies have also built on commissioned work done by other actors and the WWF has drawn on expertise from the scientific community.

The WWF report "Climate Solutions, The WWF Vision for 2050", issued in 2007 and coordinated by the WWF Global Energy Task force, highlights the role of renewables and points to great potential in wind energy (WWF 2007). The report is a thorough study on sustainable energy technologies, but also covers aspects such as investment and finance. Such studies factor in a range of critical issues from the perspective of innovation.

These studies are important in the context of the WWF seeking to influence decision-making in many intergovernmental organizations and their various conferences. Although the WWF has a much broader agenda than wind energy, and even energy policy, many of their activities center on global warming and climate change. Also, the relevance of new forms of sustainable energy is emphasised. Furthermore, the WWF has forged many relations with single firms and with certain groups of firms, and seeks to encourage a pro-active business community, although it is critical of single firm action and generally prefers binding rules covering an entire industry (Pamlin 2002).

Civil society organizations not only deliver relevant inputs for corporate strategies and for the work of intergovernmental agencies but also establish joint platforms. Drivers for innovation are found across the institutional landscape and are given a special quality when major actors are brought together. The Renewable Energy Policy Network for the twenty-first Century (REN21) is an important initiative working outside the formal framework of UN agencies – but coordinating with and including some of these – and includes a number of government agencies, industry associations and civil society groups (REN21 2010). The World Summit on Sustainable Development (WSSD) in 2002 did not bring the role of renewables to the forefront of sustainable development, but new plans soon emerged. Thus, the REN21 was under preparation in late 2004 and was stimulated by the International Conference for Renewable Energies held in the summer of that same year. A new network was officially established in 2005.

A large number of interested parties were involved in these processes, but it is interesting to note that the German government, which hosted the 2005 conference, agreed to provide a basic organizational infrastructure and, therefore, became an important catalyst. Since then, the network has developed its own capacity and is an independent forum for policy development in the area of renewables; it runs a large number of concrete projects under its International Action Program (IAP). It is not a network exclusively committed to wind energy issues, however, but it is focused on building a sustainable energy agenda as an alternative to other forums focusing on energy, environment, or development.

First, it organises its own conferences and brings different stakeholders together to formulate strategies and exchange experiences. Updating and ensuring easy flow of information to the actors in the network are key priorities. These frequent exchanges provide important inputs to the work of the involved industries and individual firms, as well as to civil society participants and government officials. Second, it provides timely inputs to other conferences organised in the UN format or in the G8 context both by preparing annual global status reports and by issuing papers that examine a wide spectrum of topics. These reports show the growing, but still largely unexhausted, role of renewables by linking these new and alternative energies to the ongoing and encompassing climate change debate. Third, it comments on various studies commissioned by other agencies on renewables, or energy in general, with the goal of influencing relevant governments, intergovernmental organizations and conferences. The goal is to stimulate policies that give higher priority to renewables.

The emergence of the REN21 shows that it has not been possible to accommodate the many interests in renewables within the context of existing organizations, such as the IEA. A new and highly specialised platform, in which the conventional energies and stakeholders are kept outside, is needed. It is also evident that traditional institutional arrangements run by governments, and with private organizations taking the role of observers, are only partially useful. In the context of this institutional renewal, civil society organizations have been granted a role which is not only helpful to funnel a civil society perspective – both an environmental and consumer one – into the network, but is also a perspective around which it is highly useful to build public policy and corporate strategies.

9.6 Conclusions

Modern interest in wind energy has evolved over the last couple of decades, and, in relation to the climate change debate, the industry has gained further momentum. Wind energy, however, is a small policy field strongly embedded in other, more encompassing policy fields. Thus, wind energy belongs to renewable energy policy, which, again, can be characterised as a subsection of energy policy. Energy policy has increasingly become an element in climate policy which has experienced a fast and fascinating career and has been vital in pointing to alternative solutions to saving the environment and halting climate change.

Although the full potential of wind energy is far from exhausted, the development of climate policy has undoubtedly been, and still is, decisive for the further development of the wind turbine industry. In other words, the innovation system is both driven by some general framework-setting strategies and by specific activities relevant to renewables and wind energy. Many of these activities are still carried out at domestic levels, which, to different extents, tap into global public policy. Indeed, both domestic and global developments have characterised these processes.

Given the key role held by pioneering firms, the wind turbine industry has entered the energy market and, through new technologies, has struggled to take market shares from coal and oil, a development sometimes actively supported by governments. In this context, public investment is of paramount importance and has consequences for national energy policy, for the competitiveness of firms and in recovering new markets. We have also seen the creation of a fledgling innovation system consisting of new institutions and policies at the global level, creating strong incentives for firms and governments alike. How can we reconcile these different figures characterizing the wind turbine industry?

Essentially, the emerging innovation system is not reducible to the behavior of firms and the properties of technological and economic domains. The creation of an innovation system very much hinges on the efforts of a range of public and private actors and on their specific abilities to mitigate conflicts and find commercially viable and socially acceptable solutions. In other words, a critical mass is produced in the form of technological, economic and political conditions that in different and complex situations stimulate each other. The rise of the wind turbine industry over the last few decades is best understood when these elements are factored in: Technological progress is essential for the industry to compete with other industries, the growth of the industry makes it politically relevant to build wind energy into an overall strategy, and political priorities will give an impetus to the industry.

It is interesting to note that wind energy is today addressed by various intergovernmental organizations, but there is no single agency specializing in this policy field and, therefore, different coordination problems arise. However, there is no doubt that these different agencies convey the message that renewables, including wind energy, should be encouraged and are important in halting climate change. In many ways, it has become easier for the wind turbine industry to leverage political institutions, but the industry is still competing with cheaper and more conventional forms of energy. The fragmentation of public institutions has, however, also complicated the concerted effort on the part of industry, as well as the inputs from civil society organizations, because several avenues of influence are available.

Whereas there is much fragmentation in public institutions and policy – a factor complicating the development of a strong innovation system – there have also been weaknesses on the side of private organizations. However, the recent collective organization of the wind turbine industry through a specialised association has given the industry a new and much clearer voice at the global level and galvanised cooperation between firms. The association is an important forum for the exchange of information and experiences. At the same time, some important civil society organizations have become strong supporters of wind energy and are active in the dissemination of knowledge and in giving the wind turbine industry some kind of moral credibility. Today, this collaboration is organised within the context of a new network, including stakeholders from the different corners of the wind energy sector, showing that various public and private actors are increasingly becoming integrated into an emerging global innovation system.

References

Bandy, J., & Smith, J. (2005). *Coalitions across border: Transnational protest and the neoliberal order*. Lanham: Rowman & Littlefield.

Bechberger, M., Mez, L., & Sohre, A. (Eds.). (2008). *Windenergie im Ländervergleich. Steuerungsimpulse, Akteure und technische Entwicklungen in Deutschland, Dänemark, Spanien und Großbritannien*. Frankfurt am Main: Peter Lang.

Betsill, M. M., & Correll, E. (2007). *NGO diplomacy: The influence of nongovernmental organizations in international environmental negotiations*. Cambridge: MIT Press.

Biermann, F., & Siebenhüner, B. (Eds.). (2009). *Managers of global change: The influence of international environmental bureaucracies*. Cambridge: MIT Press.

Braithwaite, J., & Drahos, P. (2000). *Global business regulation*. Cambridge/New York: Cambridge University Press.

Braudel, F. (1992). *Civilization and capitalism 15th–18th century. Volume 1: The structures of everyday life. The limits of the possible*. Berkeley: University of California Press.

della Porta, D., Andretta, M., Mosca, L., & Reiter, H. (2006). *Globalization from below: Transnational activists and protest networks*. Minneapolis: University of Minnesota Press.

EREC & Greenpeace. (2007). *Energy (r)evolution – a sustainable energy outlook*. Amsterdam: European Renewable Energy Council and Greenpeace.

Finnemore, M., & Barnett, M. (2004). *Rules for the world. International organizations in global politics*. Ithaca/London: Cornell University Press.
Greenpeace. (2000). *Offshore wind energy in the North Sea: Technical possibilities and ecological considerations – a study for Greenpeace*. Wilhelmshaven: German Wind Energie Institute.
Greenpeace. (2004). *Sea wind Europe*. Amsterdam: Greenpeace.
Greenpeace, & GWEC. (2006). *Global wind energy outlook*. Amsterdam/Brussels: Greenpeace and Global Wind Energy Council.
Hawkins, D. G., Lake, D. A., Nielson, D. L., & Tierney, M. J. (Eds.). (2004). *Delegation and agency in international organizations*. Cambridge/New York: Cambridge University Press.
IPCC (2002). Intergovernmental panel on climate change. Nineteenth session. Geneva, 17–20 April, Appendix g.
IPCC (2004). 16 years of scientific assessment: In support of the climate convention. December 2004. http://www.ipcc.ch/pdf/10th-anniversary/anniversary-brochure.pdf. Accessed 28 Sept 2010.
Jaccard, M. (2005). *Sustainable fossil fuels: The unusual suspect in the quest for clean and enduring energy*. Cambridge/New York: Cambridge University Press.
Keck, M. E., & Sikkink, K. (1997). *Activists beyond borders: Advocacy coalitions in international politics*. Ithaca/London: Cornell University Press.
Levy, D. L., & Newell, P. (2005). *The business of global environmental governance*. Cambridge: MIT Press.
Lewis, J. I., & Viser, R. H. (2007). Fostering a renewable energy technology industry: An international comparison of wind industry policy support mechanisms. *Energy Policy, 35*(3), 1844–1857.
Metz, B., Davidson, O., Bosch, P., Dave, R., & Meyer, L. (2007). IPCC climate change 2007. Contribution of working group III to the fourth assessment report of the IPCC. Intergovernmental panel on climate change.
Pamlin, D. (2002). WWF discussion paper. Corporate social responsibility. CSR – an overview. Gland, Switzerland: WWF.
Reinicke, W. (1998). *Global public policy*. Washington, DC: The Brookings Institution Press.
REN21 (2010). About REN 21. http://www.ren21.net/ren21/default.asp. Accessed 28 Sep 2010.
Ronit, K. (2007). Introduction: Global public policy: Yhe new policy arrangements of business and countervailing groups. In K. Ronit (Ed.), *Global public policy: Business and the countervailing powers of civil society* (pp. 1–42). London/New York: Routledge.
Ronit, K., & Schneider, V. (2000). Private organizations and their contribution to problem-solving in the global arena. In K. Ronit & V. Schneider (Eds.), *Private organizations in global politics* (pp. 1–33). London/New York: Routledge.
Steiner, A., Wälde, T., Bradbrook, A., & Schutyser, F. (2006). International institutional arrangements in support of renewable energy. In D. Assmann (Ed.), *Renewable energy: A global review of technologies, policies and market* (pp. 152–168). London: Earthscan Publications.
Szarka, J. (2007). *Wind power in Europe: Politics, business and society*. Houndsmills: Palgrave Macmillan.
Toke, D. (2007). Wind power, governance, and networks. In J. Murphy (Ed.), *Governing technology for sustainability* (pp. 168–181). London: Earthscan.
WBCSD (2010). About the WBCSD: History of the WBCSD. http://www.wbcsd.org/templates/TemplateWBCSD2/layout.asp? type=p&MenuId=NDEx&doOpen=1&ClickMenu=LeftMenu. Accessed 26 Aug 2010.
WWF. (2007). *Climate solutions: The WWF vision for 2050*. Gland/Switzerland: WWF.

Chapter 10
From Niche to Mass Markets in High Technology: The Case of Photovoltaics in Germany

Gerhard Fuchs and Sandra Wassermann

10.1 Introduction

The energy supply system in industrialised nations is changing in ways that are often conceived of as a technological and institutional regime change. Victor (2002) sees the sector in its third structural transformation. The exact outcome of this regime change is uncertain as of yet, but one element of a future governance structure will be the increased importance of decentralised forms of electric power generation and a shift towards more environmentally sustainable technologies, such as renewable energy technologies, which in the past were pushed forward by a diverse coalition of actors. This article will focus on one of the most innovative developments in the area of renewable energy technologies: photovoltaics (PV).

We will use a broad lens in order to examine the growth of PV as a source of electric power generation and as a business sector in Germany. PV can be considered an unusual success story in which political actors' ability to make a significant impact on renewable energy production and the associated economic activity looms large.

It will be argued that the growth of renewable energy takes place within networks of governance comprising formal regimes at multiple levels, informal norms and practices, as well as market structures and processes. Actors within these networks include national and sub-national authorities, multilateral institutions, firms, and NGOs. Technological development and market growth of PV are thus viewed as embedded in a broad social, economic, and political system of governance. Corporate strategies, social movements, and public policy interact within, as well as constitute, the essential elements of governance in this sector. We will

An earlier version of this chapter was published in STI Studies (Fuchs and Wassermann 2008).

G. Fuchs (✉) • S. Wassermann
University of Stuttgart, Stuttgart, Germany
e-mail: gerhard.fuchs@sowi.uni-stuttgart.de

J.M. Bauer et al. (eds.), *Innovation Policy and Governance in High-Tech Industries*, 219
DOI 10.1007/978-3-642-12563-8_10, © Springer-Verlag Berlin Heidelberg 2012

further argue that policy on PV in Germany is characterised by a specific mission orientation, the concentration of main actors, a long-term orientation, and substantive subsidies. PV is a successful, as well as "planned" innovation, something quite uncommon in the literature on innovation. Caniels and Romijn (2008, p. 246) have argued that within the literature on strategic niche management there is a shortage of analyses focussing on success stories and a lack of understanding about the processes by which policy and technological experiments culminate in viable market niches that ultimately will contribute to a regime change in a specific sector. This article attempts to fill that gap.

This chapter develops the main points in several steps. To set the stage, we will clarify our concept of innovation and describe the elements of the technological system of PV. Based on these introductory remarks, we will discuss the factors responsible for PV's breakthrough. It is too early to claim that PV will continue to be a success story in the future or that PV will eventually play a dominant role in the development of a new energy regime. PV is growing but it is still not in a settled and stable state – albeit, it is already larger than many "established" sectors. The particular technological and institutional prerequisites enabling photovoltaics' achievements have not been studied in great detail. Main lessons from this unique case are reiterated in the concluding remarks.

10.2 Innovation and Sectoral Systems of Innovation

Before discussing German innovation policy focussing on the development and market expansion of photovoltaics, we have to establish conceptual foundations for our analysis. We start with some general reflections on innovation and innovation policy, drawing from the literature on systems of innovation and strategic niche management on the one hand, and the advocacy coalition approach on the other.

10.2.1 Innovation Policy

Since the 1990s, a global shift in policies towards research and technology can be observed: the promotion of innovation has become a centre piece of official national as well as of supra- and sub-national policies. This shift in emphasis reflects discussions on the role of the state in promoting technology, as well as new ideas about how new technologies can become successful in various markets.

The traditional model in research and technology policies either centred on the support of basic research, which eventually should bring about new technologies ripe for the markets (technology push), or opted for a mission-oriented approach deciding to support a specific technology and financing its development through specific companies or research laboratories (Hiskes and Hiskes 1986).

Innovation research has shown that there is no linear development from basic research to successful technological innovation in the market. Support of basic research does not guarantee the eventual development of products that become widely accepted and thus achieve commercial success. But "market success" has become a top priority in times of increasing worldwide competition in crowded markets. The introduction of new, innovative products is considered to be a precondition for maintaining a competitive edge. In order to be commercially successful, it is of vital importance to reach a critical mass within a comparably short time frame (Rogers 1995, p. 313).

In related discussions of the state's influence on technological innovation processes, a dire picture has been painted, accentuating the conviction that the state is unable to pick technologies that will succeed in the market. Along with an increasingly prevalent attitude that markets are the best innovators and should be left alone, policy instruments worldwide seem to converge, looking increasingly alike (Holzinger et al. 2007). This neoliberal understanding, the support for markets, and "the retreat of the state," (Strange 1996) emerged in the 1990s and was accompanied by new types of policies and policy instruments which also affected the design of technology policy. Research and technology policy was transformed into innovation policy and mainly focused on funding basic research and network activities, as well as joint projects between firms and research institutes in order to stimulate knowledge flow and to ensure that the results of scientific research could be used and adopted commercially (Nooteboom 1999; Edquist 2001, p. 18). Public actors, however, were not supposed to select a certain technology in advance and would abstain from market stimulation programmes. Networks can potentially facilitate producer-customer relationships or can even result in the creation of advocacy coalitions. These are considered an important pre-condition for successful radical innovations by most experts (Weimer-Jehle and Fuchs 2007).

Although the market discourse has achieved nearly universal legitimacy, counter tendencies have always been visible as well. Complexity theory and the literature on governance have aimed at creating a new understanding on the role of politics (Kappelhoff 2000; Werle and Schimank 2000). On the one hand, social developments are unpredictable and evolutionary, but on the other hand, these evolutionary dynamics have always been accompanied by conscious planning and shaping (Czada and Schimank 2000). Thus, political actors are seen as interacting in governance networks together with other actors who also try to influence social developments. One of the measures relying more directly on the activities of public actors is the politically supported creation of niche markets. This new form of innovation policy selects a certain technology (or precursors to a technology) in advance and tries to accelerate its development. It might even help to shape the mode of its application. Such politically created niche markets work through market stimulation programmes, such as subsidies or the provision of soft loans for prospective customers, as well as through modes of legitimising the developing technology, in order to raise its public acceptance (Edler 2007). Especially in the area of environmental technologies, strategic niche management has increasingly become accepted as an instrument of innovation policy (Kemp et al. 1998; Kemp

2002; Coenen 2002) in the hopes that even the transformation of entire technological regimes is a viable option (Berkhout et al. 2003, p. 4; Caniels and Romijn 2008).

The design of national policies has to consider existing institutional frameworks and socio-cultural conditions. Work in the tradition of the Varieties of Capitalism approach claims that if national innovation policy stresses national comparative institutional advantages, it can be more successful. In other words, a system dominated by non-market coordination will have difficulties pushing new technologies dependent on a flexible and quick functioning market mechanism. On the other hand, the support of technologies which require the non-market coordination of various actors will be difficult to put into effect in liberal market economies. Based on this highly stylised interpretation we argue that the creation of (sheltered) niche markets can be a successful policy instrument especially in coordinated market economies (hypothesis 1).

Considering the fact that photovoltaics can be seen as a technological innovation that is supported in order to transform the energy sector, the existence of political and social forces strongly opposing it for ideological as well as economic reasons (e.g., rent seeking) can be assumed. As Jänicke (1997, p. 7) has shown, changes in actor constellations have resulted in improved conditions for innovation in environmentally friendly products. With regard to actor constellations and situational factors enhancing policy change, the policy analysis literature refers to the role of advocacy coalitions that are crucially important in order to spur institutional or cultural changes (Litfin 2000). We will argue that the success of innovation policy depends on its ability to create and mobilise an advocacy coalition supporting the technology in question, especially if strong incumbent actors (such as the established energy providers) exist (hypothesis 2).

10.2.2 Innovation

Innovation can be defined as artefacts, processes, ideas and strategies, which successfully change routines and are embedded in specific contexts of development and usage. Innovation as such is not just a new idea or technical system, but one which is being successfully implemented. Innovation in this sense is not a linear process but occurs by interactive relationships and feedback mechanisms between institutional and organisational elements of science, technology, learning, production, policy, firms and potential or actual market demand. Some technologies may only become innovations due to interactions between producers and users, or the specific way customers use and apply new technical artefacts (Malerba 2004, p. 24). The acceptance and use of a new technology, at any rate, plays a crucial role in the innovation process. Thus, new and better technologies in our context are only referred to as innovations if their development is embedded and accompanied by the establishment of a successful industry, and if they find their way to the market.

10.2.3 Innovation and Uncertainty

It is generally acknowledged that economic and other activities face the problem of uncertainty (Beckert 1996). This is even more so in the case of innovations, particularly if potential new products would have to cope with incumbent products and the existing infrastructures and routines supporting them. Proven ways to cope with uncertainty are the development and reliance on routines, customs, regulations, established institutions, etc.

Innovating firms may not know which application or design a new technology should be given in order for it to be successful on the market. This can lead firms to become hesitant when implementing significant changes, even as they face a volatile environment that increases pressures to introduce new products, seek new markets and introduce new technologies, practices, and organisational methods into their production processes. Uncertainty can also make it more difficult for firms to obtain external funding for their innovation projects. Customers may not trust a new and unproven technology. This leads to another blocking mechanism for the diffusion of a new technology–lack of legitimacy.

Here we are confronted with the paradox that innovation, as a routine-changing mechanism, also depends on routines, albeit currently developing ones. Innovation policy can attempt to reduce uncertainty by establishing a mix of policy instruments along with a viable support coalition. Whenever innovation policy can provide technological developments with legitimacy, the financial system will become more willing to invest in innovative firms, and potential customers may feel more secure and be more inclined to purchase new technologies (Carlsson and Jacobsson 1997, p. 285).

The role of uncertainty can be seen very clearly in the developments of the 1990s. At one point, the German PV-industry was close to extinction and production facilities were moved, since producers could not form stable expectations as to whether the institutional framework in Germany would provide favourable conditions for the further development of the PV industry or not.

As Edquist (2001, p. 17) suggests, a systemic view on innovation policy should not only analyse the role of the state but also include feedback mechanisms on how the rest of the system, social structures, routines or even discrete occurrences influence innovation policies. As German governance has always been characterised by close linkages and the reliance on common interests between government, industry, business associations, and trade unions (Hall and Soskice 2001; Harding 2000), this established form of governance has also shaped German innovation policies and will continue do so in the case of PV.

10.2.4 The Transformation of Electric Power Generation

Photovoltaics are treated as an innovation within and for the industrial sector of electric power generation. As already briefly mentioned, this sector is undergoing substantial changes in nearly all industrialised nations. The dynamics leading to these changes are also important in order to understand the case of PV, because they opened a window of opportunity which helped to push PV as a new option of energy supply.

The traditional electric power system can be looked upon as a large technical system (Mayntz and Hughes 1988), tightly coupled and run by a few, powerful incumbent actors. Energy generation is highly centralised in big power stations – open markets hardly exist. Price regulation is common and huge subsidies for the development of old and new technologies (e.g., coal, nuclear energy) make it difficult to determine "real" prices. There are suggestions that the costs of producing electricity gained out of coal or oil would double, if transparent external costs were taken into account (Milborrow 2002). Incumbent energy technologies have received direct and indirect subsidies for decades (Jacobsson and Bergek 2004, p. 210). R&D expenditures in these closed markets are nevertheless low, and innovation is slow-moving and incremental. R&D expenditures depend to a very large degree on the interpretation of political signals regarding the regulation of technology.

Two trends are transforming the traditional ways: the liberalisation of infrastructures and environmental issues, in particular, concerns about global warming. Hopes for an effective regime to address climate change have shifted from the emphasis on a mandatory multilateral agreement, the Kyoto protocol, towards a plethora of regional, national, and sub-national programmes and initiatives. Policy responses include carbon emission limits and trading systems, direct subsidies for renewable energies, and Renewable Portfolio Standards that mandate the use of specific volumes of renewable energy in electricity generation. Such policy responses are required because the market will not, by itself, respond adequately to the environmental challenge.

Given the rapid growth expected in global markets for low-emission technologies, the policy agenda is also driven by economic development goals, as countries vie for competitiveness and market share in these emerging fields. Liberalisation can have differing effects for renewable energies. If energy prices fall as a result of liberalisation and increased market competition (as economic theory would make us believe), the price target that renewables must meet becomes more challenging and liberalisation might prove to be an impediment for their further spread. On the other hand, policies and systems such as quotas and renewable energy certificates can be compatible with more competitive market structures as the experiences of the last years have shown – supported, for example, by a general increase in energy prices. In fact, many of the policies which have been implemented for the support of renewables operate within the framework of a transition to market liberalisation (OECD 2008).

Finally, beyond the problems of lacking transparency and the prevalence of risk-averse actors, there is the constraining factor of centralised energy infrastructures as they have developed and have become established over decades. National grids are mainly tailored to the operation of centralised power plants and thus cement their existence. Alternative technologies like photovoltaics follow an opposite decentralised logic that does not easily fit the established technological concepts and thus face difficulties competing with incumbent technologies (Stern 2006, p. 355).

In sum, these conditions have led to the widely accepted conviction that policy instruments which aim to create niche markets for renewable energies are needed. Even the European Commission, which traditionally favours market instruments and is rather critical towards demand side policy actions, has opted for market stimulation programmes for renewable energy technologies (Commission of the European Communities 2005; European Parliament and Council 2001). This is true in spite of the fact that until recently the European Commission and the OECD both disapproved the German model of market stimulation, and instead had favoured quota models which use market signals in order to increase the supply of renewable energy (Busch 2005, p. 235).

10.3 Photovoltaics: Technological Characteristics

Before analysing photovoltaics as a case of successful innovation, we need to provide a short introduction on the technologies and applications we are talking about. Photovoltaics use solar cells to produce electric power[1]. The most common type of solar cell consists of either mono-crystalline or poly-crystalline silicon, which is conventionally produced and used by the electronics (semiconductor) industry. Crystalline silicon technologies represent 93% of the photovoltaics world market (Solarbuzz 2007). Mono-crystalline silicon cells are characterised by their ability to convert a relatively large section of the light spectrum into electricity with an efficiency of up to 24.7% under ideal laboratory conditions (Solarserver 2007). Poly-crystalline silicon cells do not achieve such high efficiency rates, but they are less costly. The same holds for amorphous and other thin film technologies that consist of cadmium telluride (CdTe) or copper indium dieseline (CIS). Due to silicon shortages since the turn of the century, research and development on non-silicon thin film technologies has become increasingly popular, and remarkable reductions in production costs have been achieved.

The photovoltaic effect was first discovered by the French physicist Alexandre Becquerel in 1839. Albert Einstein's theoretical work on the photoelectric effect won him the Nobel Prize in 1921. Thus, basic research on photovoltaics has been conducted for quite some time. Yet the first applications did not appear until the 1950s, when Bell Laboratories invented the first solar cell and the US government started to use solar cells on satellites. "The satellite market became the first significant commercial market and annual production rose to about 0.1 MWp

[1] Photovoltaics should not to be confused with solarthermy, which is the conversion of solar energy into usable thermal energy.

[Mega Watt peak] per year in the late 1960s" (Jacobsson et al. 2002, p. 10). It is striking that the first satellite project using solar power, was under US Navy management and monitored by the Department of Defense. Some authors therefore pointed out that the case of photovoltaics was one of many technological developments in which the military played a crucial role (Clark and Juma 1987, p. 142; Jacobsson et al. 2002, p. 10). Due to US export restrictions, in the 1960s the European Space Agency (ESA) had to rely on German companies such as Siemens and Telefunken to get involved in photovoltaics research and production for space programmes (Jacobsson et al. 2002, p. 16). Since the 1970s, and largely due to the oil crises, interest in the development of various terrestrial applications grew and led to further R&D activities, mainly in the U.S. and Japan. A range of off-grid applications emerged, that were mainly used for consumer electronics like calculators and watches or as stand-alone "power stations" for SOS telephones and for remote places like buoys, yachts, mountain huts, and camping. Furthermore the idea of solar home systems to be employed in developing countries came up. Rather distinct from these off-grid photovoltaics are newer forms of applications which supply electricity to the grid just as conventional power technologies. Grid-connected applications can be found as roof-top systems, ground-mounted systems or as systems integrated into house façades. However, demonstration projects which employed photovoltaics in order to supply electricity to the grid were not implemented before the 1990s. Grid-connected photovoltaics are, therefore, a rather new development and it is striking that since 1999 they have rapidly outpaced other forms of applications in International Energy Agency (IEA) reporting countries (IEA 2005).

10.4 Success Indicators

In this section we wish to highlight the successful development of PV with the help of quantitative indicators. In order to measure "success" we will use the indicators installed PV power, production, export sales, employees, and patents. As Fig. 10.1 impressively shows, installed PV power was at a relatively low level, then doubled for the first time in 2000 and has grown continuously since then. These findings demonstrate the correlation between policy instruments that were applied by the federal Red-Green coalition government, the regulatory instrument EEG and the 100,000 roofs programme, and the expansion of the market (see below).

In 2005 "...Germany accounted for more than 93% of the EU 25" (Jäger-Waldau 2002, p. 75) installations. Stable political and socio-economic conditions not only convinced private households to install photovoltaic power, but solid markets also stimulated the investment in new production capacities for solar cells and modules.

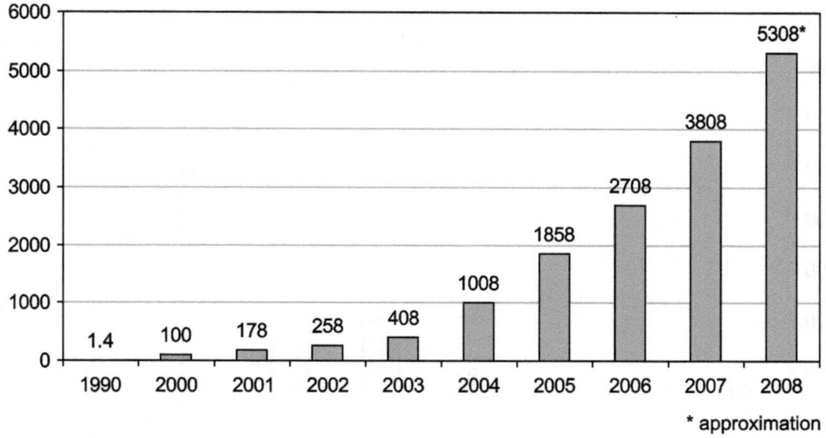

Fig. 10.1 PV power installed in Germany (MWp) (Source: Bundesverband Solarwirtschaft e.V. (BSW-Solar 2010))

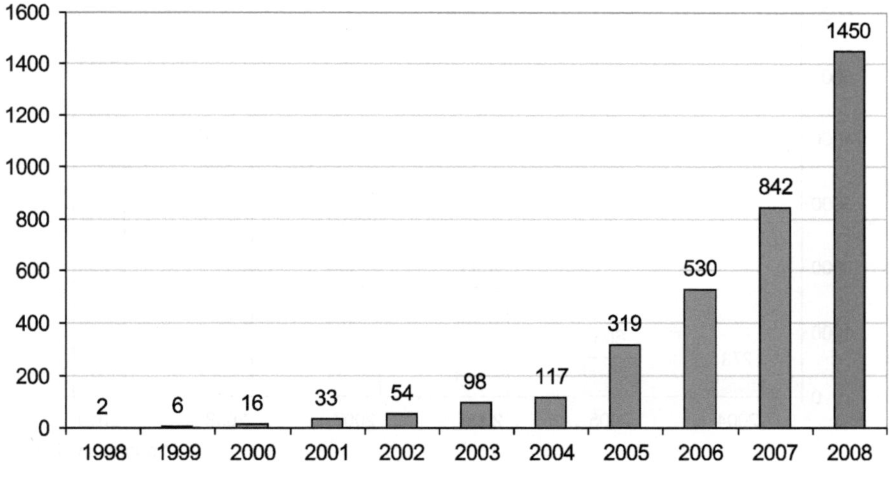

Fig. 10.2 Solar cell production in Germany (Mw) (Source: Bundesverband Solarwirtschaft e.V. (BSW-Solar, multiple years))

As Fig. 10.2 shows, cell production has grown to almost 1,500 MW annually. Sales as well as export shipments of the German photovoltaics industry have been rising with a comparable rate, as can be seen in Figs. 10.3 and 10.4.

Sales figures and numbers of photovoltaics power installed clearly show its market success. An even more common way of measuring innovation is patent data, since "…patents provide a uniquely detailed source of information on inventive activity" (see OECD 1994, p. 9). As Fig. 10.5 shows, Japan is by far the most

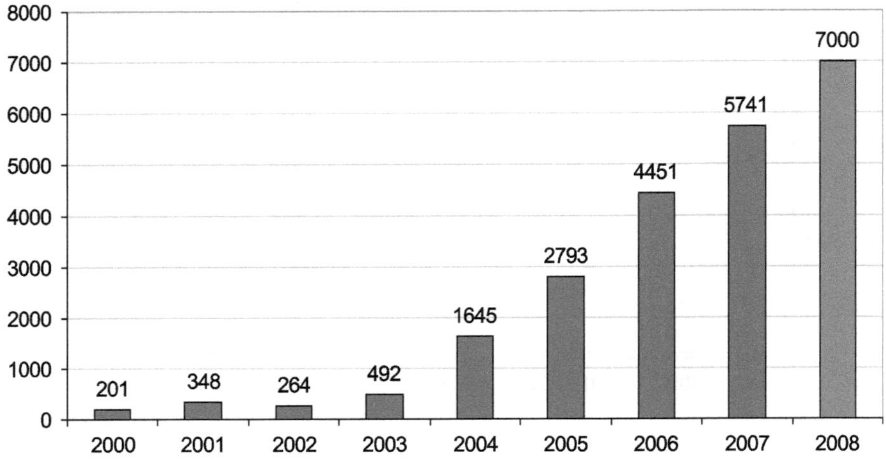

Fig. 10.3 Sales of the German photovoltaics industry (Source: Bundesverband Solarwirtschaft e.V. (BSW-Solar 2010))

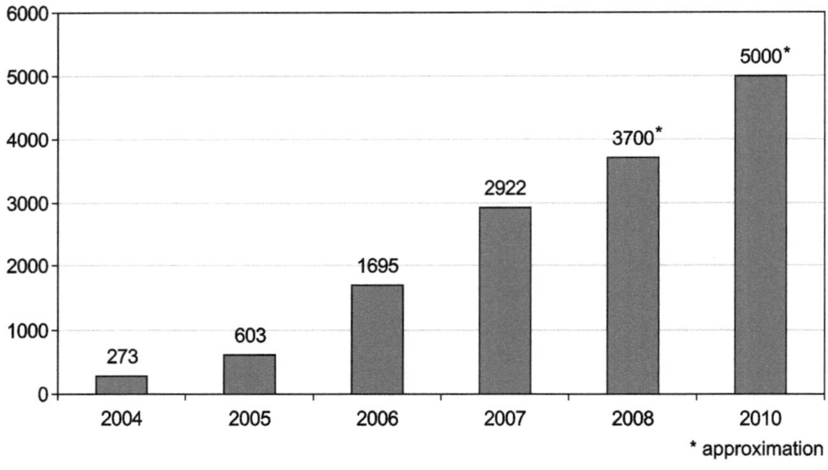

Fig. 10.4 Export sales of the German photovoltaics industry (Source: Bundesverband Solarwirtschaft e.V. (BSW-Solar 2010))

active nation in patent applications, followed by the U.S. and Germany[2]. As can be seen from Fig. 10.6, German patent activities reflect quite well the global increase

[2] It is striking that Japan accounts for 74% of all patent applications, but this is mainly due to characteristics of the Japanese patent law system, which makes the process of applying for a patent easier and cheaper than in the U.S. and Germany. Furthermore, in Japan, normally one invention is divided into small elements and for each a patent application is filed (Siemer 2005, p. 66). Therefore, comparing German patent activities with the Japanese would be biasing.

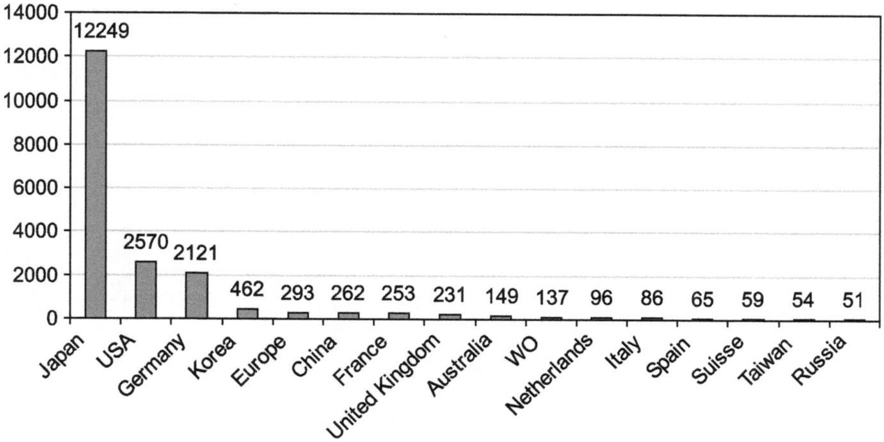

Fig. 10.5 Global patent applications in photovoltaics

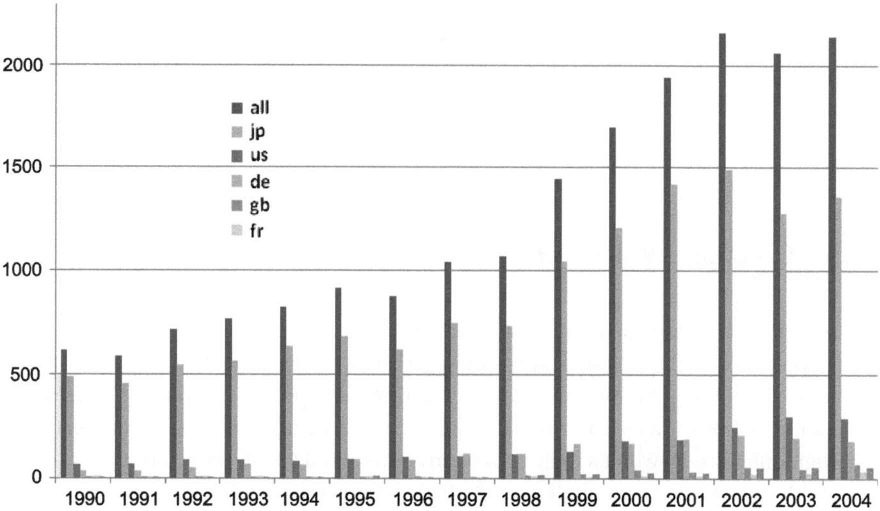

Fig. 10.6 First filing of photovoltaic patents (Source: Visentin et al. 2005)

of photovoltaics patents from around 500 in the early 1990s up to around 2,000 in 2002. The numbers for Germany do not differ significantly from those for the U.S., and Germany is far ahead of other industrialised countries, such as its European neighbours. The data seems to suggest that rather than being a precondition for the further development of PV, the economic success of PV spurred hectic activities to protect intellectual property.

These figures demonstrate the (at least short term) success of the PV industry. It is expanding production in Germany and off shore, it is increasing the export ratio of its production, it is employing ever more people, it is operating profitably and

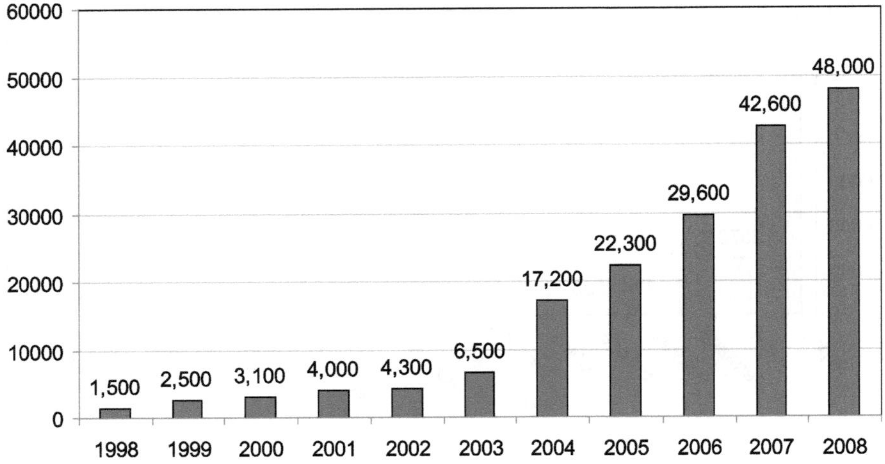

Fig. 10.7 Employees in the sector (Source: Bundesverband Solarwirtschaft e.V. (BSW-Solar 2010))

continually accumulates intellectual capital. Meanwhile more corporations are active in this sector and more people are employed in the sector than in many other established economic sectors (Fig. 10.7).

10.5 Characteristics and Development of the Industry

In the following, we will analyse the development of photovoltaics based on the hypothesis that an advocacy coalition is a crucial factor for the formulation and implementation of successful innovation policies. So-called advocacy coalitions supporting environmental policies consist of administrative and academic environmentalists, as well as members of environmental social movements who cooperate with industrial actors, such as manufacturers of renewable energy technologies (Jänicke 2007, p. 140). Lobbyism is often a conservative mechanism as it requires that the lobbyists be in a position of economic power. Therefore, one would not assume that environmentalists are able to form an effective advocacy coalition since interest groups that support emerging technologies normally are neither well positioned financially nor do they have the ability to influence powerful political actors.

Although the photovoltaics advocacy coalition was not formed by very powerful actors and groups, it has intelligently managed to use external events to gain strong social backing for its ideas. Such support was needed as it faced powerful opponents in the incumbent energy providers. "Substituting established technologies implies, (. . .), that new interest groups will challenge existing ones, and a realignment of the institutional framework, and a transformation of the energy provision system cannot be expected to be achieved without overcoming considerable opposition from vested interests involved with the incumbent technologies" (Jacobsson et al. 2002, p. 3). In the formative stage, the PV advocacy coalition aimed to support the diffusion of the technology in order to reach the critical mass needed to achieve substantial change in the energy sector. Once this critical mass had been reached, self-stabilizing effects occurred. Consequently, the critical mass itself accounted for a further consolidation of the advocacy coalition and contributed to its success.

10.5.1 The Formative Stage in the 1980s

The story of PV began like many other cases in German research policy. From the early 1980s on, common instruments of public research and development funding, such as financing research departments conducting basic research on PV, were employed. The external trigger for early research had been the oil crisis in the 1970s. At that time, the ministry of research and technology (BMFT) was in charge of photovoltaics policy programmes. Initially, the support for new technology had been integrated into the unit for non-nuclear energy technologies. In 1976, a unit of its own was created (Ristau 1998, p. 40). Interestingly, many of the programmes financing photovoltaics projects were carried out by the ministry of economic cooperation and development, since during the 1970s the future of photovoltaics applications was seen in solar home systems for developing countries, i.e., the focus was on off-grid applications. When oil prices dropped again and the conservative-liberal coalition under Chancellor Kohl assumed power, policy actions promoting photovoltaics declined severely. In 1985 public funding of photovoltaics related research and development projects accounted to less than 53 Mio DM. However, institutional actors involved in research on photovoltaics had been established. When external events such as the Chernobyl accident and the discussions on environmental problems and climate change emerged, these actors, together with environmentalist groups, managed to set the agenda for photovoltaics. When political actors started to attribute a higher priority on environmental problems, the Green party, on the one hand, and highly motivated researchers on the other, acted as transmission belts between external events and political and social discourses.

In the 1980s, specialised photovoltaics departments and research institutes were being created, such as the Fraunhofer Institute for Solar Energy Systems ISE in Freiburg (in 1982), the Centre for Solar Energy and Hydrogen Research Baden-Württemberg in Stuttgart/Ulm (in 1988), or specialised physics departments, such as at the Carl von Ossietzky University in Oldenburg. The latter can be seen as a typical example of how the formation of the photovoltaics advocacy coalition depended on highly committed individual actors. They were influenced by the experiences of early anti-nuclear power activists, who were criticised for their lack of reasonable alternatives for energy provision (Gabler 2007). The formation of research groups and departments dedicated to the development of alternatives to nuclear power became the first strategic step towards the formation of an advocacy coalition supporting photovoltaics. Furthermore, the creation of specialised departments and institutes attracted environmentally committed scientists. On this foundation, local networks consisting of environmentalists and researchers emerged. This was especially the case in Freiburg, where the Fraunhofer Institute for Solar Energy Systems ISE merged with a vivid environmental scene that positively influenced network activities and enabled local strategies of niche management (Niewienda 2006).

Federal innovation policy at that time was mainly carried out in the form of direct project funding. The main recipients were the Fraunhofer Institute for Solar Energy Systems ISE, the Hahn-Meitner-Institute, the Institute for Solar Energy Supply Techniques, and two industrial actors: AEG-Telefunken and Siemens Solar. The early photovoltaics programmes "provided opportunities for universities, institutes and firms to search in many directions, which was sensible given the underlying uncertainties with respect to technologies and markets" (Jacobsson and Lauber 2006, p. 262). Research funding was not only given to one technology, but competing technologies, such as crystalline silicon and thin-film technologies. Additionally, research and development of inverters (to make grid-connected applications work) had begun.

Interestingly, these research projects on the one hand, and the absence of market stimulation programmes on the other, led to an odd situation: whereas the big two German companies engaged in photovoltaics production were able to develop internationally competitive products and German research on photovoltaics achieved a leading position in the world, the technologies developed could not be sold at home due to a lack of domestic demand (Ristau 1998, p. 45). Actually, photovoltaic technologies developed in Germany were ready for testing. However, owing to the characteristics of the energy sector, coupled with the difficulties of creating private demand and the absence of political interest and financial support, at that time it looked very unlikely that photovoltaics could succeed in the German market. The supporting advocacy coalition was in its infancy, consisting only of highly committed scientists, environmental groups (Gabler 2007) and the newly founded German association for Solar Energy. In these early days the advocacy coalition was too weak, particularly as it had not yet incorporated more powerful industrial lobbies. On the other hand, very influential lobby groups supporting fossil fuels and nuclear power worked hard to prevent competition from renewable energies. They joined forces with the ministry of economics (Ristau 1998, p. 46) and heavily relied on old research and development contacts and networks within the ministry of research (Ristau 1998, p. 44).

Eventually, external events, such as the nuclear accident in Chernobyl in 1986, changed public opinion and the attitudes towards nuclear power substantially, and opened a window of opportunity for a general discussion on a transformation of the energy sector. Within 2 years, opposition against nuclear energy increased from 50% to over 70% of the population (Jahn 1992). Whereas before, only the Green party had argued against nuclear power, this position was now also adopted by the Social Democrats, who opted for phasing out nuclear power plants. In addition to the national antipathy towards nuclear energy the influence of a growing Green party as well as powerful environmental movements were important factors. Considering all these "external events", the German government – compared to other European governments relatively early – felt compelled to support research and development, as well as diffusion, of renewable energy technologies, including photovoltaics.

10.5.2 First Attempts at Market Stimulation in the Early 1990s

Market stimulation programmes traditionally are among the policy instruments of the ministry of economics but they were not employed until 1991. As we have mentioned before, the ministry of economics deliberately refused to support the photovoltaics research and development projects of the ministry of research. Since the new technology was definitely not economically competitive in Germany, it either had to fail, find its markets abroad (in Southern regions, as off grid applications in the developing world), or find support domestically via an artificial niche market. Finally in 1991, the situation changed when the first law regulating energy feed-ins was developed and passed. The law had been initiated by Green Party and CDU/CSU parliamentarians and it could finally pass due to cross-factional support (Ohlhorst et al. 2008, p. 16). Leading up to the adoption, lobbying activities for a range of different associations had been of vital importance. Besides the newly founded renewable energy associations, the incumbent association of hydropower plants was active, so that especially Bavarian parliamentarians supported the law. In retrospect, analysts assume that at the time the future impact of the law was underestimated, allowing its passage without greater difficulties (Ohlhorst et al. 2008, p. 17). The law described a mechanism which required utilities to remunerate energy from renewable sources that was fed into the grid. Producers of renewable electric power received 90% of the average revenue per kilowatt hour from the utilities. Even though the first feed-in law was like a market stimulation programme, it contained a market mechanism, which, at the beginning, was not seen as critical. With declining energy prices throughout the 1990s (mainly due to European deregulation policies), this policy instrument was too weak to trigger a market expansion for photovoltaics.

The law was accompanied by the 1,000-roofs programme in the early 1990s, which enabled first experiences with grid-connected photovoltaics applications, an initiative that can be seen as a typical instrument of strategic niche management. The 1,000-roof programme, starting in 1991 and ending in 1995, was a mixture of demonstration and a market stimulation programme. It offered soft loans for private households who were interested in participating in the grid-connected photovoltaics test stage. The programme was not only accompanied by electro-technical and physical tests on inverters, cell duration, etc. (Grochowski et al. 1997), but also by social research which studied customers' motives and social affiliations (Gennenig and Hoffmann 1996). This first niche programme became crucial for institutional capacity building and symbolised an initial step towards a transformation of the energy sector. Routines and motives of first movers could be revealed, and thus enabled the advocacy coalition to improve its diffusion strategy by better taking into account special needs of potential users. The accompanying social research revealed that 75% of the participants were academics, and 22% were teachers. The majority declared environmental reasons as the main motive to participate in the programme. Interestingly, only 15% of the participants could be characterised as real energy savers; instead the majority did not intend to abstain from comfort, for example, by changing routines. On the other hand, 38% of the participants were extremely interested in technical aspects of their PV application and carried through technical implementations on their own. Fifteen percent of the participants admitted status reasons as their main motive when purchasing their PV application. For them it was extremely important that the technology was widely visible (Genennig and Hoffmann 1996, p. 111).

When the 1,000-roofs programme ended and the German government did not immediately develop follow-up programmes, "one could observe a shift in the investment activities of the big European PV-companies from Europe towards the US" (Jäger-Waldau 2002, p. 40). The ministry of economics created a market launch programme for renewable energy technologies in 1995. But since it only provided 4.5 Mio. DM for photovoltaics, it did not meet the expectations of the photovoltaics industry (Ristau 1998). This is a striking example of the relationship between uncertainty and innovation. Throughout the 1990s, German policy did not systematically aim to reduce uncertainty, as the programmes were inadequately financed and were not based on long-term considerations. The result was that the development of technical innovations and marketable products came to a halt. This only changed when the Green party together with the Social Democrats came into power at the federal level in 1998.

Despite the identified shortcomings, it has to be acknowledged that the 1990s can be characterised by early (successful) investments. Publicly funded R&D, as well as the first market stimulation programmes and the first feed-in law not only led to the establishment of an initial knowledge base, but also to the creation of an embryonic advocacy coalition consisting of scientists, an infant industry and its interest groups, as well as highly committed environmentalists. Some of them appeared as first movers in the market, i.e., they were the first costumers, taking part in the 1,000-roof programme. Even though the programme offered soft loans and the power produced was remunerated, these first users did not benefit in a monetary sense. They did not make a return on their investment, nor did they earn money. Instead they appeared as "the hard core" of the advocacy coalition, mainly acting out of ideological reasons. But there was positive feedback from the early investment, which, for example, resulted in the ability of the coalition to shape further institutional change and to initiate sectoral transformation. Taken together, these first political programmes had significant effects. For one, public awareness of the new technology rose and photovoltaics received legitimacy. Political support in the form of subsidies found broad approval in public opinion. Furthermore, a number of new, often small, firms entered the market, "among these, we find both module manufacturers and integrators of solar cells into facades and roofs, the latter moving the market for solar cells into new applications" (Jacobsson and Lauber 2006, p. 266). Before these developments, the market had been dominated by the two big players, Siemens and AEG Telefunken. In 1991, when the 1,000-roofs programme was initiated 99.5% of market demand was satisfied by these two companies. Even in 1993, once the programme was opened for European competitors like BP-Solar and the Italian firm Helios, Siemens and ASE still held a market share of 70% (Ristau 1998, p. 48).

10.5.3 Strategic Niche Management in the Mid-1990s

Throughout the 1990s, industrial solar associations were gradually founded, that aimed to improve and enhance political support of the infant technology and its commercialisation. Additionally, (local) groups and societies, like the Solar Group Aachen e.V., Eurosolar (European Association for Renewable Energies), and the German Association for the Promotion of Solar Power were founded and tried to build up political momentum. These groups discussed the suitability of political instruments, including blue prints for a new feed-in law or another roof-programme. They were joined by local politicians that strongly favoured the idea of renewable energy and opted for more decentralised energy systems. For them, grid-connected photovoltaics applications met both of these aims. So it was a coalition of local politicians, the Green party, researchers, environmental societies, and business associations that managed to influence the federal government to improve and enhance its innovation policy for photovoltaics. Especially when the 1,000-roofs programme ended, strategic niche management appeared at the local level: protagonists of the solar scene were successful in implementing local feed-in laws, inspired by the Solar Group Aachen e. V. In contrast to the federal law, which only regulated the remuneration of photovoltaics power at arm's length, the concept of the Solar Group Aachen e. V. worked with cost-covering prices. The development of a policy instrument that aimed to convince users to purchase PV for return on investment reasons changed secondary aspects of the programme. Still adhering to its policy core, the PV coalition learned new ways to achieve its goal. The new mechanism paved the way for the wider diffusion of photovoltaics by making them a financially attractive investment for more than just ideologically motivated environmentalists.

These initiatives were strongly supported by the infant photovoltaics industry and its associations. The solar industry intensified its lobbying. With some of the global players that were also involved in cell production, such as Siemens and ASE, becoming part of the advocacy coalition, political pressure became more effective. Siemens was already producing in the US, complaining that due to the lack of domestic demand in Germany, it would not make sense to come back to Germany. ASE threatened to follow Siemens, in the case that no follow-up programme would be started. In response, the federal government started debating a 100,000-roofs programme. This long-term-perspective for public funding to create a niche market incentivised ASE to stay in Germany and even build new production plants. It increased its capacity from 20 to 50 MW by the end of 2002 under the name of RWE-Schott Solar (Jacobsson and Lauber 2006, p. 268).

During the PV coalition's formative stage significant opposition arose. Industrial organizations, especially German utilities had strongly opposed political instruments to support photovoltaics, such as the early energy feed-in law from 1991 (Wong 2005,

p. 135). In 1994, Preussen Elektra lodged a complaint against this law at the European and the federal level. Opposition formed not only due to general criticism towards subsidising renewable energy technologies, but also because of the specific design of the feed-in law, which disadvantaged some of the utilities. Since renewable energy is mainly produced in the windy regions near the coast (wind power) and photovoltaics applications are concentrated in the sunny South, this bias meant that some Northern utilities or their customers had to finance subsidies for renewable energy technologies. The case was dismissed in the courts but the discussion did not recede.

10.5.4 Reaching a Critical Mass (1998–2009)

Sabatier (1998) argued that policy change can only be achieved following external perturbations, such as changes in the government coalition or impacts from other subsystems. This also seems to be true in the case of PV. When, in 1998, the Green party, together with the Social Democrats, formed the federal government, the photovoltaics advocacy coalition took its chance. Now, it no longer had to be content with merely influencing the rebuilding of institutional frames and policy programmes from the outside of political institutions. The Greens took over the ministry of the environment and this initiated the institutionalisation of the photovoltaics advocacy coalition within the centre of political power. The situation in the late 1990s was accompanied by international and European developments, such as the liberalisation and deregulation of the energy sector. In the wake of the Kyoto protocol, international organisations as well as the European Commission made CO_2 reduction a top-priority political goal.

As a consequence, the change in political power constellations was linked to a beginning of the restructuring of the energy sector. Institutional settings and the infrastructure of the energy sector started to become more open and fluid. Corporate structures were being reorganised and replaced by more competitive management and governance structures. Thus, innovation in photovoltaics was accompanied by the re-structuring of the energy sector and social innovations, like new management concepts, new user routines, "new roles and identities of electricity customers, new policy problems, regulatory concepts, institutions and governance arrangements" (Voß et al. 2003, p. 4). It can be assumed that these changes and transformation processes in the sector not only shaped the background, but, more fundamentally, have been crucial factors in triggering innovation in photovoltaics. Institutional changes, such as deregulation in the energy sector and objectives formulated by the European Union concerning the transformation of the energy sector, opened up a policy window of opportunity for the success of an advocacy coalition working against the resistance of powerful advocates of traditional energy sources. Two policy instruments were designed and implemented, which are widely believed as having been decisive for the German photovoltaics success story. The actual design of the instruments was prepared and debated by solar groups, societies and associations. Groups like Eurosolar, the German Association for the

Promotion of Solar Power, and Greenpeace were extremely important for an adjusted "relaunch" of the 1,000-roofs programme and the first feed-in law of 1991. The locally prevailing feed-in tariffs could later serve as blueprints for a new feed-in system on the federal level. Furthermore, the lobbying activities of associations and environmental groups helped to shape a novel roofs-programme on a far larger scale.

In 1999, the 100,000 roof programme was created. It was a market stimulation programme, which offered soft loans with 10 years duration and 2 years free of redemption. In 2000, the Renewable Energy Law was passed. It set a fixed feed-in tariff of around 50 cents[3] per kWh for 20 years, with a 5% decrease annually for installations after 2002. Compared to the first feed-in law, which had been heavily opposed by the utilities, the additional costs of renewable energies were now shared and only 5% the financial charges had to be paid by the utilities. The law was inspired by the local feed-in laws for solar power. The skills that had been achieved on the local level helped the Green Party to move the concept to the federal level. For this process, it was extremely helpful that one of the main protagonists of the local groups, who had organised local feed-in tariffs, was elected as a federal deputy in 1998 and thus could bring in experiences he had on the local level (Rosenbaum et al. 2005, p. 79). He was among the Green deputies who initiated a discursive process involving various actors, such as environmental groups, solar industry associations, the association of the machinery and equipment producers (VDMA), the metal workers trade union, solar cell producers, and politicians from some *Länder*.

This institutionalisation of an intermediate level of conflict can be interpreted along the lines of the notion of policy learning advanced by Sabatier (Sabatier and Jenkins-Smith 1993). The panel did not intend to conduct a general discussion on the future of the German energy supply system (the policy core, still separating the coalitions). Instead, it only discussed the issue of financial support for renewable energy technologies. Hence, in 1998, the Green party acted as a policy broker, searching for compromises in secondary aspects that could be supported by the majority of actors and thus enlarge and finally stabilise the advocacy coalition in a way that it would survive even without institutional backing in the future. "The unorthodox coalition even included a major utility (...); as a result the big utilities were not united in their opposition" (Jacobsson and Lauber 2006, p. 267). Further innovation in PV was still funded by public research grants – albeit at a decreasing rate. Public funds were concentrated more on network and cluster projects, many of which were embedded in structural policies in order to help the economically underdeveloped regions in the East of Germany. Regional cluster and network policy is a rather new instrument that aims to create an innovation-friendly environment by fostering collective identities and trust in order to support the formation and development of local networks (Dohse 2007). Within the past few years, the

[3] The exact amount is subject to size and application: electricity from roof-top systems is reimbursed higher than electricity sourced from ground mounted systems.

solar industry has figured out where to settle down in order to receive subsidies. What we can see nowadays, are photovoltaics clusters in East Germany, predominantly located near the small town of Thalheim, in the vicinity of Bitterfeld, Saxony-Anhalt. In particular, small start-ups, which have emerged after 2000, have settled down in the East. One of the world leaders in cell production is Q-Cells, a firm, founded in Berlin in 1999, which soon moved to Thalheim in order to start cell production in 2001. Q-Cells is an example of a German success story, i.e., it perfectly reflects the effectiveness of the 100,000-roofs programme and the Renewable Energy Law. By the end of 2002, it employed 82 persons; at the end of 2004 it already had 484 employees, a number which has grown to 1,700 by the year 2007.

Q-Cells is also an example of the photovoltaics industry's increasing ability to acquire financing and venture capital from the private sector and the equity market. Since October 2005, Q-Cells has been listed on the Frankfurt stock exchange, and, since December 2005, in its technology index TecDax. The first German PV firm to be listed on the stock exchange was the Solon AG in 1998. It was soon followed by Solar World AG in 1999, Sunways AG in 2001, Solar-Fabrik AG in 2002 and many others. All these companies were young start-ups, small and medium-sized companies, which differed considerably from the multinational firms, such as Siemens and ASE, which had been dominating the early PV industry.

The development and success of these new firms is evidence that the industry has left the formative stage (i.e., the niche market) and has been entering the take-off stage (i.e., is ready for market expansion). Market expansion and the activities of new actors in the sector have been accompanied by a significant enlargement and diversification of the photovoltaics advocacy coalition. This applies to producers as well as to users. Whereas first producers like the Freiburg Solar-Fabrik, founded in 1996 by the environmentalist Georg Salvamoser, were embedded in local solar networks and were not solely led by return on investment thinking, motives and behaviours of producers like Q-Cells, Solar World or Solon do not differ from producers in other sectors.

Additionally, due to the Renewable Energy Law (EEG), users of photovoltaics are no longer necessarily led by "green" motives, as it has increasingly become profitable to purchase solar modules, especially for farmers, who have plenty of space on their barn roofs, which can be used as a building ground for the rather cheap thin film technology (Rosenbaum et al. 2005, p. 85).[4] Furthermore, this development is supported by the wide acceptance of solar energy by the German public. This trend is vividly reflected in the Christian Democratic Party, which has now firmly accepted the policy of supporting photovoltaics. So, when in 2005 the Red-Green government ended and was replaced by the grand coalition of Social

[4] The literature on strategic niche management sees the prevalence of economic motives as an impediment to the success of policies (Hoogma et al. 2002). We are arguing that exactly the opposite mechanism (addressing economic motives) has been essential for the success of PV policies.

Democrats and Christian Democrats, the new government did not opt to take a new path. The Renewable Energy Law was not abolished and a recent amendment to the law does not entail comprehensive changes for PV support.

The take-off stage has been accompanied by organizational changes that have helped to consolidate the chosen path. In 2002, after the re-election of the Red-Green government, coalition talks assigned the ministry of the environment full responsibility for renewable energies. Whereas the beginning of the formative stage had been characterised by conflicts between the ministry of economics and the ministry of research, both being rather averse to substantially supporting photovoltaics, in 2002 the situation changed completely. The ministry of the environment is now responsible for the Renewable Energy Law as well as the public financing of photovoltaics related R&D.[5]

Meanwhile the photovoltaics industry in Germany is highly differentiated, thanks to its ability to employ diverse methods of production and in its ability to build up important links to related industries. Therefore photovoltaics related R&D is not just research on new materials and cell efficiencies.

The German machine building industry and photovoltaics are in a mutually beneficial relationship. Machine building has benefited from the emergence of the photovoltaics industry. At the same time, German solar producers gained advantages from the expertise of the machine building industry as innovations in photovoltaics happen mainly through cost reductions in production processes. For the German machine building sector, a strategic orientation to PV manufacturing equipment can be observed. The development of "turn-key" facilities helped to enable mass production and facilitated the standardisation process (Dewald 2007, p. 132). These are crucial preconditions to achieving economies of scale and making PV applications more competitive (Auer 2008, p. 12).

Furthermore, architects and craftsmen, especially electricians, have adapted well to the new technology as a growth option for their businesses, and associated institutions of vocational education have managed to adjust their curricula. Thus, well-known bottlenecks that often constrain the diffusion of new technologies have been overcome. The specific dynamics of the advocacy coalition described in this article can be illustrated with an examination of the machine building industry. This industry is an actor which cannot be considered to be part of the energy policy subsystem proper but is strongly supporting the PV coalition by now. At the beginning of the formative stage, there existed a single-minded coalition supporting renewable energy technologies. At that time, it shared a joint policy core, which was the transformation of the energy sector, substituting nuclear and fossil power plants for renewable energy technologies. Learning processes during the course of

[5] Another form of institutionalisation is the so called "Glottertal talks", which are strategic talks on photovoltaics-related R&D. These talks originated in 1987 but have gained importance particularly during the last couple of years. Researchers and representatives of the leading institutes and companies meet with members of the ministry of the environment in order to discuss future public R&D activities for PV.

this stage helped to develop new policy instruments. Radical opposition against the traditional energy sector, based on theories and visions highlighting worst case scenarios on the one hand and demonstrations and blockade actions on the other, gave way to more pragmatic considerations and helped the coalition to gain political power. The new PV policy core of the transformed coalition is now characterised not as purely oppositional to traditional forms of energy supply, but as supporting PV. Its formation has been accompanied by new theories, visions and ideas on generating demand for PV by reducing costs, increasing returns, spreading information, and eventually by finding ways to enlarge the coalition. These dynamics have resulted in the integration of actors like the machine building industry and even some of the utilities, who either do not belong to the policy subsystem or explicitly share another policy core and representatives of various parties from the Eastern part of Germany. At this time, with the original policy core changed to support PV, the ground has been prepared for the integration of a very heterogeneous set of actors.

10.6 A Future for Photovoltaics? Conclusions and Lessons

At the beginning of this chapter, we claimed that the creation of niche markets can be a successful policy instrument in coordinated market economies (hypothesis 1) if a powerful advocacy coalition can be mobilised (hypothesis 2). Our analysis has shown that the support of PV after 1998 was successful in establishing a growing and profitable economic activity. The PV industry can produce and sell its products both in Germany and abroad. The story, however, also demonstrates that the success of such a policy depends on many favourable circumstances. It does not only need broad political and public support that goes beyond the initial policy core, but also a delicate architecture of instruments that are geared towards the special characteristics of the system to be supported. The policy instruments are mostly not generic but geared towards the specific problems of the PV industry.

The success of PV is also dependent on general conditions that offer a window of opportunity for change. The electric power sector over the last years has faced new challenges. These challenges have come from market liberalization, the expectation that the sector should contribute to environmental aims, and the development of new technologies (e.g., forms of renewable energy) that are difficult to integrate into the dominant regime of the sector. PV successfully exploited the fact that it was a decentralised, small technology which could be connected to the grid without severe difficulties and compatibility problems. It could rely on existing scientific strengths in this area, as well as the expertise of suppliers (e.g., machine building industry). Some elements of path dependency are therefore present in the development of PV.

The political instruments that were developed offered long-term security for the industry as well as incentives to build new production units in the disadvantaged regions of the new German *Länder*. The users of PV modules were guaranteed a

20 year security on their investments. Consequently, PV could serve many masters. The strength of the coalition was recently demonstrated when the federal government amended the Renewable Energy Law without implementing important changes. It achieved nearly unanimous support by a public in favour of clean technologies and was supported by an advocacy coalition comprised of scientists, politicians, environmentalists, and an increasingly economic group of actors.

Taken together, the many beneficial factors and the very specific composition of the advocacy coalition also point to the difficulties in imitating this successful experiment in other areas. The lesson is not that the same policy should be and can be pursued in other cases as well. The general message, rather, is that customised innovation policies need to reflect the specific conditions and opportunities in the targeted areas.

References

Auer, J. (2008). Deutscher Maschinenbau macht Wirtschaft fit für die Zeit nach dem Öl, Deutsche Bank Research, Energie und Klimawandel, Aktuelle Themen 435. Frankfurt am Main: Deutsche Bank.

Beckert, J. (1996). Was ist soziologisch an der Wirtschaftssoziologie? *Zeitschrift für Soziologie, 25*(2), 125–146.

Berkhout, F., Smith, A., & Stirling, A. (2003). Socio-technological regimes and transition contexts. *Working Paper Series 2003/3*. Brighton: SPRU Science and Technology Policy Research.

BSW-Solar (2010). Statistic data on the German photovoltaic industry, June 2010. Available at http://en.solarwirtschaft.de/fileadmin/content_files/factsheet_pv_engl.pdf. Retrieved 31 May 2011. German Solar Industry Association.

Busch, P. O. (2005). Institutionalisierter Politiktransfer mit Nebenwirkungen: Die Ausbreitung von Quoten und Einspeisevergütungen. In K. Tews & M. Jänicke (Eds.), *Die Diffusion umweltpolitischer Innovationen im internationalen System* (pp. 233–255). Wiesbaden: VS-Verlag.

Caniels, M. C. J., & Romijn, H. A. (2008). Strategic niche management: Towards a policy tool for sustainable development. *Technology Analysis & Strategic Management, 20*(2), 245–266.

Carlsson, B., & Jacobsson, S. (1997). Diversity creation and technological systems: A technology policy perspective. In C. Edquist (Ed.), *Systems of innovation: Technologies, institutions and organizations* (pp. 266–294). London: Pinter.

Clark, N., & Juma, C. (1987). *Long-run economics: An evolutionary approach to economic growth*. London: Pinter.

Coenen, R. (2002). Umlenken auf nachhaltige Technologiepfade. In A. Grunwald (Ed.), *Technikgestaltung für eine nachhaltige Entwicklung* (pp. 389–405). Berlin: Edition sigma.

Commission of the European Communities (2005). Communication from the Commission of 7 December 2005 "The support of electricity from renewable energy sources", COM(2005) 627 final. Official Journal C 49 of 28 Feb 2006.

European Parliament and Council (2001). Directive 2001/77/EC of the European Parliament and of the Council of 27 September 2001 on the promotion of electricity produced from renewable energy sources in the internal electricity market. Official Journal of the European Communities, 27 Oct 2001, L283/33.

Czada, R., & Schimank, U. (2000). Institutionendynamik und politische Institutionengestaltung: Die zwei Gesichter sozialer Ordnungsbildung. In R. Werle & U. Schimank (Eds.),

Gesellschaftliche Komplexität und kollektive Handlungsfähigkeit (pp. 23–45). Frankfurt am Main: Campus.

Dewald, U. (2007). Innovationssystem Photovoltaik in Deutschland. FVS BSW-Solar Themen 130–135.

Dohse, D. (2007). Cluster-based technology policy – The German experience. *Industry and Innovation, 14*(1), 69–94.

Edler, J. (Ed.). (2007). *Bedürfnisse als Innovationsmotor. Konzepte und Instrumente nachfrageorientierter Innovationspolitik*. Berlin: Edition sigma.

Edquist, C. (2001). The system of innovation approach and innovation policy: An account of the state of the art. Lead paper presented at the DRUID conference, Aalborg, 12–15 June 2001.

Fuchs, G., & Wassermann, S. (2008). Picking a winner? Innovation in photovoltaics and the political creation of niche markets. *Science, Technology & Innovation Studies, 4*(2), 93–113.

Gabler, H. (2007). Interview at the Zentrum für Sonnenenergie- und Wasserstoff-Forschung, 9 Jan 2007.

Genennig, B., & Hoffmann, V. U. (1996). *Sozialwissenschaftliche Begleituntersuchung zum Bund-Länder-1000 Dächer Photovoltaik-Programm, Abschlussbericht*. Leipzig: Umweltinstitut Leipzig e.V.

Grochowski, J., Decker, B., Kiefer, K., & Rössler, E. (1997). Mindererträgsanalysen und Optimierungspotentiale an netzgekoppelten Photovoltaikanlagen des 1000-Dächer-Programms. In Forschungsverbund Sonnenenergie (Ed.), *Themen 96/97* (pp. 32–38).

Hall, P., & Soskice, D. (Eds.). (2001). *Varieties of capitalism: The institutional foundations of comparative advantage*. New York: Oxford University Press.

Harding, R. (2000). Resilience in German technology policy: Innovation through institutional symbiotic tension. *Industry and Innovation, 7*(2), 223–243.

Hiskes, A. L., & Hiskes, R. P. (1986). *Science, technology, and policy decisions*. Boulder/London: Westview Press.

Holzinger, K., Jörgens, H., & Knill, C. (2007). Transfer, Diffusion und Konvergenz von Politiken. *Politische Vierteljahresschrift, Sonderheft 38*. Wiesbaden: VS-Verlag.

Hoogma, R., Kemp, R., Schot, J., & Truffer, B. (2002). *Experimenting for sustainable transport: The approaches of strategic niche management*. London: Spon Press.

IEA (2005). Total photovoltaic power installed in IEA PVPS countries. http://www.iea-pvps.org/isr/22.htm. Accessed 12 June 2007.

Jacobsson, S., & Bergek, A. (2004). Transforming the energy sector: The evolution of technological systems in renewable energy technology. In K. Jacob, M. Binder, & A. Wieczorek (Eds.), *Proceedings of the 2003 Berlin conference on the human dimensions of global environmental change* (pp. 208–236). Berlin: Environmental Policy Research Centre.

Jacobsson, S., & Lauber, V. (2006). The politics and policy of energy system transformation – Explaining the German diffusion of renewable energy technology. *Energy Policy, 34*, 256–276.

Jacobsson, S., Andersson, B. A., & Bangens, L. (2002). Transforming the energy system – the evolution of the German technological system for solar cells. *SPRU Paper No 84*. Brigthon, Sussex: SPRU Science Policy Research Unit.

Jäger-Waldau, A. (2002). *Status of PV research, solar cell production and market implementation in Japan, USA and the European Union*. Ispra: Institute for Environment and Sustainability, Renewable Energy Unit.

Jahn, D. (1992). Nuclear power, energy policy and new politics in Sweden and Germany. *Environmental Politics, 1*(3), 383–417.

Jänicke, M. (1997). Umweltinnovationen aus der Sicht der Policy-Analyse: Vom instrumentellen zum strategischen Ansatz der Umweltpolitik, Papier vorgelegt im Rahmen des BMBF-Projektverbundes "Abschätzung der innovativen Wirkungen umweltpolitischer Instrumente" am 9.12.1996 in Königswinter. Berlin: Freie Universität Berlin, Forschungstelle für Umweltpolitik.

Jänicke, M. (2007). Trendsetter im "regulativen Kapitalismus": Das Beispiel umweltpolitischer Pionierländer. In K. Holzinger, H. Jörgens, & C. Knill (Eds.), *Transfer, Diffusion und Konvergenz von Politiken. Politische Vierteljahresschrift, Sonderheft 38* (pp. 131–149). Wiesbaden: VS-Verlag.

Kappelhoff, P. (2000). Komplexitätstheorie und Steuerung von Netzwerken. In J. Sydow & A. Windeler (Eds.), *Steuerung von Netzwerken* (pp. 347–389). Opladen: Westdeutscher Verlag.

Kemp, R. (2002). Environmental protection through technological regime shifts. In A. Jamison & H. Rohracher (Eds.), *Technology studies and sustainable development* (pp. 95–126). München/ Wien: Profil.

Kemp, R., Schot, J., & Hoogma, R. (1998). Regime shifts to sustainability through processes of niche formation: The approach of strategic niche management. *Technology Analysis & Strategic Management, 10*(2), 175–195.

Litfin, K. T. (2000). Advocacy coalitions along the domestic-foreign frontier: Globalization and Canadian climate change policy. *Policy Studies Journal, 28*(1), 236–252.

Malerba, F. (2004). *Sectoral systems of innovation: Concepts, issues and analyses of six major sectors in Europe*. Cambridge: Cambridge University Press.

Mayntz, R., & Hughes, T. P. (Eds.). (1988). *The development of large technical systems*. Frankfurt am Main: Campus.

Milborrow, D. (2002). External costs and the real truth. *Windpower Monthly, 18*(1), 32.

Niewienda, A. (2006). Interview at DS-Plan. Stuttgart, 17 Dec 2006.

Nooteboom, B. (1999). Innovation and inter-firm linkages: New implications for policy. *Research Policy, 28*(8), 793–805.

OECD (1994). The measurement of scientific and technological activities: Using patent data as science and technology indicators. *Patent manual*. Paris: Organisation for Economic Co-operation and Development.

OECD. (2008). *Environmental policy, technological innovation and patents*. Paris: Organisation for Economic Co-operation and Development.

Ohlhorst, D., Bruns, E., Schön, S., & Köppel, J. (2008). Windenergieboom in Deutschland: Eine Erfolgsstory. In M. Bechberger, L. Mez, & A. Sohre (Eds.), *Windenergie im Ländervergleich. Steuerungsimpulse, Akteure und technische Entwicklungen in Deutschland, Dänemark, Spanien und Großbritannien*. Frankfurt am Main: Peter Lang.

Ristau, O. (1998). *Die solare Standortfrage. Der technologische Wettstreit zwischen den USA, Japan und Deutschland*. Bad Oeynhausen: Bröer-und-Witt-Solarthemen.

Rogers, E. M. (1995). *Diffusion of innovation* (4th ed.). New York: The Free Press.

Rosenbaum, W., Mautz, R., & Byzio, A. (2005). *Die soziale Dynamik der regenerativen Energien – am Beispiel der Fotovoltaik, der Biogasverstromung und der Windenergie*. Göttingen: Soziologisches Forschungsinstitut Göttingen (SOFI).

Sabatier, P. A. Jenkins-Smith, H. (eds.) (1993): Policy change and learning: An advocacy coalition approach, Boulder, CO: Westview Press.

Sabatier, P. A. (1998). The advocacy coalition framework: Revisions and relevance for Europe. *Journal of European Public Policy, 5*(1), 98–130.

Siemer, J. (2005). Das Orakel vom Amt. Patentanmeldungen für Photovoltaik. *Photon*(2005–10), 66–70.

Solarbuzz (2007). Solar Cell Technologies. http://www.solarbuzz.com/Technologies.htm. Accessed 6 Aug 2007.

Solarserver (2007). Dünnschicht-Solartechnik: Neue Technologien zur Kostensenkung der Photovoltaik. http://www.solarserver.de/solarmagazin/artikeljuni2006.html. Accessed 6 Aug 2007.

Stern, N. (2006). *Review on the economics of climate change*. London: HM Treasury.

Strange, S. (1996). *The retreat of the state: The diffusion of power in the world economy*. Cambridge: Cambridge University Press.

Victor, D. G. (2002). Electric power. In B. Steil, D. G. Victor, & R. R. Nelson (Eds.), *Technological innovation and economic performance* (pp. 385–415). Princeton: Princeton University Press.

Visentin, A., Voignier, V., & Königstein, C. (2005). Technology trends in photovoltaics for the last decade: An analysis on the basis of patent application filing data. Unpublished manuscript. European Patent Office.

Voß, J-P., Fischer, C., Schumacher, K., Cames, M., Pehnt, M., Praetorius, B., et al. (2003). Innovation: An integrated concept for the study of transformation in electricity systems. *TIPS Discussion Paper 3*. Frankfurt am Main: Sozialökologische Forschung.

Weimer-Jehle, W., & Fuchs, G. (2007). Generierung von Innovationsszenarien mit der Cross-Impact Methode. In Forum für Energiemodelle und Energiewirtschaftliche Systemanalysen in Deutschland (Ed.), *Energiemodelle zu Innovation und moderner Energietechnik* (pp. 37–90). Berlin: Lit.

Werle, R., & Schimank, U. (Eds.). (2000). *Gesellschaftliche Komplexität und kollektive Handlungsfähigkeit*. Frankfurt am Main: Campus.

Wong, S.-F. (2005). Obliging institutions and industry evolution: A comparative study of the German and UK wind energy industries. *Industry and Innovation, 12*(1), 117–145.

Chapter 11
Governance of Large Innovation Projects: The Implementation of the Electronic Health Card in Germany

Achim Lang and Alexander Mertes

11.1 Introduction

The introduction of the electronic health card (eHC) in Germany, replacing an older version with less functional aspects, marks the largest telematic project in the history of the German health care system. It provides comprehensive information to all participating organizations linking 127,000 physicians, 65,000 dentists, 21,000 pharmacies, 2,200 hospitals, and around 200 compulsory and private health insurance companies. The implementation of the eHC is expected to cost more than one billion Euros. The eHC contains a memory chip and microprocessor that store administrative, personal as well as treatment related data from the insured.

In January 2009, 3 years after the originally envisioned date, reading devices were dispatched to health care providers in selected areas. Due to numerous delays, the date for the trial launch had to be postponed several times. Even after the first reading devices had been dispatched, some organizations attempted to thwart the whole project. Less than a year earlier, in May 2008, the 111th German Medical Assembly had rejected the implementation of an eHC. Initially, the eHC had been selected as a prestigious project within the German high technology and innovation policy. The implementation of the eHC fanned high hopes and aspirations on the side of policy makers and public administration. The project aims at realizing objectives relevant to health care as well as industry and innovation related policies (BMG 2007).

The conception of largely autonomous policy subsystems is a central assumption in prominent policy theories such as the advocacy coalition framework and

A. Lang (✉) • A. Mertes
University of Konstanz, Konstanz, Germany
e-mail: achim.lang@uni-konstanz.de

J.M. Bauer et al. (eds.), *Innovation Policy and Governance in High-Tech Industries*, DOI 10.1007/978-3-642-12563-8_11, © Springer-Verlag Berlin Heidelberg 2012 punctuated equilibrium theory (Sabatier and Weible 2007; Adam and Kriesi 2007; True et al. 2007). Compared with the recognition of interdependencies in research on federal policy-making (Scharpf 1985) the assumption that policy takes place in closed issue networks (Heclo 1978) has obstructed a clearer view of multifaceted interdependencies between policy subsystems, which have so far been widely neglected.

This essay takes up this gap in policy theorizing and explores, how policy and administrative coordination (Braun 2008) is adjusted in a temporary policy network transcending domain boundaries, such as the eHC. This will be done by analyzing the emerging policy network with a focus on structural barriers to coordination and on inconsistencies in goals and task settings.

The main hypothesis is that actor constellations and institutional logics ascribed to different policy domains – such as innovation and health care policy – can hardly be accommodated without causing frictions between actors.

The chapter is subdivided into four parts. In the first part the guiding principles of German high technology and innovation policy are scrutinised. Then the concept of policy networks is introduced and its relevance for governance and policy implementation is highlighted. In the third section, actor structures and interest positions regarding the implementation of the eHC are analysed. The chapter concludes with a short summary of the main findings and gives some recommendations for future implementations of large technical systems.

11.2 Innovation Policy: Governing Large Technical Projects

To get an understanding of the coordination structures of the eHC, it is necessary to describe the driving forces and protagonists behind German innovation policy and compare them with governance arrangements in the health care domain. Two examples, the "fast breeder"[1] project and the German tolling system can illustrate German innovation policy.

After World War II, large technical innovations were promoted by the Ministry of Nuclear Issues which pursued an active industrial and innovation policy. The Ministry of Nuclear Issues strived for new technological innovations of high political and symbolic value (Weyer 2004). The use of nuclear energy for non-military purposes was the first such project. The breeder reactor, which later became known as fast breeder, was one of the first high-tech innovations in a large scale project that was headed by the German Federal Ministry of Nuclear Issues (today Federal Ministry of Education and Research). By 1959, the plan for constructing a fast breeder had emerged at the nuclear research center in Karlsruhe, Germany. Between 1971 and 1974, the breeder reactor was built. The fast breeder was to become a project of national prestige and a showcase of German industrial policy. The Federal government proclaimed it to be the solution to a secure national

[1] A fast breeder is a nuclear reactor with an advanced technology.

energy supply. Yet after investing immense sums in nuclear research and a long construction phase, the project was abandoned in 1991 (Keck 1984; Marth and Koehler 1998).

In 2002, Toll Collect Ltd. was established as a joint venture between German Telecom, Daimler PLC and the French Cofiroute. In September 2002, the newly founded company was won the tender and was awarded a contract for deploying a new German tolling system. The system was supposed to set new standards and become the most advanced tolling system in the world. The system operates via satellite, using for the time being, the American GPS and when available the European Galileo-System. The tolling system allows automatic charging of tolling fees via lorries′ on-board-units. However, technical difficulties delayed the introduction of the highly complex system by 16 months until the beginning of 2005. Johannes Weyer (2005) concluded that the Ministry of Traffic, Construction and Urban Planning had blindly trusted Toll Collect Ltd. and not established control mechanisms to oversee system development.

Both examples illustrate a general feature of German innovation policy: that government stipulates and conceptually co-designs new innovations but lacks direct control of the innovation and implementation process. The state limits itself to create a framework for self-organised and inherent dynamic non-governmental projects (Dolata 2004, p. 23). But the government also acts as the driving force behind technology developments with direct state intervention and project support (Weyer 2004).

Compared to innovation governance the Federal Government pursues a more restrained approach in health policy. The relationships between state and associations of the German health system are generally described as corporatist (Alemann and Heinze 1979; Döhler and Manow 1995; Bandelow 1998). Corporatist means, in this case, the integration of organised interests and their participation in the formulation and execution of political decisions (Schmitter and Streeck 1999; Streeck and Kenworthy 2005). As organised interests, associations take on an intermediate position by representing the interests of their members to state ministries and agencies and by justifying political agreements with their members. Unlike in pluralistic arrangements, it is assumed that associations do not only influence government policy but also take part in the formulation and execution of state measures. German health policy reveals clear signs of corporatism (Döhler and Manow 1995; Rosenbrock and Gerlinger 2006).

In the following sections we therefore discuss how the institutional amalgamation of policy domains is implemented and how institutional logics merge.

11.3 Governance in and through Policy Networks

We draw on the typology of van Waarden (1992) and Adam and Kriesi (2007) for guidance on classifying the coordination structures. These authors have developed a multi-dimensional grid for the analysis of coordination forms that facilitates comparison of empirical configurations with theoretical ideal types. Essentially there are three dimensions: actors, structure and distribution of power (Adam and Kriesi 2007; van Waarden 1992):

- The actor dimension refers to the number of actors and their interest positions.
- Structure includes features such as the type and density of interaction and the centrality of actors and networks.
- The power distribution dimension refers to power asymmetries between societal and state actors.

Using these typologies, it is possible to stipulate ideal types of coordination and control, and verify them empirically. Different types of coordination form a continuum between two extreme points. On the one hand, a main actor may determine the policies by himself without the participation of other actors (hierarchy); on the other hand, many players may decide spontaneously and independently (market). Networks align themselves between these two end points and represent a recombination of market and hierarchy (Börzel 1998; Thatcher 1998; Schneider and Janning 2006). They are characterised by a limited number of autonomous actors and by the ability to pursue objectives through coordinated action, a trait typical of hierarchies (Mayntz 1997). In addition, further ideal types can be distinguished. Frans van Waarden (1992) defines "macro-corporatism" as a network characterised by cooperation between state actors and selected interest groups. It is assumed that in a macro-corporatist network there is no state actor dominance.

The coordination and control structures for the eHC were examined using a quantitative network analysis. The data on which the calculations are based were collected through a standardised questionnaire between December 2007 and March 2008. Questionnaires were sent to all 42 members[2] of the advisory board of the *Society for Telematics Applications in Health Care* (gematik), thus covering all the political actors that influence the implementation of the eHC. A total of 29 were returned, which amounts to a response rate of 69%.[3]

The standardised questionnaire contains questions on the substantive positions of the actors (such as expected costs and data privacy), communication styles, negotiation in decision-making bodies, as well as information exchange and influence reputation. The calculation and visualization of network-analytical

[2] Overall, 43 actors participate in the eHC. The health insurance associations Arbeiter-Ersatzkassen-Verband (AEV) und Verband der Angestellten-Krankenkassen (VdAK) submitted a joint representative to the boards so a total of 42 actors were contacted. The standardised questionnaire was completed by seven insurance companies, four service providers, a state actor, five actors from industry, three actors from the area of patients and consumers, three scientific actors, and representatives of six test regions.

[3] In a pre-test, a questionnaire with 12 questions was sent to nine actors. On the basis of their feedback, eight questions were selected for the final version of the questionnaire. In the first stage of the survey in December 2007, this version of the questionnaire was sent to all stakeholders via e-mail. Reminder messages were sent every other week. In February 2008, a reminder letter including the questionnaire was sent to all respondents who had not yet responded.

measures were performed with the computer programmes UCINET, VISONE and NETSCAPE.[4]

11.4 The Introduction of the eHC: Institutional Governance in Temporary Policy Fields

In 2003, legislation was passed to modernise public health insurance. The legislative initiative to introduce telematic devices in the health care system (*Gesetz zur Organisationsstruktur der Telematik im Gesundheitswesen*) set the institutional frame for the implementation of eHC. The compulsory administrative and medical issues of the eHC are detailed in Section 291a of the fifth chapter of the Social Security Code (SGB V). The statutory provisions for the field test regions were adjusted by the federal ministry of health (BMG) in 2005 and 2006. Also, new specifications related to the eHC were laid out by the BMG (BMG 2005, 2006).

The legal guidelines for the eHC are modeled after prior large scale innovation policy projects, in which the federal government was the driving force behind technology innovation, but left the management of the projects to private companies. The new legislation introduced gematik, which is in charge of the technical realization of the project. It also lists the entire structure and members of gematik: peak associations of the German health system as members, technical committee, advisory board, the peak associations' share of the company, voting procedure and the status of the BMG. Formally the BMG does not have a right to vote in the general meeting, but it is permanent observer during meetings. The BMG is also present during advisory board meetings and chairs the "architecture board" which is supposed to accelerate the decision making process.

The funding of the operating company after the introduction of the eHC is also established in the new legislation. Contributions to meet the needed funding are allocated according to the number of members of the peak associations. The refunding is done by the health insurance funds after the average PC and IT equipment of the care providers has been valued. The operating expenses of the gematik are reimbursed by the health insurances. The surcharges for care provider are quoted according to the usage of the telematics. Costs for the introduction of the eHC for project offices and project management lie with the health insurance funds as well. Health care providers receive funds from a surcharge from the health insurances for each billing case. The legal guidelines demand that the insurers and health care providers have to take on the budget of the gematik and the costs for the introduction and further development of the eHC (FES 2005). Additionally, investments by the individual doctors, dentists, hospitals, pharmacists and other care provider are needed. It is estimated that investments reach up to

[4] Further explanations with regards to the measures can be found in the respective figures.

€3,500 by each user for the new data processing system. Hospitals will have to expect costs of up to €500 for each data processing terminal (FES 2005).

11.4.1 Actors: Categories and Items of Interest

The implementation of the eHC is a newly created policy domain. It is made up of different actors and actor groups in the fields from health and innovation policy that jointly participate in the introduction of the eHC (see Table 11.1).

The Federal Ministry of Health (BMG) spearheads the implementation of the eHC. Gematik's shareholder agreement could only be decided with consent of the BMG. The shareholders of gematik are composed of the top associations in the German health system, i.e., representatives of providers, the Statutory Health Insurance (GKV) and the Private Health Insurance (PKV). Among these leading associations of health care are also the Federal Chamber of Dentists (BZÄK) and the Federal Medical Council (BÄK). The health insurance providers of the GKV and PKV make up a block that holds 50% of the voting shares at the shareholder meeting. The representatives of health service providers such as doctors and hospitals command the other 50%. Additionally, the shareholders of gematik form a technical committee consisting of experts for technical solutions in health care. The technical committee brings together technical experts of the shareholders, whose judgments are essential for the decisions in the shareholder meeting. The

Table 11.1 Actors and their institutional imbedding

Sectors	Shareholders of gematik	Advisory board	Architectural board
Public and private health insurers AOK, BKK, KNAPP, See-KK, BLK, IKK, AEV/VdAK, PKV	X	X	X
Service providers DAV, DKG, BZÄK, KZBV, BÄK, KBV, BPtK	X	X	X
State BMG, BMF, BSI, BMWi, EU KOM	–	X	X
Industry BITKOM, ZVEI, VHitG, ADAS, D21	–	X	–
Consumers/patients Federal Commissioner of Data Protection and Freedom of Information (Peter Schaar), Government Commissioner for the Concerns of Patients (Helga Kühn-Mengel), BAG, VZBV, VdK	–	X	X
Science UniF, FHD, TmF, GVG	–	X	X
Test regions	–	X	–

EU KOM = European Commission, Directorate General Information Society. BITKOM, VHitG, ADAS, D21, and ZVEI are business associations in the information and communications technology domain. BAG, VZBV, and VdK are consumer or patient organizations. UniF, FHD, TmF and GVG are public or private research centers.

technical committee can be seen as a link between the IT company gematik and its shareholders. Therefore, there is always a representative of gematik participating in the technical committee.

In addition to the technical committee, a 42-member advisory board was created. It makes recommendations on legislative, organizational and technical measures for the implementation of the eHC. Recommendations of the advisory board are non-binding for the shareholders. Shareholders have the right to participate in the meetings of the board, in order to enable tight interlocking. Industry associations represent major companies involved in the eHC (see gematik 2008). In the context of the introduction of the eHC, gematik published various public tenders for the manufacture of the eHC: a health professional card (HPC), the connector and the card terminals (reading devices for the eHC and HPC). Large companies such as T-Systems, Siemens and Sagem had submitted bids and received contracts. In addition to the industry associations, three associations from the social sector are represented on the advisory board. The Federal Commissioner for Data Protection and Information Technology and the Federal Commissioner for the Interests of Patients were thematically assigned to this group of social organizations. Actors in the field of consumer protection and data privacy are rather critical towards the introduction of eHC. Science organizations and the seven test regions are also members of the advisory board. Representatives of federal ministries (e.g., the Ministry of Labour and Social Affairs in Baden-Württemberg) stand in for the test regions.

Another key body is the architectural board, founded in 2005 as a consequence of a regulation of the Federal Ministry of Health and Social Security (since 2005 BMG). It is designed to contribute to a more efficient resolution of contentious issues regarding architectural decisions during the test phases. The architectural board consists of seven members, five of whom with right to vote: a representative of all service providers, a representative of health insurers, a representative for all scientific organizations and a representative of the BMG. Among the permanent, non-voting participants are a representative of the Federal Commissioner for Data Protection and Freedom of Information, a representative of the Federal Office for Information Security (BSI), and the Managing Director of gematik. The official aim of this panel is to achieve consensus-oriented decision-making among members. Before an extensive technical debate ensues in the technical committee, the implementation can be discussed in a 'small round' in the architectural board and delegated to the Technical Committee if necessary. The board may be seen as a body of institutional influence of the BMG on the decision-making process of the top organizations. The architectural board and the technical committee meet regularly and discuss the latest technological solutions. The Board meets quarterly to state overall objectives and convey the views of affected groups (BMG 2005).

The main objective of the new health card, as articulated by the German Federal Ministry of Health, is to improve communication with all healthcare stakeholders. The BMG has stated the increased involvement of patients as yet another target. Patients may obtain full insight into the stored data (e.g. vaccination status, allergies, history of disease). In an upgraded version of the eHC, medical letters,

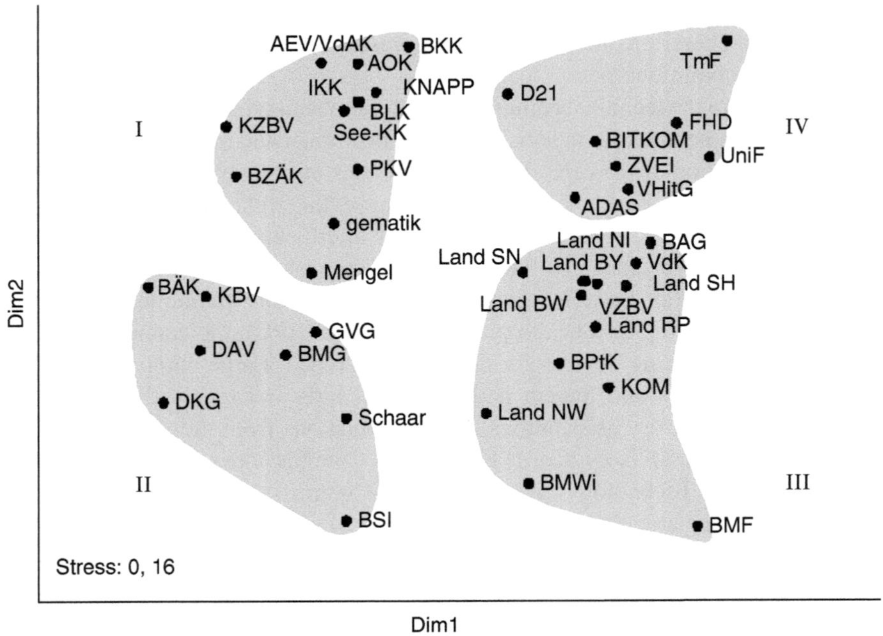

Fig. 11.1 Interest positions (multidimensional scaling and cluster analysis)
Annotation: MDS calculates the paired distances between the positions of interest and displays the positions of all organizations to each other in a two-dimensional space - which is derived from the data matrix. The calculation of the distances between the actors' positions of interests abides by the Euclidean metric, which states the distance of two object profiles as their shortest gap. The stress-value shows how well we displayed thedistances.The paired distances also form the basis for the hierarchical cluster analysis.They become the starting point for cluster algorithms, which combine the objects to clusters. In this study, Ward's Algorithm is employed, as this constitutes a conservative procedure that does not have any contracting or dilating properties, i.e., Ward's Algorithm has the tendency to form neither a single large nor many small clusters. Rather,tends to form groups equal in size.

reports and picture files, such as X-rays are to be made accessible in the form of an electronic patient record in a data-protective environment (data server). Likewise, the eHC should lead to more efficiency in the health sector (BMG 2007).

However, the interests of actors involved in the implementation of the eHC differ in some respects. To determine similarities and differences in their positions more precisely, actors were asked to name the other ones with whom they share the most ideas regarding the eHC. The resulting estimates indicate the distance between the positions of each stakeholder as they are perceived by the interviewees. With the help of multidimensional scaling (MDS) and hierarchical cluster analysis, these assessments can be illustrated (Fig. 11.1). The graphical representation shows that the positions of interest can be assigned to four groups.

Group I comprises almost all the sponsors and two organizations of service providers, namely the Federal Chamber of Dentists (BZÄK) and the Federal

Table 11.2 Evaluation of the eHC (% approval)

	Cluster I	Cluster II	Cluster III	Cluster IV	Chi^2	Total
The eHC does not pose a security risk for patients	77.8	33.3	90.0	85.7	4.75	79.3
Costs of the eHC are equitably distributed	0.0	33.3	10.0	42.9	6.01*	17.2
The eHC reduces labour costs in the health sector	44.4	33.3	20.0	57.1	2.65	37.9
The eHC reduces costs	44.4	0.0	50.0	42.9	2.47	41.4

* Significant on the level $p <= 0.1$ (Fisher exact test)

Note: Responses to some questions were not provided by all organizations. Of the Cluster I organizations, 9 (out of 12) responded, in Cluster II 3 (out of 8), in Cluster III 7 (out of 8) and in Cluster IV 10 (out of 14)

Association of Dental Insurance (KZBV). Gematik and the Commissioner of Data Protection complete the first cluster. The sponsors are more positive towards the eHC. They expect cost cuts ranging in the millions thanks to reductions in the number of duplicate investigations and the use of e-prescriptions (Zeit Online 2007). Respondents of the other organizations, however, are more critical towards the cost argument (see Table 11.2). Also, the presumption that the eHC will lower the workload in the health sector is only shared by a minority of respondents (although the percentage is higher than in other groups). The outstanding characteristic of this group is, however, that no organization has agreed with the statement that the costs of the eHC are evenly distributed.

The second cluster is formed - apart from four organizations of service providers – by the BMG, the BSI, Data Protection Officer Schaar and the Society for Insurance Science (GVG). Of all groups, this one is mainly concerned that the eHC will fail to reduce costs and might even lead to massive cost increases. Surprisingly, the BMG is also in this group. Thus it seems that the public rhetoric deviates from the perceived position of interest. In particular, service providers are skeptical of the eHC. There are concerns of longer treatment times due to the creation of e-prescriptions as experienced during tests (Frankfurter Allgemeine Zeitung 2008). A second point of criticism, put forward by the service providers, is data insecurity as a result of long-lasting storage of information in external central computers. Similarly, there are concerns regarding PIN protection of access to the data on the eHC. Critics refer to older patients who forget their PIN or have difficulties entering the PIN into the reader. For these reasons, the eHC in its present form was rejected by the 111th German Medical Assembly. The doctors voiced the threat not to participate in the roll-out of the eHC (Frankfurter Allgemeine Sonntagszeitung 2007).

Cluster III consists of the test regions, the consumer protection and patients' associations as well as the Federal Ministry of the Economy and the Ministry of Science. These organizations regard the eHC rather positive in reference to cost-cutting and do not perceive any data security concerns. In the fourth cluster we find the organised business interests together with university and science organizations. In this cluster there is a high approval rate of the eHC.

11.5 Distribution of Power: Voting Rights, Influence Reputation and Information Control

According to the new legislation a group of shareholders needs at least 67% of the votes in order to win a decision. If none of the groups succeeds reaching this threshold for its suggestions and the actors cannot find unanimous consent, the BMG ultimately decides upon the issue. Examples for such measures are the founding of the architectural board and an ordinance in 2005 providing direction to gematik. According to the initial plan, the ministry was supposed to receive the so-called "fourth milestone" of the architectural solution[5] for the introduction of the eHC by September 2005. However, gematik could not bring forward the necessary system architecture for components in due time. The BMG reacted with an ordinance in 2005, which instructed gematik on further steps. Consequently, the architectural board was implemented to review architectural decisions and make decisions on controversial questions. The architectural board has four voting members. A BMG representative chairs the board. In case of ties, the BMG representative's vote is decisive. Through the establishment of the architectural board and the possibility to set up substitute performances the health ministry can make crucial decisions. The judicial basis for this is article 80 of the Constitution of the Federal Republic of Germany (*Grundgesetz*), which empowers the ministry to enact ordinances. In another ordinance in October 2006 related to testing measures for the introduction of eHC, the BMG, for instance, defined the development of telematic infrastructure and the requirements for components and services.

The law does not prescribe any specific voting procedures for the technical committee and advisory board. The technical committee regularly submits suggestions to the shareholders, which are then decided upon in the shareholder's meeting. These recommendations include current solutions, which are voted on among telematic experts of top associations at the technical committee. In contrast, recommendations by the advisory board provide general directions and are less specific. Members of the technical committee have one vote each; recommendations to be presented at the shareholders' meeting require a majority of 67%.

In order to examine informal aspects of power, respondents were asked to assess the influence of every actor on the implementation of eHC. The reported influence assessments were summarised (in the network analytical unit "indegree"). Figure 11.2 clearly shows the central position of the BMG. The ministry received an indegree of 134. This is the highest number of all actors and amounts to about 95% of the highest possible indegree. Gematik, which was founded as a state enterprise and led by the top associations in the health sector, has the second most central position in the network (indegree = 121), right after the BMG. The most central shareholders are the Federal Association of Panel Doctors (KBV), the BÄK, the Federal Association of Local Healt Insurance Funds (AOK) and the AEV/VdAK (indegree of 114 to 103).

[5] This is what the concrete technical implementation of the eHC is called.

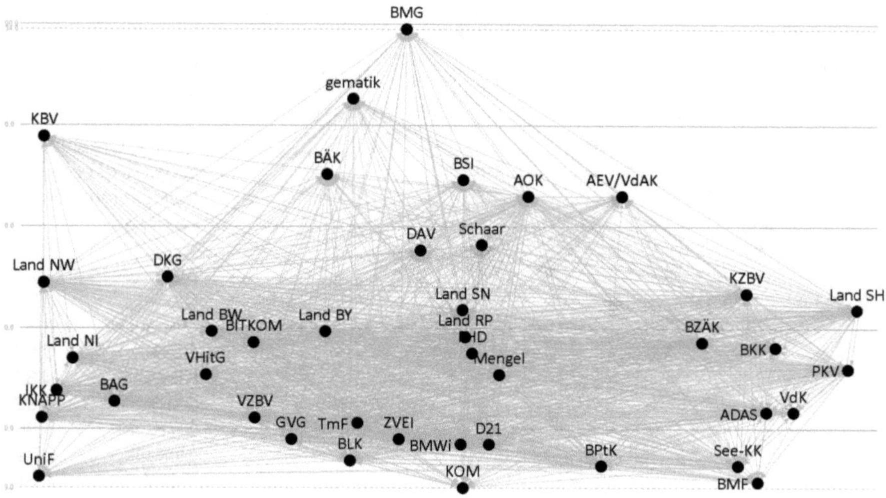

Fig. 11.2 Influence reputation
Annotation: Lines indicate influence-attribution. Organizations located at the top are attributed higher influence.

The least central shareholders of the gematik are the BLK (indegree = 54) and the health insurance See-Krankenkasse (indegree = 53). Also, various members of the advisory board often hold central positions – above all the BSI with an indegree of 106 and Peter Schaar, at the time of the empirical study serving as the Federal Commissioner for Data Protection and Freedom of Information with an indegree of 94. The seven test regions hold moderately central positions with indegrees between 73 and 87.

Two thirds of the respondents viewed the BMG as the central actor in the network, which makes decisions unilaterally or under consideration of the negotiating members. 31% saw the BMG as an actor that specifies goals or the frame of negotiation. Only one of the 29 respondents considered the BMG as an equal actor. It is quite obvious that the majority of the interviewed actors attributed the decisive, central role in the network to the BMG. Thus, the majority stated that it conceived the negotiating style surrounding the eHC as cooperative or consultative.

The information exchange network, as shown in Fig. 11.3, is composed of all 42 members[6] of the advisory board of gematik. The density of interaction amounts

[6] It must be mentioned at this point that the data matrix was made symmetric in order to acquire statements on actors who did not participate in the survey. Additionally, betweenness centrality scores were calculated for the data matrix, which only contains the organizations that have taken part in the survey. The comparison of both calculations showed that the distortions due to this transformation were minimal. Actors, who were graded as central in the first data matrix, similarly obtained central positions in the second data matrix.

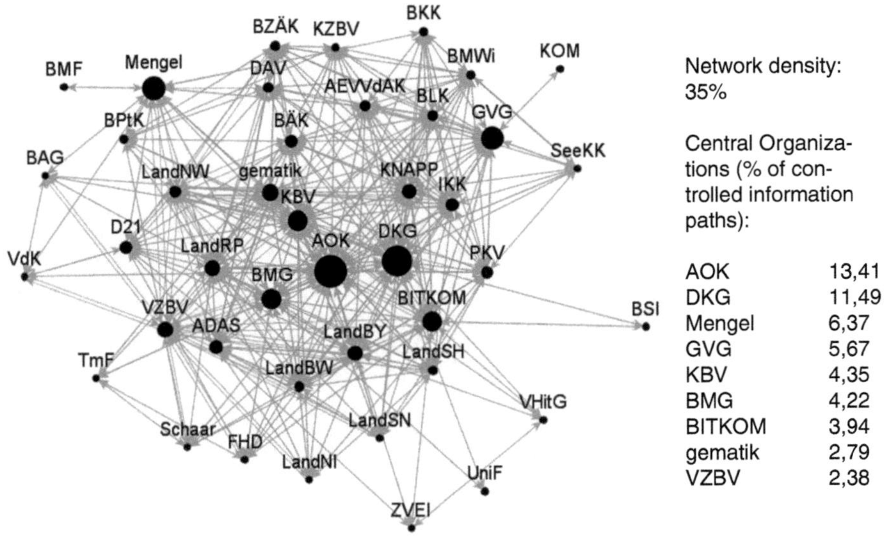

Fig. 11.3 Information exchange and control
Annotation: Lines indicate information flow. The size of the nodes indicates the scope of information control (Betweenness Centrality).

to 35%[7] and is thus relatively high. Therefore, no actor can obtain a relationally prominent position and dominate the information flow entirely or at least predominately.

The betweenness centrality, which states the number of information paths between two network actors that an actor can control, offers an appropriate measure of actor centrality (Wasserman and Faust 1994, pp. 188–191; Scott 2001, pp. 86–88). Two shareholders of gematik are the most central actors in the communication network in which the BMG only plays a secondary role. AOK as the most central actor in the eHC network can control around 13.5% of all information flows in the network. Apart from AOK, the German Hospital Society (DKG) holds a very central role. It is responsible for the distribution of about 11.5% of all information used in the network. Apart from these two central actors, Helga Kühn-Mengel (the Commissioner for the Concerns of Patients), the GVG, BMG, Federal Association for Information Technology, Telecommunications and New Media (BITKOM) and KBV hold moderately central roles in the network. Four to seven percent of information runs through each of these actors. Figure 11.3 also shows clearly that the organizations of Clusters I and II, in other words the sponsors and service providers, are foremost involved in the information exchange, while scientific organizations and the test regions hold peripheral positions.

[7] Network density measures relate the actual number of ties in a network to the maximum possible number of ties in a network. Thirty-five percent means that 35% of maximally possible ties between actors have been realised.

Table 11.3 Evaluation of the position of the BMG and the negotiating style (% approval)

		Cluster I	Cluster II	Cluster III	Cluster IV	Chi^2	Total
Role of BMG	Equal member	0.0	0.0	0.0	14.3	3.26	3.4
	Decides frame for negotiation	0.0	66.7	0.0	14.3	12.57*	10.4
	Decides negotiating goals	33.3	0.0	20.0	14.3	1.84	20.7
	BMG decides in the common interest	33.3	0.0	60.0	57.1	4.28	44.8
	BMG decides alone	33.3	33.3	20.0	0.0	3.00	20.7
Negotiating style	Consultative	55.6	0.0	40.0	42.9	2.88	41.4
	Negotiation-intensive	11.1	33.3	10.0	0.0	2.52	10.3
	Coordinating	33.3	33.3	20.0	0.0	3.00	20.7
	Cooperative	11.1	100.0	50.0	71.4	9.71*	48.3

*Significant at the level p <= 0.05 (Fisher exact test)

Note: Responses to some questions were not provided by all organizations. Of the Cluster I organizations, 9 (out of 12) responded, in Cluster II 3 (out of 8), in Cluster III 7 (out of 8) and in Cluster IV 10 (out of 14)

A comparison between reputation and information control reveals significant differences for individual actors in the implementation network. The influence reputation of the BMG, for instance, does not correspond with its control possibilities in the area of information exchange. Thus, the BMG is credited with the highest influence value, while it has low capacities to control of information exchange. Likewise, gematik, the KBV, BÄK, BSI and AEV/VdAK are assessed as influential actors who do not, however, command corresponding shares in information control. The BSI is an exemplary case of the commonly low correlation between influence on implementation and actual control possibilities, as it holds a high influence reputation (70%), but cannot execute any information control. The Federal Agency prescribes for the telematic infrastructure safety standards in accordance with SGB V § 291b. This probably explains its high reputation. Its low embedding into the network however means that the conflict of interest in the area of data security cannot be sufficiently moderated by the BSI, the decisive actor in this question. Apart from the BSI, the AEV/VdAK and BÄK also display low information exchange with the other actors in relation to their reputations. In contrast, AOK and the DKG command distinctive control resources that do not necessarily correlate with reputation values. This is one of the reasons for securing the dominance of actors in healthcare policy over those in innovation policy.

Table 11.4 presents a systematic overview of the previously mentioned factors on the formation of the implementation network (information exchange). It can be seen that the formation of the coordination network results basically from actual and perceived agreement in opinion and from sector and advisory board membership. All factors have a positive effect on the occurrence of a tie between two actors. The positive effect of actual and perceived agreement on the several aspects of

Table 11.4 Effects on network formation (QAP-correlation coefficients)

	Perceived influence	Financial burden	Agreement in opinion	Perceived agreement in opinion	Advisory board member	Sector
Information exchange	−0.024	0.037	0.119**	0.429***	0.351***	0.330***
Perceived influence	–	−0.005	0.100	0.041	0.108	0.081*
Financial burden	–	–	0.125*	−0.057	−0.214**	−0.181***
Agreement in opinion	–	–	–	0.029	−0.037	−0.031
Perceived agreement in opinion	–	–	–	–	0.286***	0.381***
Advisory board member						0.468***

*$p \leq 0.1$, **$p \leq 0.05$, ***$p \leq 0.01$

Notes: The quadratic assignment procedure QAP is a two-step algorithm used to test the association between networks. In the first step, correlation coefficients between corresponding cells of the two data matrices are computed. In the second step, the algorithm randomly permutes rows and columns of one matrix several 100 times and each time recalculates the correlation measures. Then the algorithm assesses the proportion of times that the permuted correlation measure is larger than or equal to the real measure calculated in step 1. A low proportion (< 0.05) is an indicator of a strong association between the matrices (Ucinet IV 2009, p. 92)

the eHC indicates that traditional sectoral interest positions play a dominant role in the emergence of the implementation network. Similarly, sector and board membership invigorate the sectoral interest perspective leading to a network core of health care organizations while information and communication technology organizations are seated at the spectator porch.

11.6 Conclusions

This essay focused on the policy and administrative coordination in a temporary and domain boundaries transcending policy network that was established to implement the eHC in Germany. The main focus has been on structural barriers to coordination and on inconsistencies in goals and task settings that resulted from power asymmetries. The analysis revealed that different policy domains can hardly be accommodated without causing frictions among participating actors. These frictions result from different institutional logics inherent in different policy domains.

The introduction of the eHC has been announced as a prestigious project within the German high technology and innovation policy. The implementation of the eHC cherished high hopes and aspirations on the side of policy makers and public administration that aim to realise objectives relevant to health care as well as industry and innovation related policies. Accordingly, the legal guidelines of the eHC have been inspired by existing large scale projects in innovation policy, in which the federal government was the driving force behind technology innovation, but left the management of the projects to private companies.

However, the actual implementation is still dominated by cleavages inherent in German health politics, namely cost transfer and cost reduction. Therefore information exchange centers around actors from the health policy domain. In particular the representatives of physicians and hospitals (Deutsche Krankenhausgesellschaft) and of statutory health insurance (AOK) have created support coalitions and dominate information flows while information technology associations and research institutes are located at the periphery of the network.

Network formation results basically from actual and perceived agreement in opinion and from sectoral and institutional embeddedness. This indicates that traditional sectoral interest positions play a dominant role in the structuration of the implementation network and it also explains the slow and cumbersome process of implementation. Especially the health care providers have turned out to be the stalling element in the whole process.

In sum, the merging of different policy domains resulted in an incoherent coordination structure in which interest positions, power and information control are not matched. These functional and structural inconsistencies have increased the administrative complexities leading to the non-intended problems in the implementation of the eHC.

A recommendation for future telematic projects, especially when involving large socio-technical systems, involves a better management of complex interdependencies. Therefore it should be considered to install boundary-spanning responsibilities, that are fostered by the institutional settings, in particular memberships of all stakeholders (not just shareholders) in advisory or other boards, eventually leading to emergent coordination structure that matches with the functional requirements.

References

Adam, S., & Kriesi, H. (2007). The network approach. In P. A. Sabatier (Ed.), *Theories of the policy process* (pp. 129–154). Boulder: Westview Press.

Bandelow, N. C. (1998). *Gesundheitspolitik. Der Staat in der Hand einzelner Interessensgruppen?* Opladen: Leske+Budrich.

BMG (2005). Verordnung über Testmaßnahmen für die Einführung der elektronischen Gesundheitskarte vom 2. November 2005. Berlin: Bundesministerium für Gesundheit, Referat Öffentlichkeitsarbeit.

BMG (2006). Verordnung zur Änderung der Verordnung über Testmaßnahmen für die Einführung der elektronischen Gesundheitskarte vom 2. Oktober 2006. Berlin: Bundesministerium für Gesundheit, Referat Öffentlichkeitsarbeit.

BMG. (2007). *Die elektronische Gesundheitskarte*. Berlin: Bundesministerium für Gesundheit, Referat Öffentlichkeitsarbeit.

Börzel, T. A. (1998). Organizing Babylon: on the different conceptions of policy networks. *Public Administration, 76*(2), 253–273.

Braun, D. (2008). Organising the political coordination of knowledge and innovation policies. *Science and Public Policy, 35*(4), 227–239.

Döhler, M., & Manow, P. (1995). Staatliche Reformpolitik und die Rolle der Verbände im Gesundheitssektor. In R. Mayntz & F. W. Scharpf (Eds.), *Gesellschaftliche Selbstregulation und politische Steuerung* (pp. 140–168). Frankfurt am Main, New York: Campus.

Dolata, U. (2004). Unfassbare Technologien, internationale Innovationsverläufe und ausdifferenzierte Politikregime. Perspektiven nationaler Technologie- und Innovationspolitiken. artec-paper Nr. 110. Bremen: University of Bremen, artec-Forschungszentrum Nachhaltigkeit

FES (2005). Die elektronische Gesundheitskarte kommt. Nutzen und Risiken der Telematik im Gesundheitswesen für Patienten und Gesellschaft. Berlin: Friedrich-Ebert-Stiftung.

Frankfurter Allgemeine Sonntagszeitung (2007). Die Ärzte jammern. 13 May 2007, p. 30.

Frankfurter Allgemeine Zeitung (2008). Die Mängel der elektronischen Gesundheitskarte. 16 Mai 2008, p. 38.

gematik (2008). Gesellschaft für Telematikanwendungen der Gesundheitskarte mbH. Berlin: gematik.

Heclo, H. (1978). Issue networks and the executive establishment. In A. King (Ed.), *The new American political system* (pp. 87–124). Washington, DC: American Enterprise Institute.

Keck, O. (1984). *Der Schnelle Brüter. Eine Fallstudie über Entscheidungsprozesse in der Großtechnik*. Frankfurt am Main: Campus.

Marth, W., & Koehler, M. (1998). The German Fast Breeder program (a historical overview). *Energy, 23*(7/8), 593–608.

Mayntz, R. (1997). *Soziale Dynamik und politische Steuerung. Theoretische und methodologische Überlegungen*. Frankfurt am Main: Campus.

Rosenbrock, R., & Gerlinger, T. (2006). *Gesundheitspolitik. Eine systematische Einführung* (2nd ed.). Bern: Hans Huber.

Sabatier, P. A., & Weible, C. M. (2007). The advocacy coalition framework. In P. A. Sabatier (Ed.), *Theories of the policy process* (pp. 189–220). Boulder: Westview Press.

Scharpf, F. W. (1985). Die Politikverflechtungs-Falle: Europäische Intergration und deutscher Föderalismus im Vergleich. *Politische Vierteljahresschrift, 26*(4), 323–356.

Schmitter, P. C., & Streeck, W. (1999). The organization of business interests. Max Planck Institut für Gesellschaftsforschung Discussion Paper 99/1. Cologne.

Schneider, V., & Janning, F. (2006). *Politikfeldanalyse. Akteure, Diskurse und Netzwerke in der öffentlichen Politik*. Wiesbaden: VS Verlag.

Scott, J. (2001). *Social network analysis: A handbook*. London: Sage.

Streeck, W., & Kenworthy, L. (2005). Theories and practices of Neocorporatism. In T. Janoski, R. R. Alford, A. M. Hicks, & M. A. Schwartz (Eds.), *The handbook of political sociology* (pp. 441–460). Cambridge: Cambridge University Press.

Thatcher, M. (1998). The development of policy network analysis: from modest origins to overarching frameworks. *Journal of Theoretical Politics, 10*(4), 389–416.

True, J. L., Jones, B. D., & Baumgartner, F. R. (2007). Punctuated equilibrium theory: Explaining stability and change in American policymaking. In P. A. Sabatier (Ed.), *Theories of the policy process* (pp. 155–187). Boulder: Westview Press.

Ucinet, I. V. (2009). *Reference guide 6*. Harvard: Analytic Technologies.

van Waarden, F. (1992). Dimensions and types of policy networks. *European Journal of Political Research, 21*(1–2), 29–52.

Von Alemann, U., & Heinze, R. G. (1979). *Verbände und Staat. Vom Pluralismus zum Korporatismus*. Opladen: Westdeutscher Verlag.

Wasserman, S., & Faust, K. (1994). *Social network analysis: methods and applications*. Cambridge: Cambridge University Press.

Weyer, J. (2004). Innovationen fördern-aber wie? Zur Rolle des Staates in der Innovationspolitik. In M. Rasch & D. Bleidick (Eds.), *Technikgeschichte im Ruhrgebiet-Technikgeschichte für das Ruhrgebiet* (pp. 278–294). Essen: Klartext Verlag.

Weyer, J. (2005). Staatliche Förderung von Großtechnikprojekten. *Ein dysfunktionaler Anachronismus im Zeitalter der Globalisierung? Technikfolgenabschätzung-Theorie und Praxis, 14*(1), 18–25.

Zeit Online (2007). Albtraum Gesundheitskarte. ZEIT Online, 17 October 2007.

Chapter 12
A Paradigm Change in Innovation Policies? Assessing the Causes and Consequences of Embryo Research Laws

Simon Fink

12.1 Introduction

The discovery of the therapeutic potential of human embryonic stem cells in 1998 was a major breakthrough for biomedicine (Gearhart 1998; Thomson et al. 1998). Proponents of embryonic stem cell research argue that effective cures for many grave diseases can be developed using human embryonic stem cells. However, this line of research is ethically highly contentious. In public debate, embryo research is often depicted as incommensurate with basic ethical standards of human dignity (Lauritzen 2001).

Despite the fundamental ethical conflicts, most nations have passed laws regulating embryo research. However, there is no consensus on the strictness of embryo research laws. Some nations have passed liberal laws and encourage stem cell research while others interdict embryo research.

A question of considerable theoretical, as well as practical, importance is whether the considerable variation of embryo research laws has led to a corresponding variation in the innovative ability of national economies. Researchers, politicians, and representatives of the pharmaceutical industry claim that stem cell research is the most promising branch of medical biotechnology. Exceedingly strict regulation of this research is thought to impede innovation in the medical biotechnology sector (Standing Committee on Legal and Constitutional Affairs 2000; Deutsche Forschungsgemeinschaft 2001; Association of the British Pharmaceutical Industry (ABPI) 2004). However, opponents of stem cell research argue that a ban of embryonic stem cell research is ethically necessary and does not inhibit innovation (Spaemann 2001; Campbell 2001; Lilge 2001). The actors in this debate have mostly used abstract reasoning, hypothetical examples, or anecdotal evidence.

S. Fink (✉)
University of Bamberg, Bamberg, Germany
e-mail: simon.fink@uni-bamberg.de

J.M. Bauer et al. (eds.), *Innovation Policy and Governance in High-Tech Industries*, DOI 10.1007/978-3-642-12563-8_12, © Springer-Verlag Berlin Heidelberg 2012

However, studies on the consequences of permissive or strict embryo research laws on the innovative ability of national research systems are rare. This chapter closes this research gap. It proceeds in three steps.

First, the chapter establishes that human embryonic stem cell research has indeed been a major critical juncture for the biotech sector. With the discovery of human embryonic stem cell research, embryo research transformed from a branch of fertility medicine into a line of research of considerable commercial interest (Spar 2004). Left parties and the pharmaceutical industry were the major political and economic actors to realise the economic potential of embryo research. Accordingly, they tried to liberalise existing embryo research laws.

Second, the chapter demonstrates the considerable variation of embryo research laws. The disparity of embryo research laws is largely due to the Christian democratic parties and the Catholic Church resisting liberalizations. The content of embryo research laws is the outcome of a struggle between secular proponents and religious opponents of embryo research.

Third, the chapter argues that the relationship between embryo research laws and the innovative ability of the medical biotech sector does not correspond to the views of either proponents or opponents. No systematic short-term effect of the strictness of embryo research laws on the innovativeness in medical biotechnology could be detected. Strict laws do not directly lead to a decline of innovations. Permissive laws do not directly lead to an increase in innovations. Rather, the field is characterised by long-term structural differences. In a few cases, the increase of the innovativeness of the sector may be the result of a political strategy, but is equally often the by-product of regulatory inactivity. The argument that strict regulation impedes the innovative ability and permissive regulation leads to an increase in innovation is not supported. However, this does not rule out that increasing return effects of different embryo research laws will play out in the long-run.

The chapter is structured in seven sections. The next section introduces the field of embryo research and outlines the basic ethical and medical problems and prospects. Methodological considerations concerning the quantitative comparison of embryo research laws are discussed in the third section followed by analyses of the impact of stem cell research on embryo research policies. The fifth section analyses the still existing variety of embryo research laws and the strategic interaction of left parties, the pharmaceutical industry, Christian democrats, and the Catholic Church. The sixth section examines the impact of different embryo research laws on the innovative ability of national economies. The last section summarises the findings.

12.2 The Problem

Human embryos appeared on the political agenda when a baby was first created using in-vitro-fertilization (IVF). For IVF, embryos are created and implanted into the womb of a woman in order to bypass certain forms of infertility. Additionally, the successful conduct of IVF demonstrated that human embryos can be cultivated in a laboratory. For technical reasons, the number of embryos created for an IVF is usually larger than the number used for implantation. These surplus embryos are kept in a frozen state, but are almost never used for fertility treatment. Medical researchers have proposed to use these surplus embryos for research purposes.

Another implication of IVF is that the embryos can be screened before implantation. The so-called pre-implantation diagnosis is still in its infancy. However, by selecting the sex of the embryo it becomes possible to avoid hereditary diseases located on the sex-determining chromosomes. This selection is ethically highly contentious. The attempt to actually change the genetic endowment of the embryo – germline therapy – is still in its infancy (Stock and Campbell 2000). The thin line between cure and research is clearly crossed with non-therapeutic research – research that uses the embryo as a raw material, without the intention of creating a child. The derivation of human embryonic stem cells is one form of non-therapeutic research. Stem cells are a very promising object for research, as they can differentiate into any mature cell type. They can be used to create, for example, brain cells, as replacements for decayed ones (Thomson et al. 1998). The source for stem cell lines can be the aforementioned surplus embryos. Scientists often demand that creation of embryos be allowed for research purposes only. Particularly, the creation of embryos using therapeutic cloning is often seen as desirable. Therapeutic cloning is the creation of an embryo with the same genetic characteristics as a mature human being. The genetic identity is especially desirable for the creation of replacement tissue or organs, as the risk of rejection is much lower if the replacement tissue has the same genetic information as the recipient. If the same cloning technique is used to create a child, it is called reproductive cloning.

The ethical problems with the diverse techniques of embryo research are complex. The clearest judgement can be made about therapeutic research. It is morally justifiable to observe the development of an embryo. The other techniques are more contentious. The most obvious problems arise with the techniques that imply the destruction of the embryo. If one conceives of the embryo as a human being, all these techniques are fundamentally wrong and should not be carried out. However, as some ethicists see the embryo as equivalent to a human being (Ryan 2001), and others do not (Steinbock 2001), the question is by no means settled. Similarly complex issues arise with germline therapy and reproductive cloning. No human being is killed using these techniques, but Habermas argues that these techniques violate the bodily and moral integrity of the cloned and/or genetically modified child (Mendieta 2004).

The embryo research policy field overlaps considerably with policy fields like artificial reproduction and abortion, as some of the core techniques and ethical questions are the same. However, embryo research poses a set of new and complex ethical problems. Additionally, the scientific basis is in a state of dynamic development, and stem cell research is the latest development. Although the research frontier is in constant movement, policy makers have to strike a balance between research ambitions and moral concerns.

12.3 Operationalizing National Embryo Research Regulatory Systems

To allow a comparative survey of national embryo research laws, this chapter proposes to operationalise embryo research laws on the dimension of their strictness. Nine basic techniques of embryo research were identified. Data on embryo research laws in 21 OECD countries (the Western OECD world) were gathered, documenting whether the nine basic techniques were allowed (coded 0) or forbidden (coded 1). In some cases, qualifications were added; these cases were coded 0.5. The techniques and their coding, as well as a list of the coded laws and details on the coding decisions are shown in Table 12.1. Added up, these binary variables constitute the Embryo Research Index (ERINDEX), which ranges from 0 to 9, with 9 indicating that every technique is forbidden. The main sources for the data are Gratton (2002), the Council of Europe (1998), and UNESCO (2004). The data were cross-checked using legal studies, case studies and e-mail correspondence with ethics councils.

Table 12.1 Composition of the embryo research index

Variable	Description of Procedure	Coding
THR	Therapeutic research: non-harming research.	0 (allowed)//1 (forbidden)
TSS	Therapeutic sex selection: the selection of the child's sex after genetic testing in order to avoid hereditary diseases.	0/1
GLTH	Germ line therapy: the manipulation of the human germ line in order to influence genetically determined characteristics.	0/1
NTHR	Non-therapeutic research: research that destroys the embryo	0/1
NTHRAG	The age or stage of development until which non-therapeutic research may be done.	0 (no time limit) 0.5 (up to 14 days after fertilization) 1 (forbidden in principle)
EPRES	Embryo production for research purposes: the production of embryos solely for the purpose of research.	0/1
ESCR	Embryonic stem cell research: research on human embryonic stem cells (which must necessarily have been created using human embryos).	0 (use and production of stem cells allowed) 0.5 (use of imported stem cell lines allowed, but no production) 1 (completely forbidden)
THERCL	Therapeutic cloning: cloning human embryos to obtain research embryos.	0/1
REPCL	Reproductive cloning: cloning human embryos to obtain a human child (the "Dolly the sheep" procedure)	0/1
ERINDEX	Sum of the aforementioned variables.	0 to 9

12.4 Stem Cell Research as a "Critical Juncture"?

Sociological literature often claims that the development of human embryonic stem cell research in 1998 can be considered a "critical juncture" that transformed the regulatory systems of embryo research (Walters 2004; Bauer and Gaskell 2001, 2002). An analysis of quantitative data on the strictness of embryo research laws contributes to an assessment of whether this thesis holds true.

A graphical analysis of the first embryo research laws in the countries under study demonstrates that the year 1998 can indeed be considered a critical juncture. In Fig. 12.1, a vertical line marks the year 1998. With the exception of Italy and Switzerland, all countries that have passed their first embryo research law after 1998 have moderate to liberal laws. All these laws – including the Swiss law – allow the creation of stem cells using surplus embryos; the laws in the Netherlands and Belgium additionally allow the so-called therapeutic cloning.

The analysis of legal changes offers additional evidence for the critical juncture thesis. Many countries had adopted embryo research laws prior to 1998 and revised those laws after 1998. An analysis of these law changes demonstrates the importance of the year 1998. In Fig. 12.2, degree and direction of the law change are plotted against time. The horizontal line marks the baseline – countries above the line have tightened their law on embryo research, while countries below the line

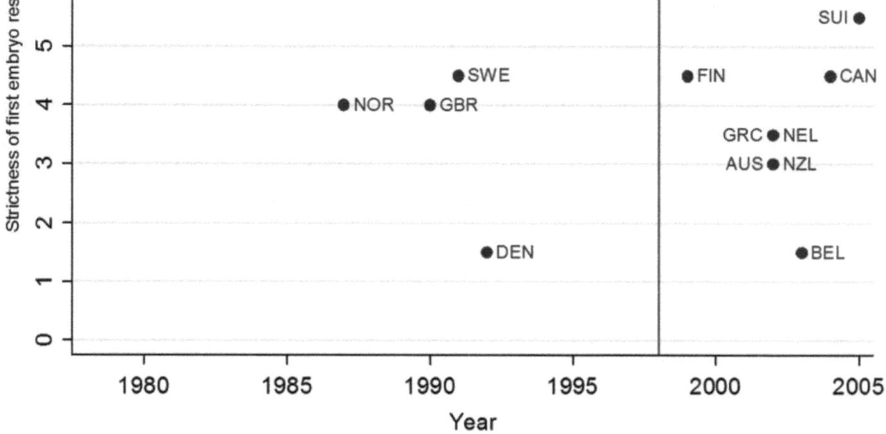

Fig. 12.1 Relation between year and strictness of the first embryo research law

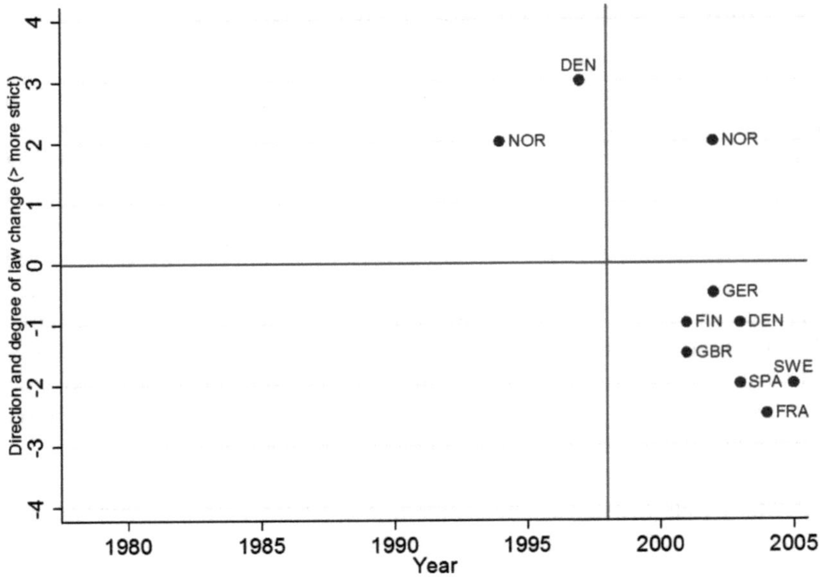

Fig. 12.2 Relation between year of law change and direction and degree of law change

have liberalised their law. The vertical line marks the year 1998, the year of the stem cell research breakthrough.

Figure 12.2 demonstrates that more existing embryo research laws were liberalised than tightened and that all liberalization measures occurred after 1998. All liberalizations concerned stem cell research. Germany allowed the import of stem cell lines; Denmark, Finland, France, and Spain allowed the production of stem cell lines using surplus embryos; Sweden and the United Kingdom allowed therapeutic cloning to create stem cells. Only Norway introduced additional restrictions after 1998.

However, the wave of liberalizations has not resulted in a universal regulatory model for embryo research (Table 12.2).

Table 12.2 demonstrates that no dominant regulatory model has emerged. A rough classification may distinguish three groups of regulators. First, a group of permissive regulators: This group is characterised by the fact that its members allow for therapeutic cloning. It is comprised of Belgium, the United Kingdom, New Zealand, and Sweden. Second, a group of restrictive regulators: This group is characterised by the fact that its members forbid all forms non-therapeutic research. Austria, Germany, Italy, and Norway belong to this group. The other countries can be termed intermediate regulators and try to steer a middle course between the other groups.

The descriptive analysis of embryo research laws allows two conclusions. First, the development of human embryonic stem cell research constitutes a critical juncture for the embryo research policy field. After 1998, a wave of liberalizations of existing embryo research laws started. New laws passed after 1998 are

Table 12.2 Embryo research laws as of 2006

Country	THR	TSS	GLTH	NTHR	NTHRAG	EPRES	ESCR	REPCL	THERCL	INDEX LAW
Belgium	0	0	0	0	0.5	0	0	1	0	1.5
UK	0	0	1	0	0.5	0	0	1	0	2.5
Sweden	0	0	1	0	0.5	0	0	1	0	2.5
N. Zealand	0	1	1	0	0	0	0	1	0	3
Australia	0	0	0	0	0	1	0	1	1	3
Denmark	0	0	0	0	0.5	1	0	1	1	3.5
Finland	0	0	0	0	0.5	1	0	1	1	3.5
Greece	0	0	0	0	0.5	1	0	1	1	3.5
Netherlands	0	0	1	0	0.5	0	0	1	1	3.5
Canada	0	0	1	0	0.5	1	0	1	1	4.5
France	0	0	1	0	0.5	1	0	1	1	4.5
Spain	0	0	1	0	0.5	1	0	1	1	4.5
Switzerland	0	1	1	0	0.5	1	0	1	1	5.5
Austria	0	1	1	1	1	1	0.5	1	1	7.5
Germany	0	1	1	1	1	1	0.5	1	1	7.5
Italy	0	1	1	1	1	1	1	1	1	8
Norway	1	0	1	1	1	1	1	1	1	8

significantly more liberal than pre-1998 laws. Second, no convergence of regulatory models can be observed. Different countries still have very distinct embryo research laws.

12.5 Heterogeneous Liberalization?

This section has two main aims. First, to explain the variation in embryo research laws over time, specifically, the wave of liberalizations that started after 1998. In order to do this, the impact of stem cell research on political actors' beliefs and preferences needs to be elucidated. Following actor-centred institutionalism (Scharpf 1997), we cannot suppose that embryo research laws adapt to research progress in a functional logic. Instead, the content of the policies is determined by interactions between self-interested actors. Thus, we must search for actors who benefit from liberalization of embryo research laws. Second, the wave of liberalization was by no means universal. Therefore, the second phenomenon in need of explanation is the variation between countries.

A theory-driven search for actors that presumably shape embryo research policies may start with the classical theories of policymaking (Schmidt 1993) and concentrate on the influence of parties and interest groups. For the purpose of this chapter, this choice of theories has to be extended. As the policy field has major ethical implications, cultural theories that conceptualise the influence of religious interests on public policy have to complement the classical theories of policymaking (Castles 1994; Minkenberg 2002).

Applying these theoretical concepts to the embryo research policy field, the first finding is that left parties were important actors in the field. Left parties changed their perception of embryo research after the stem cell breakthrough. Before stem cell research claimed to turn embryo research into an economically beneficial enterprise, embryo research ranked low on the political agendas of left parties. However, after 1998, left parties were amongst the foremost proponents of stem cell research and initiated most of the liberalizations of embryo research laws.

Figure 12.3 illustrates the preference change of left parties. If all embryo research laws from 1978 to 1998 are considered, no clear correlation between the strength of left parties and the strictness of embryo research laws emerges (see the nearly horizontal regression line in Fig. 12.3). However, upon closer inspection of the data, an intriguing pattern emerges. Austria and Spain can be considered outliers with regard to a partisan difference thesis. Both have strict embryo research laws in spite of left governments. However, the laws in Austria and Spain were passed before 1998, before the breakthroughs in stem cell research. At that time, the potential economic benefits of embryo research could not be anticipated. Left parties did not rank the issue high on their political agenda. After 1998, left parties only passed permissive laws (the dashed regression line in Fig. 12.3). The same is true for the liberalization of existing laws. After 1998, left governments only introduced liberal new laws or liberal amendments. This pattern suggests that embryo research has not been a salient issue for left parties in the entire period examined in this chapter. After embryo research promised to be economically beneficial, left parties enacted permissive laws.

The preference change of left parties can also be observed by tracing the lawmaking processes in some countries. The British Labour Party is the most

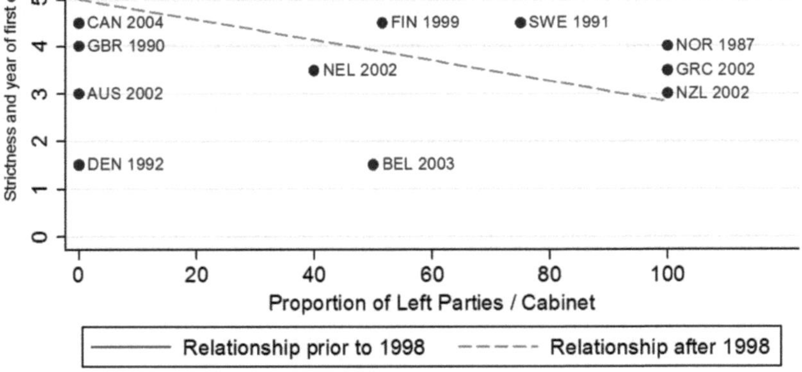

Fig. 12.3 Relationship between the strength of left parties and the strictness of embryo research laws

noticeable example. In the 1980s, the Labour Party had no clear preference on embryo research. A small fraction supported embryo research, another small fraction was opposed, and a large group lacked an interest in the issue (Mulkay 1997). However, this changed rapidly after 1998. Leading actors in the Labour Party recognised that embryo research, and stem cell research in particular, could form a major facet of a knowledge society, and support economic growth (Blair 2000b). Using strategic framing of the issue, the government convinced the disinterested majority of the Labour party of the economic benefits and outmanoeuvred the remaining opponents (Banchoff 2005; Fink 2009). The Prime Minister's office centrally coordinated decision making on biotechnology, and devoted vast resources to the promotion of stem cell research. At the end of the transformation process that had started in the minds of a few Labour leaders, the United Kingdom had one of the most liberal embryo research laws in the world.

Other country cases that support this conclusion are Belgium, Finland, and Sweden (Burrell 2005; Schiffino and Varone 2004). In these three countries, left parties were the driving force behind liberal embryo research laws – but only after the issue became economically relevant.

The second group of actors that changed their preferences concerning embryo research was the pharmaceutical industry. Before 1998, embryo research was seen as basic research with little economic implications, or as a commercially uninteresting branch of reproductive medicine. However, with the discovery of the therapeutic potential of human embryonic stem cells, the industry changed its evaluation of embryo research.

The United Kingdom is the most salient example. The potential commercial benefits of embryo research were not apparent in the 1980s, and the pharmaceutical industry refrained from lobbying activities and did not take part in the legislative process that led to the Human Fertilisation Act of 1991 (Mulkay 1997). After 1998, the British pharmaceutical industry and the Royal Society lobbied for the admission of therapeutic cloning and depicted stem cell research as a strategically important line of research (Sleator 2000). The British pharmaceutical industry association (ABPI) replied to a government inquiry: "In what areas are there opportunities for the UK research base to excel and contribute to the economy and society, which might form the basis of future strategic research programmes over the next 10 years? Stem Cells" (Association of the British Pharmaceutical Industry (ABPI) 2004). The economic argument was first adopted by the governing Labour Party, but representatives of the Conservatives also reasoned that the scientific and economic success of the United Kingdom was strongly dependent on stem cell research: "The question is not whether the research will be done but whether it will be done in the United Kingdom: whether we, within the limited scope of our jurisdiction, wish to sanction such activity in order to promote our pharmaceutical research base..." (Philip Hammond, Conservatives, see Sleator (2000)) These arguments of the pharmaceutical industry and the research lobby for a liberal law were very successful in the United Kingdom. The Human Embryo and Fertilisation Regulations and the Human Reproductive Cloning Act of 2001 were at the time of their introduction the most liberal embryo research laws in Europe.

Other cases that demonstrate how stem cell research was seen as an economic asset are Australia (Dodson and Gray 2002; Standing Committee on Legal and Constitutional Affairs 2001), Denmark, Finland, and Sweden (Burgermeister 2003; Burrell 2005). All these cases show considerable policy change. Before 1998, the debates were primarily framed in terms of ethics, basic research and reproductive medicine. After 1998, economic considerations entered the debate as equally important arguments.

Thus, the new framing of the policy field in economic terms by left parties and the pharmaceutical industry explains the transformation of the regulatory landscape after 1998. The embryo research policy field came into the focus of economic interests, and left parties and the pharmaceutical industry tried to achieve economic success by demanding and passing liberal embryo research laws.

Nevertheless, we still observe a large variation in embryo research laws. This variation cannot be explained by the power of left parties and the pharmaceutical industry. Instead, two other actors merit our attention. The proponents of liberal embryo research laws met with two equally powerful opponents in political and societal arenas: Christian democratic parties and the Catholic Church. The result of the ensuing political struggle determined the content of embryo research laws.

There is a strong correlation between the strength of Christian democratic parties and the strictness of embryo research laws. Christian democrats have only passed restrictive laws (see Table 12.3), and never liberalised embryo research laws when they held government power. Italy and Norway appear as outlier cases in Figs. 12.1 and 12.2. They are the only countries that have passed restrictive embryo research laws or tightened existing embryo research laws after 1998. These country cases

Table 12.3 Embryo research laws in force as of 2006, and political constellation in the year the law was passed

Country	Year	ERINDEX	Christian democracy	Left parties	Veto players	Catholics
BEL	2003	1.5	0	50	5	75%
GBR	2001	2.5	0	100	1	9%
SWE	2005	2.5	0	100	1	1%
AUS	2002	3	0	0	2	26%
NZL	2004	3	0	100	1	15%
FIN	2001	3.5	0	50	5	0%
GRC	2002	3.5	0	100	1	n.a.
NEL	2002	3.5	0	40	3	31%
DEN	2003	3.5	0	0	2 (3)	3%
SPA	2003	4.5	0	0	1	94%
CAN	2004	4.5	0	0	1	46%
FRA	2004	4.5	0	0	2	85%
SUI	2005	5.5	28.56	28.56	4	46%
AUT	1992	7.5	44	50	2	78%
GER	2002	7.5	0	93.75	2	34%
NOR	2003	8	79	0	3 (4)	3%
ITA	2004	8	25	0	4	97%

can now be explained by referring to the strong role of Christian democrats in both countries (Kallerud 2004; Ramjoué and Klöti 2004).

Viewed over all countries and the whole period under study, the consistency of Christian democrats' preferences is remarkable. They were sceptical towards embryo research in the 1980s, and have mostly kept this preference. Theoretically, this corroborates the assumption that Christian democratic parties have a distinctly Christian heritage and try to pass corresponding policies (Hanley 1994). According to the empirical analysis of embryo research policies, Christian democrats are not "Conservatives with another name" (Broughton 1988; Duverger 1966).

The second actor that strongly opposed the liberalization of embryo research laws was the Catholic Church. We observe a persistent influence of religious norms and actors on embryo research policy. This is corroborated by multivariate regression models (Fink 2008b). If all other factors are held constant, religious countries passed stricter laws. The pattern can also be discerned in a graphical analysis (Fig. 12.4). the more religious the population of a country, the stricter the embryo research laws in that country.

Qualitative analyses suggest that this pattern can be explained by the power of the Catholic Church as a societal veto player (Fink 2009). As a political actor, the Catholic Church combines a clear and non-negotiable position against embryo research (Vatican Congregation for the Doctrine of Faith 1987) with high

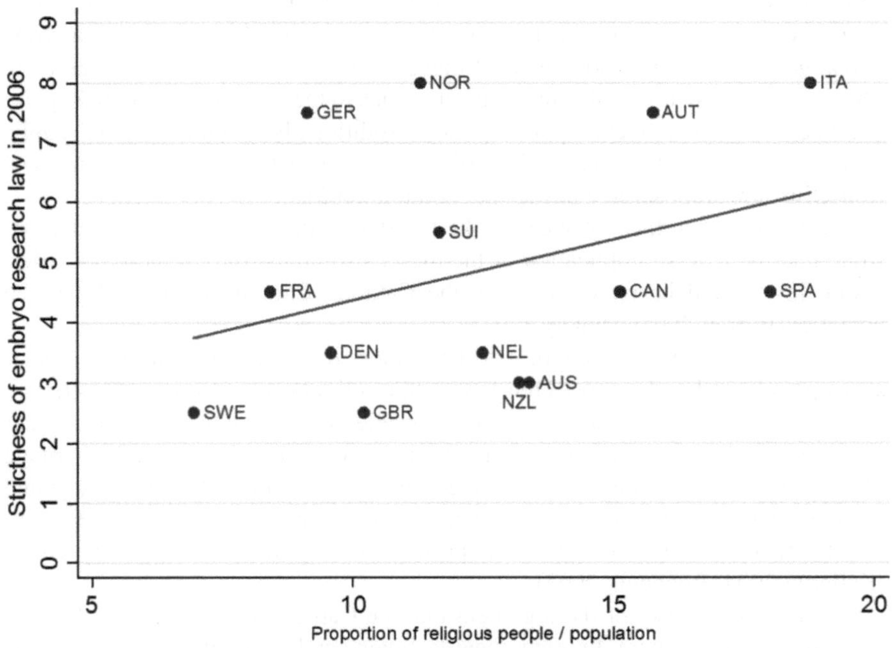

Fig. 12.4 Relationship between the proportion of religious people and the strictness of embryo research laws

mobilization potential in religious societies. When these factors interacted, the Church succeeded in pressing for restrictive embryo research policies.

The prime example to support the thesis of Catholic Church influence is Italy. During the 1990s, secular coalition governments ruled Italy. The Church saw that the chances of passing a restrictive law that coincided with its religious preferences were low. Thus, the Church blocked the passage of a law altogether (Ramjoué and Klöti 2004). The Church's influence was transmitted primarily via the Christian Democratic Party in parliament. However, as the religious-secular cleavage runs orthogonally to the labour-capital cleavage in Italy, many political actors of the centre-left governments were also devout Catholics. Left parties could not rely on their parliamentary majority on morality issues, and a blockade ensued. The blockade was overcome only when a centre-right coalition with the participation of Christian democrats won the elections in 2001. Following the 2001 elections, Italy introduced one of the strictest embryo research laws in the world (Lorenzi 2003). The test of the Church's power came when secular groups challenged the new law in a referendum. The Catholic Church mobilised its adherents and achieved a victory (Arie 2005). Thus, the Italian policymaking process demonstrates the power of the Catholic Church to block legislation on moral issues in religious countries until a friendly government takes over and the "right" law can be passed (Fink 2009).

Another example for the considerable influence of the Catholic Church is Austria. The Catholic Church played a major role in the design of a very restrictive embryo research law (Stranzinger 1992; Hadolt 2005). The position of the Catholic Church was very coherent and strongly influenced by the official catholic doctrine of *donum vitae* (Körtner 2002), which forbids any non-therapeutic embryo research. The Catholic Church was able to use its close ties to the Christian democratic party (Grabner 1999) to influence public policy. Additionally, it mobilised the public using its media outlets, up to the point where the Church published its own restrictive law proposal in a newspaper (Hadolt 2005).

On the other hand, country cases like Sweden or the United Kingdom demonstrate that the reverse implication of the thesis also holds true. In these countries, liberal laws could be passed because the Catholic Church has no mobilization potential (Burrell 2005; Mulkay 1997).

In Fig. 12.4, some cases appear to be outliers to the general rule that secular states pass liberal embryo research laws while religious countries pass restrictive laws. Germany and Norway are secular societies that – relative to their secular background – have passed surprisingly restrictive laws. For Germany, the heritage of national socialism seems to have played a major role (Fink 2007a; Rothmayr and Ramjoué 2004. Norway, on the other hand, is an example for a tremendous success of Christian democrats that tried to push forward policies on morality issues (Kallerud 2004; Bondevik 2003).

A problematic case for the argument that the Catholic Church is an influential actor in the field is Spain. In many regards, Spain can be regarded as a case that is most similar to Italy. The Catholic Church traditionally has a major role in public life and can mobilise a religious population. However, this constellation seems to be eroding. Spain liberalised its embryo research law in 2003, despite resistance of the Church. Viewed in a larger frame, the liberalization of the embryo research law is only a facet in a major struggle between the Spanish Catholic Church and secular forces. At the beginning of the 21st century, the Catholic Church in Spain not only had to mobilise against embryo research, but also against gay marriage and the introduction of Islamic religious education in schools (Haines 2005; Simons 2004). Thus, the case of Spain hints at the fact that the power of the Catholic Church to block policies that are not according to its preferences may be eroding. At the moment, it cannot be determined whether this is an idiosyncrasy of the Spanish case or a facet of a general erosion of Church power.

To summarise, two groups of actors are responsible for the major transformations of the regulatory systems for embryo research after 1998. Left parties and the pharmaceutical industry saw the commercial potential of embryo research and tried to liberalise embryo research laws. Whether they succeeded was contingent on the power of two other actor groups who opposed most applications of embryo research: Christian democratic parties and the Catholic Church. If Christian democratic parties held government power or the Catholic Church had broad mobilization potential, they succeeded in passing strict embryo research laws or preventing liberalizations.

12.6 The Impact of Embryo Research Laws on the Innovative Ability of National Economies

The previous sections have demonstrated that we witness a heterogeneous trend towards liberal embryo research laws. Considerable differences in the strictness of embryo research laws still persist.

The major question posed at the outset was whether these differences in the strictness of embryo research laws have an impact on the innovative ability of national economies.

To operationalise the innovative ability of the medical biotech sector, this chapter proposes to measure patents in microbiology as a share of the total number of patents. This measure can be loosely termed "biotech innovation quota". There are three rationales for this operationalization. First, a growing number of patents in microbiology and genetic engineering should be the first sign of an improved research environment. Second, the proportion of patents in the sector is an inter-subjective measure that allows comparisons between countries and over time. Third, from a policy maker's viewpoint, if one considers the biotech sector to be strategically important, the quota of biotech patents is a good benchmark to assess whether the sector prospers. Given the scarce supply of cross-country and time series data about biotechnology research performance (Van Beuzekom 2001; Arundel 2003), using the proportion of medical biotech patents is a reasonable proxy for the innovative ability of the sector (Fink 2007b).

An analysis of the relationship between the strictness of embryo research laws and innovations may start with a simple cross-sectional plot. Figure 12.5 plots the mean proportion of biotech patents (1998–2005) against the strictness of national embryo research laws (2006).

Figure 12.5 shows that restrictive embryo research laws are correlated with a lower innovative quota in stem cell research. The countries with the most restrictive embryo research laws – Austria, Germany, Norway and Italy – have the lowest biotech innovation quota. This finding resonates with country case studies (Burrell 2005; Körtner 2002). However, a permissive law does not guarantee innovations in the medical biotech sector. A comparison of Denmark and Finland demonstrates that countries with very similar embryo research laws exhibit considerable differences in their biotech innovation quota.

However, a cross-sectional perspective may simply reflect stable long-term level differences. From a political and strategic point of view, the more interesting question is whether a correlation between permissive embryo research laws and a high innovative ability can also be shown over time. If the introduction of a liberal embryo research law is followed by an increase in the biotech innovation quota – or the introduction of a restrictive law is followed by a decline of the biotech innovation quota – the case for the strategic importance of embryo research laws – and the incommensurability of innovation and ethics – would be strengthened.

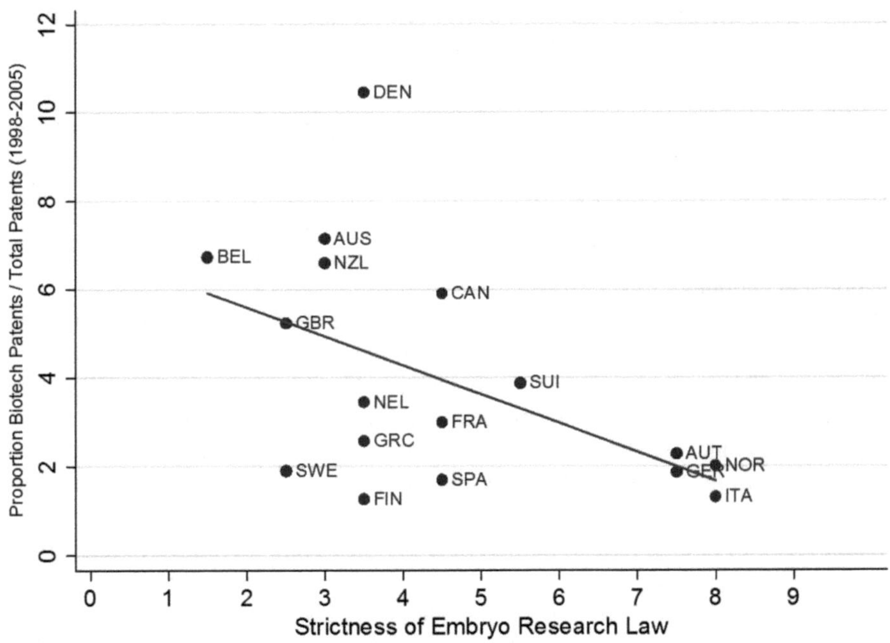

Fig. 12.5 Relationship between strictness of embryo research laws and innovative ability – cross sectional perspective

Figure 12.6 offers a longitudinal perspective to complement the cross-sectional picture. The data show the development of the biotech innovation quota in three country groups (permissive, average and strict regulators according to the reasoning in section 3, Table 12.2).

Figure 12.6 demonstrates that the findings in Fig. 12.5 partly reflect structural differences. The group of restrictive regulators has a lower biotech innovation quota than the group of permissive regulators from the outset. On the other hand, Fig. 12.6 shows that these level differences have increased. After the breakthroughs in stem cell research the biotech innovation quota has increased in all countries, but most markedly in the group of permissive regulators.

In conjunction, Figs. 12.5 and 12.6 suggest that permissive embryo research laws might be a necessary condition for a high biotech innovation quota, but are no sufficient condition (see for example Sweden or Finland in Fig. 12.5).

Figure 12.7 plots the biotech quota for the group of permissive regulators: Belgium, New Zealand, the United Kingdom and Sweden. In all other countries under study, the quota remains more or less stable. This implies that there is no sharp decline in the biotech innovation quota in the countries that have passed strict laws on embryo research. This finding cautions us further against the hypothesis that permissive embryo research laws have a short-term effect on the innovative ability of the sector.

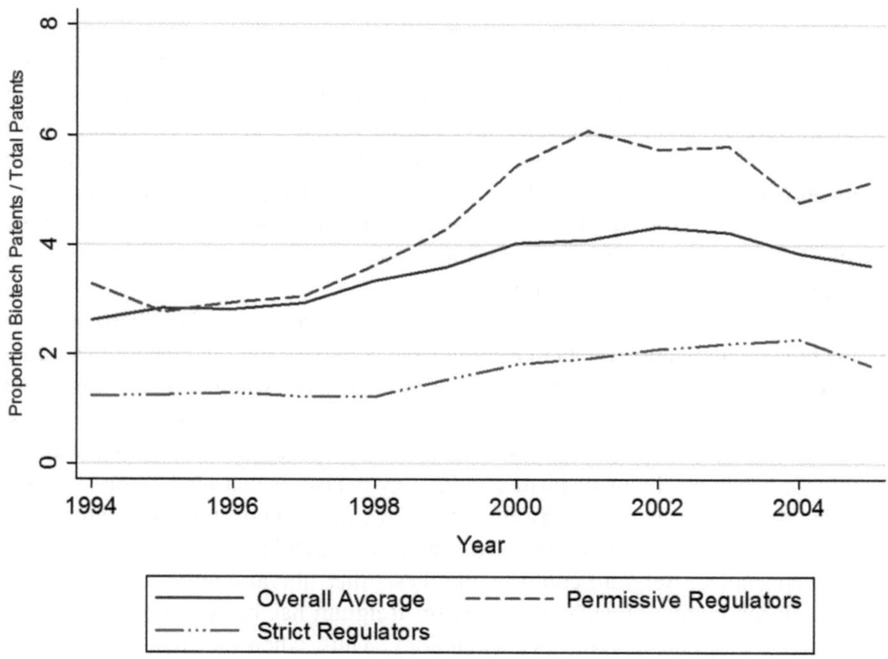

Fig. 12.6 Development of the biotech innovation quota over time for permissive and strict regulators, compared to the overall mean

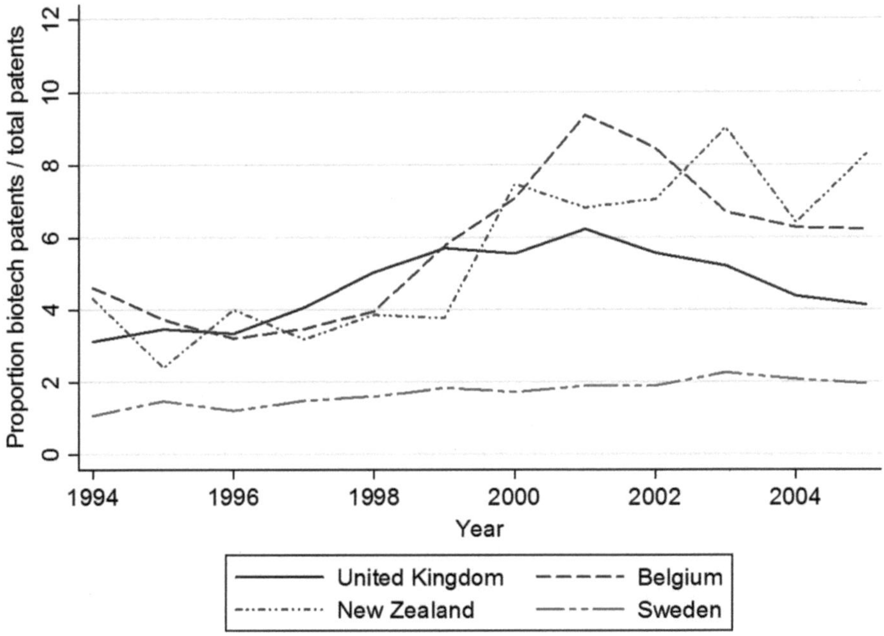

Fig. 12.7 Development of the biotech quota in the group of permissive regulators

The country trajectories depicted in Fig. 12.7 suggest that caution is necessary when assessing the ability of governments to steer biotech innovation. There are only three possible examples of "heroic innovation policy" in the country sample studied: Belgium and New Zealand have passed very liberal embryo research laws (see Table 12.2). Both countries have seen a considerable rise of their biotech innovation quota (see Fig. 12.7). The United Kingdom has passed a very liberal law, but its increase in biotech innovation quota is not as marked as in Belgium or New Zealand. The fourth country in the permissive group, Sweden, has not witnessed a change in the biotech innovation quota.

The four countries presented in Fig. 12.7 merit our attention. The sharp increase of the biotech innovation quota in Belgium and New Zealand (and the more smooth increase in the United Kingdom) raises the question whether these cases are evidence for the thesis that permissive embryo research laws lead to innovations, or if we see only statistical artefacts. And if these countries prove to be evidence for the thesis, what are the mechanisms and policy measures that lead to the success? On the other hand, the case of Sweden raises the question of why the country was not able to capitalise on its permissive embryo research law.

At first glance, Belgium confirms the thesis that liberal embryo research laws lead to an innovative biotechnology sector. Belgium has been one of the leading countries in artificial reproductive technology. Thus, when the stem cell research breakthroughs occurred, Belgium already had an established research base in applied medical biotechnology (Varone and Schiffino 2004). Belgium had no special law regulating embryo research up to 2003, and was considered a "bioethical paradise" (Varone and Schiffino 2004). Public opinion was very positive towards biotechnology (Schiffino and Varone 2004). Together with the United Kingdom, Belgium was considered to be one of the most research-friendly environments for stem cell research in Europe, and is either coordinator or project partner in a large share of EU-funded research projects involving stem cells (European Commission 2005). However, the Belgian success story is a by-product of political struggle and not the result of a political strategy to promote life sciences. The boom in biotech patents in Belgium occurred from 1998 to 2001. However, the very permissive law, the *Loi relative à la recherche sur les embryons in vitro*, was passed only in 2003. Up to this time, the lack of a law in Belgium cannot be considered as a part of a coherent political strategy. Rather, intense political struggle within a coalition comprising Christian democrats prevented the passage of a law on embryo research (Varone and Schiffino 2007; Schiffino and Varone 2004). Only when the Christian democrats left the coalition due to electoral defeat, a law could be passed. Therefore, the biotech boom from 1998 on occurred to some extent "behind the backs" of the political actors.

Thus, Belgium confirms the thesis that liberal embryo research laws are associated with a prospering and innovative medical biotech research sector, although this cannot be attributed to a political strategy, and is rather the unintended consequence of policy deadlock.

New Zealand's success story is similarly equivocal. New Zealand has had an ethics committee regulating embryo research since 1993. As early as 1996, a bill regulating embryo research – the Human Assisted Reproductive Technology (HART) bill – came into the parliamentary arena, but lay dormant for a long time in the Health Committee (Barr 2003b). Following the stem cell research breakthroughs, the Labour government reanimated the bill in 2003. Due to the Westminster system with few veto points, the government was able to push the liberal bill through (Barr 2003a), and the HART act was passed in 2004. Public opinion towards medical biotechnology and stem cell research was generally positive (Warren and Osborne 2006) and supported by headlines like "Stem cells could end need for heart transplants," "Blind could see again with new medical breakthrough," "'Incurable' illness falls to gene therapy," or "World on the edge of a new era of drug discovery" in the New Zealand Herald. New Zealand universities are amongst the leading research institutions in stem cell research, with a particular record in neurological research (Futurewatch 2006). What makes the increase in stem cell-related patents even more intriguing is that the amount of state funding is comparatively low. Only NZ$2.3 million per annum are allocated to stem cell projects (Futurewatch 2006). However, similar to the Belgian case, the increase in innovations in stem cell research occurred before the permissive law was passed. In the New Zealand case, the delay of the law was not due to coalition struggles, but rather to conflicts and hesitation within the governing party. However, the conclusion remains the same: The success of the sector seems to have been an unintended consequence rather than the result of a political strategy.

Thus, the case of New Zealand leads to a similar conclusion like the Belgian case. On the one hand, the liberal regulatory situation seems to have been supportive for the increase in innovations in the stem cell field. On the other hand, this does not reflect the intended consequences of a political strategy.

The United Kingdom was the first country to liberalise its embryo research law after the breakthroughs in stem cell research. The Human Fertilisation and Embryology Regulations from 2001 allowed therapeutic cloning, and were part of an explicit strategy to promote biotechnology as an integral element of the knowledge society (Blair 2000a; Banchoff 2005). As a traditional leader in biotechnology (Gottweis 1998), with the Royal Society playing a strong role as policy advisor (Krönig 2001), and an already established overview and licensing system (the Human Fertilisation and Embryology Authority (HFEA)), the United Kingdom was in an optimal position to build on its experience and strengthen its innovativeness in the biotech sector. However, as Fig. 12.7 shows, the bulk of the increase in biotech innovations occurred from 1996 to 2001, under the old Human Fertilisation and Embryology Act, dating from 1990. This act was permissive from the outset, and introduced with the explicit aim to strengthen the United Kingdom's research base in biotechnology (Mulkay 1997). Hence, the United Kingdom could capitalise on the stem cell research breakthroughs because the regulatory framework that was already in place was liberal enough to keep researchers in the country.

The case of the United Kingdom thus fully confirms the thesis that permissive embryo research laws lead to an increase in the innovative ability of the biotech sector; although, the case suggests that the effects are to be assessed on a long-term time frame.

Sweden at first glance seems to contradict the thesis that liberal embryo research laws are associated with a high innovation quota. Sweden has had a relatively liberal embryo research law since 1991. This law was changed in 2005 to allow therapeutic cloning, and with the explicit aim to strengthen the Swedish research position in biomedical applications (Kulawik 2003). However, as can be seen in Fig. 12.7, the relatively liberal law of 1991 was not accompanied by an increase in the biotech innovation quota. The Swedish case illuminates the limits of a quantitative approach to innovativeness. The quota of patents may not have increased, but according to all observers, Sweden is a world leader in stem cell research (Burrell 2005; Torgersen et al. 2002; Kulawik 2003). The funding of SKR257.3 Mio (€27 Mio) from 2003 to 2008 expresses the high priority that stem cell research enjoys in the Swedish innovation system (Hague 2006). The Karolinska Institute in Stockholm and the Sahlgrenska Academy in Gothenburg are amongst the leading suppliers of stem cell lines. Thus, the quality of the Swedish innovations in the biomedical sector is high, though its proportion compared to total patents is low. This may reflect a distinct "patenting culture" (Packer and Webster 1996), focusing more on quality than on quantity.

The case of Sweden is illustrative for two reasons. First, it confirms the thesis that permissive embryo research laws can lead to innovations. Second, it illustrates the limits of a purely quantitative approach to the field and the usefulness of qualitative information.

To sum up: There seems to be an association between strict embryo research laws and a low innovation quota in stem cell research in a cross-country comparison, which would confirm the thesis that strict embryo research laws inhibit innovation. However, this association has to be interpreted very cautiously. First, the reverse does not automatically hold true. Permissive embryo research laws are not consistently associated with a high innovation quota in stem cell research. The variation within the innovation quota increases as the embryo research laws get more permissive, but there are countries with permissive or intermediate embryo research laws and a low innovation quota in stem cell research. Second, the disaggregation of the data and the study of country trajectories reveal that there are only very few countries in which the innovation quota in stem cell research has changed substantially in the last 13 years. This also means that the countries that have passed strict laws have not experienced a decline of their innovation quota. Third, in the countries that have experienced a sharp increase of the innovation quota in the stem cell area, there is some evidence that this increase is causally linked to a permissive regulatory situation. However, there is less evidence that this is due to a conscious political strategy. If we consider the temporal dimension, the increase of innovations in the medical biotech sector often occurred before political actors had decided on how to regulate the sector. Only in two countries under study – Sweden and the United Kingdom – can the prospering of the biotech sector be attributed to a political strategy. Fourth, the time frame of the analysis is still rather short. All we can safely conclude is that embryo research laws have no significant systematic effect in the short term. What the long-term effects are – possibly in the form of path-dependent or self-reinforcing dynamics (Pierson 2000) – is open to speculation. Finally, all the results must be interpreted in light of the used indicator. The quota of microbiology/stem cell patents is only a proxy measure for the innovativeness of the sector. It does not say anything about the total number of patents in the sector – a metric in which, for instance, Germany can easily outshine Belgium. And it does not say anything about the importance or quality of the patents (as the case of Sweden has demonstrated). Thus, all the conclusions from this analysis must be taken with some caution as to their generalizability.

12.7 Summary

This chapter made three interrelated arguments concerning the interaction of political factors, embryo research laws and the innovative ability of national economies in fields related to stem cell research.

First, the chapter demonstrated that the development of human embryonic stem cell research constituted a critical juncture for the field. Powerful actors like left parties and the pharmaceutical industry lacked an interest in the issue before 1998. However, after stem cell research promised to turn embryo research into a commercially beneficial enterprise, these two actors mobilised to liberalise existing regulatory systems. Thus, a wave of liberalizations ensued.

Second, the chapter demonstrated that religious interests still play an important role in the field. The Catholic Church and Christian democratic parties were major players and were able to resist liberalization pressures. This is surprising, as it is often claimed that scientific and economic interests dominate the field.

Essentially, the chapter made a pluralist argument. The most important factor determining the content of embryo research laws is the balance of power between secular and religious actors. As has been argued in more detail elsewhere (Fink 2008a, b), institutional factors do not add much additional explanatory power. The power struggles between actors can explain most of the variation.

Third, the resulting variation of embryo research laws did not have a clear-cut impact on the innovative ability of national economies in fields related to stem cell research. Stable, long-term differences in innovativeness were dominant, and government interventions in the form of permissive laws did not have a predictable and stable effect in the short term. Innovativeness of the medical biotech sector seemed to be, in considerable parts, determined by stable structural differences. Policy measures, like permissive or strict embryo research laws, seldom had a short-term impact on the innovativeness of the sector.

This finding cautions the hopes – and promises – of many actors that claim to introduce permissive embryo research policies in order to reap short-term gains in innovative ability. This strategy may work, but more often, changes in the innovativeness of the sector cannot be attributed to strategic political decisions. This finding also casts doubts on the ability of states to steer scientific developments and sectors, and to force innovations by policy measures.

However, proponents of strict bioethics laws should not draw the conclusion that embryo research laws do not matter at all for the innovative ability of a national economy. None of the countries that have passed strict regulations was able to raise its biotechnology innovation quota, while at least some of the countries with permissive regulations were able to increase their biotech innovation quota. Second, due to the relative youth of the research and policy field, this article could only examine a relatively short time frame. What the long-term consequences of different embryo research laws are remains an open question. Recent theorizing about the self-reinforcing nature and nonlinear dynamics of social processes (Pierson 2004; Mayntz and Nedelmann 1987) suggests that small differences in innovative ability may add up at an increasing rate, thereby generating path-dependent developments. Maybe the question of how to regulate stem cell research will prove to be a critical juncture, and 20 years from now, the countries that chose a permissive law today will have a lead in the sector that none of the other countries can catch up on.

Theoretically, this finding illuminates the fact that the thesis, "Strict laws lead to a decline of innovative ability whereas permissive laws lead to an increase in innovative ability" is not appropriate to grasp the complexities of the social world. From a social scientific viewpoint, this thesis – which is often heard even from representatives of science – reflects an overly simplistic worldview. If a national economy is seen as a complex adaptive system (Schneider and Bauer

2007), the notion that only one factor determines the ability of this system to generate innovations is ruled out from the beginning. In a system whose properties are constantly shaped by the complex interactions of a variety of social actors, laws are only one amongst many factors. Laws may roughly channel the properties of the whole system in the long term – but only seldom change the innovativeness of the system in the short run.

References

Arie, S. (2005). In Europe, Italy now a guardian of embryo rights. Christian Science Monitor, 14 June 2005.

Arundel, A. (2003). Biotechnology indicators and public policy. *STI Working Papers, 2003*(5).

Association of the British Pharmaceutical Industry (ABPI) (2004). Science and innovation: Working towards a ten-year investment framework. A response by the ABPI. http://www.abpi.org.uk/information/pdfs/S&I_consultation.doc. Accessed 20 April 2006.

Banchoff, T. (2005). Path dependence and value-driven issues: The comparative politics of stem cell research. *World Politics, 57*(2), 200–230.

Barr, T. (2003a). GE free NZ Inc. Submission on the supplementary order paper on the human assisted reproductive technology bill. http://www.biopolitics-berlin2003.org/doc_rt.asp?p=1&id=51. Accessed 28 February 2006.

Barr, T. (2003b). Made to order – the government buys into GE babies. *OrganicNZ, 62*(4), 20.

Bauer, M. W., & Gaskell, G. (Eds.). (2001). *Biotechnology 1996–2000: The years of controversy*. London: Science Museum.

Bauer, M. W., & Gaskell, G. (Eds.). (2002). *Biotechnology: The making of a global controversy*. Cambridge: Cambridge University Press.

Blair, T. (2000a). Biotechnology: Investing in the future. Speech held on 17 November 2000, London, http://webarchive.nationalarchives.gov.uk/20060715135117/number10.gov.uk/page1548.

Blair, T. (2000b). Wir werden Europa führen. Frankfurter Allgemeine Zeitung, 8 December 2000.

Bondevik, K. M. (2003). Speech at Trans-Atlantic Dialogue on GenEthics and Health. Centre for Strategic & International Studies, Washington, DC.

Broughton, D. (1988). The social bases of Western European conservative parties. In B. Girvin (Ed.), *The transformation of contemporary conservatism* (pp. 193–224). London: Sage.

Burgermeister, J. (2003). Danes begin stem cell work. The Scientist Daily News, 28 October 2003.

Burrell, R. (2005). Assisted reproduction in the Nordic countries: A comparative study of policies and regulation. Nordic Committee on Bioethics.

Campbell, C. S. (2001). Source or resource? Human embryo research as an ethical issue. In P. Lauritzen (Ed.), *Cloning and the future of human embryo research* (pp. 34–49). Oxford: Oxford University Press.

Castles, F. G. (1994). On religion and public policy: Does Catholicism make a difference? *European Journal of Political Research, 25*(1), 19–40.

European Commission. (2005). *European research projects involving stem cells in the 6th Framework Programme*. Brussels: European Commission.

Council of Europe. (1998). *Medically assisted procreation and the protection of the human embryo: Comparative study on the situation in 39 states*. Strasbourg: Council of Europe.

Deutsche Forschungsgemeinschaft (2001). Stellungnahme zur Forschung mit embryonalen Stammzellen.

Dodson, L., & Gray, D. (2002). Embryo ban hits research. http://www.theage.com.au/articles/2002/02/25/1014471629519.html.

Duverger, M. (1966). *Sociologie politique*. Paris: PUF.

Fink, S. (2007a). Ein deutscher Sonderweg? Die deutsche Embryonenforschungspolitik im Licht international vergleichender Daten. *Leviathan, 1*(2007), 107–128.

Fink, S. (2007b). Ethics vs. innovation? The impact of embryo research laws on the innovative ability of national economies. *Science, Technology & Innovation Studies, 3*(2), 133–149.

Fink, S. (2008a). *Forschungspolitik zwischen Innovation und Lebensschutz. Die Determinanten von Embryonenforschungspolitiken im internationalen Vergleich* (Staatslehre und politische Verwaltung). Baden-Baden: Nomos.

Fink, S. (2008b). Politics as usual or bringing religion back in? The influence of parties, institutions, economic interests and religion on embryo research laws. *Comparative Political Studies, 41*(12), 1631–1656.

Fink, S. (2009). Churches as societal veto players: Religious influence in actor-centered theories of policy making. *West European Politics, 32*(1), 77–96.

Futurewatch. (2006). *Stem cell research in New Zealand.* Wellington: Ministry of Research, Science and Technology.

Gearhart, J. (1998). New potential for human embryonic stem cells. *Science, 282*(5391), 1061–1062.

Gottweis, H. (1998). *Governing molecules: The discursive politics of genetic engineering in Europe and the United States.* Cambridge, MA: MIT Press.

Grabner, P. (1999). *Technik, Politik und Gesellschaft: Eine Untersuchung am Beispiel des österreichischen Gentechnikgesetzes.* Wien: Peter Lang.

Gratton, B. (2002). Survey on the national regulations in the European Union regarding research on human embryos. Report for The European Group on Ethics in Science and New Technologies (EGE).

Hadolt, B. (2005). *Reproduktionstechnologiepolitik in Österreich: Die Genese des Fortpflanzungsmedizingesetzes 1992 und die Rolle von ExpertInnen.* Reihe Soziologie: Institut für höhere Studien Wien. 74.

Hague, A. (2006). *Stem cell research in Sweden: An overview.* Stockholm: British Embassy.

Haines, L. (2005). Spain greenlights therapeutic cloning. The Register, 11 July 2005.

Hanley, D. (1994). Introduction. Christian Democracy as a political phenomenon. In D. Hanley (Ed.), *Christian democracy in Europe: A comparative perspective.* London: Pinter.

Kallerud, E. (2004). The ambiguity of progress: Biotechnology in Norway. In M. Häyrinen-Alestalo, & E. Kallerud (Eds.), *Mediating public concern in biotechnology a map of sites, actors and issues in Denmark, Finland, Norway and Sweden. Norwegian Institute for Studies in Research and Higher Education Rapport 2/2004* (pp. 83–112).

Körtner, U. (2002). Ethische Fragen der Biotechnologie. In C. Kopetzki, & H. Mayer (Eds.), *Biotechnologie und Recht* (pp. 1–14). Wien: Manz´sche Verlags- und Universitätsbuchhandlung.

Kulawik, T. (2003). *Gender representation and the politics of biotechnology in Sweden: Explaining liberal regulations in a social democratic state.* Paper presented at the The ECPR General Conference, Marburg, September 19–21

Lauritzen, P. (Ed.). (2001). *Cloning and the future of human embryo research.* Oxford: Oxford University Press.

Lilge, W. (2001). The case for adult stem cell research. *21st Century Science and Technology Magazine, Winter 2001–2002.*

Lorenzi, R. (2003). Italy approves embryo law. *The Scientist Daily News.* 12 December 2003.

Mayntz, R., & Nedelmann, B. (1987). Eigendynamische soziale Prozesse: Anmerkungen zu einem analytischen Paradigma. *Kölner Zeitschrift für Soziologie und Sozialpsychologie, 4*(39), 648–669.

Mendieta, E. (2004). Habermas on human cloning. *Philosophy and Social Criticism, 30*(5–6), 721–743.

Minkenberg, M. (2002). Religion and public policy: Institutional, cultural, and political impact on the shaping of abortion policies in Western Democracies. *Comparative Political Studies, 35*(2), 221–247.

Mulkay, M. (1997). *The embryo research debates: Science and the politics of reproduction.* Cambridge: Cambridge University Press.
Packer, K., & Webster, A. (1996). Patenting culture in science: Reinventing the scientific wheel of credibility. *Science, Technology & Human Values, 21*(4), 427–453.
Pierson, P. (2000). Increasing returns, path dependence, and the study of politics. *American Political Science Review, 94*(2), 251–267.
Pierson, P. (2004). *Politics in time: History, institutions, and social analysis.* Princeton: Princeton University Press.
Ramjoué, C., & Klöti, U. (2004). ART policy in Italy: Explaining the lack of comprehensive regulation. In I. Bleiklie, M. L. Goggin, & C. Rothmayr (Eds.), *Comparative biomedical policy: Governing assisted reproductive technologies* (pp. 42–63). London: Routledge.
Rothmayr, C., & Ramjoué, C. (2004). Germany. ART policies as embryo protection. In I. Bleiklie, M. L. Goggin, & C. Rothmayr (Eds.), *Comparative biomedical policy: Governing assisted reproductive technologies* (pp. 174–190). London: Routledge.
Ryan, M. A. (2001). Creating embryos for research: On weighing symbolic costs. In P. Lauritzen (Ed.), *Cloning and the future of human embryo research* (pp. 50–66). Oxford: Oxford University Press.
Scharpf, F. W. (1997). *Games real actors play: Actor-centered institutionalism in policy research.* Boulder: Westview Press.
Schiffino, N., & Varone, F. (2004). ART policy in Belgium: A bioethical paradise? In I. Bleiklie, M. L. Goggin, & C. Rothmayr (Eds.), *Comparative biomedical policy: Governing assisted reproductive technologies* (pp. 21–41). London: Routledge.
Schmidt, M. G. (1993). Theorien in der international vergleichenden Staatstätigkeitsforschung. In A. Héritier (Ed.), *Policy-Analyse. Kritik und Neuorientierung* (pp. 371–393). Opladen: Westdeutscher Verlag.
Schneider, V., & Bauer, J. (2007). Governance: Prospects of complexity theory in revisiting system theory. Quello Center Working Papers, 07–01. East Lansing, MI: Michigan State University.
Simons, M. (2004). Church and state clash, noisily, in Spain. NYTimes on the Web, 4 October 2004.
Sleator, A. (2000). Stem cell research under the Human Fertilisation and Embryology Act 1990. *House of Commons Library Research Paper 00/93.* London.
Spaemann, R. (2001). Gezeugt, nicht gemacht. Die verbrauchende Embryonenforschung ist ein Anschlag auf die Menschenwürde. In C. Geyer (Ed.), *Biopolitik* (pp. 41–50). Frankfurt/M: Suhrkamp.
Spar, D. (2004). The business of stem cells. *New England Journal of Medicine, 351*, 211–213.
Standing Committee on Legal and Constitutional Affairs (2000). Scientific, ethical and regulatory aspects of human cloning and stem cell research. *Public hearing. House of Representatives, Parliament of Australia.*
Standing Committee on Legal and Constitutional Affairs. (2001). *Scientific, ethical and regulatory aspects of human cloning and stem cell research.* Parliament of Australia: House of Representatives.
Steinbock, B. (2001). Respect for human embryos. In P. Lauritzen (Ed.), *Cloning and the future of human embryo research* (pp. 21–33). Oxford: Oxford University Press.
Stock, G., & Campbell, J. (Eds.). (2000). *Engineering the human germline: An exploration of the science and ethics of altering the genes we pass on to our children.* Oxford: Oxford University Press.
Stranzinger, P. (1992). *Die artifizielle menschliche Fortpflanzung aus medizinischer, ethischer und juristischer Sicht: Staatliche und kirchliche Standpunkte zur artifiziellen Insemination, zur In-vitro-Fertilisation und zum Umgang mit menschlichen Embryonen - ein Vergleich.* Salzburg, Salzburg: Univ.

Thomson, J. A., Itskovitz-Eldor, J., Shapiro, S. S., Waknitz, M. A., Swiergiel, J. J., Marshall, V. S., et al. (1998). Embryonic stem cell lines derived from human blastocysts. *Science, 282*(5391), 1145–1147.

Torgersen, H., Hampel, J., Jelsøe, E., et al. (2002). Promise, problems and proxies: Twenty-five years of debate and regulation in Europe. In M. W. Bauer & G. Gaskell (Eds.), *Biotechnology: The making of a global controversy* (pp. 21–94). Cambridge: Cambridge University Press.

UNESCO. (2004). *National legislation concerning human reproductive and therapeutic cloning.* Paris: UNESCO.

Van Beuzekom, B. (2001). Biotechnology statistics in OECD member countries: Compendium of existing national statistics. *STI Working Papers, 2001*(6).

Varone, F., & Schiffino, N. (2004). Regulating red and green biotechnologies in Belgium: Diverging designs of biopolicies. *Archives of Public Health, 62*(1–2), 83–106.

Varone, F., & Schiffino, N. (2007). Conflict and consensus in Belgian biopolicies: GMO controversy versus biomedical self-regulation. In É. Montpetit, C. Rothmayr, & F. Varone (Eds.), *The politics of biotechnology in North America and Europe: Policy networks, institutions, and internationalization* (pp. 193–213). Lanham, MD: Lexington Books.

Vatican Congregation for the Doctrine of Faith. (1987). Donum vitae: Instruction on respect for human life in its origin and on the dignity of the procreation. *Origins, 40*, 697–711.

Walters, L. (2004). Human embyronic stem cell research: An intercultural perspective. *Kennedy Institute of Ethics Journal, 14*(1), 3–38.

Warren, J., & Osborne, R. (2006). *Attitudes to embryonic stem cell research in New Zealand.* Wellington: Bioethics Council of New Zealand.

Part IV
Conclusions

Chapter 13
Innovation Policy and High-Tech Development: Conclusions

Achim Lang, Volker Schneider, and Johannes M. Bauer

13.1 Introduction

The contributions to this book have examined innovation systems and innovation policy in high-tech sectors. At the center of inquiry has been the adaptation of these sectors to the increasing complexity of economic and technological conditions, actor constellations, and policy interdependencies. Chapters have presented systematic descriptions and analyses of a variety of actors, issue cleavages, and institutional constellations in national and sectoral innovation systems. Authors have offered descriptions of political power struggles, technical and economic coordination problems, as well as the formation of advocacy coalitions. They have examined how these conditions facilitate or thwart the development of effective governance. Chapters have also documented major changes in governance during the recent past.

13.2 A Complex Systems Approach to Innovation Policy

The conceptual lens informing many of the contributions to this book is rooted in the theory of complex adaptive systems. Although multiple definitions of complexity coexist (Mitchell 2009), we generally consider the number of actors in a system, the internal density of links between these actors, and links to the external environment (exostructure) as basic systemic features that influence the complexity of systems (Bunge 1996; 2000). The number of actors reflects the totality of subsystems that make up the larger system. From a political science and governance

A. Lang (✉) • V. Schneider
University of Konstanz, Konstanz, Germany
e-mail: achim.lang@uni-konstanz.de

J.M. Bauer
Michigan State University, Michigan, USA

J.M. Bauer et al. (eds.), *Innovation Policy and Governance in High-Tech Industries*,
DOI 10.1007/978-3-642-12563-8_13, © Springer-Verlag Berlin Heidelberg 2012

perspective, this might be the number and variety of political actors involved in a certain policy domain or governance setting, or the number of advocacy coalitions that participate in a policy discourse.

The internal density of links between these actors measures their interaction, interdependencies and embeddedness. Such links are sometimes called "epistatic" (Kauffman 1993; Frenken 2006) as they capture the effects of changes in the behavior of one actor on others. Viewed in political terms, they can be interpreted as influence tactics or power games. In a governance framework, epistatic links between actors denote mechanisms by which actors mutually coordinate their behaviors. Basic coordination mechanisms include observation, influence and bargaining.

The third element – the exostructure linking actors and subsystems to higher-order systems (e.g., policy domains, national and global political systems) – describes the interdependence between systems. Changes in one system may trigger or inhibit changes in other, co-evolving systems. Intersystemic interdependence has not been systematically addressed in the policy sciences. Policy domains are commonly considered to be largely separated, independent of each other, and dominated by policy domain experts. For the purpose of this volume a distinction was made between *pooled, sequential and reciprocal* forms of interdependence (Saavedra et al. 1993; Thompson 1967). This provides a conceptual tool to account for relationships between policy domains such as innovation, technology and health policy.

Dutton, Schneider and Vedel expanded this basic theoretical perspective and highlighted the complexity and nestedness of games across issue areas. This "ecology of games" (EOG) emphasises social and political conflicts within nested decision-making processes that involve public as well as private actors and that are related to complex social and technical interdependencies. The EOG approach is aimed at an actor-centred, dynamic reconstruction of social interactions, where social processes are the consequences of interactions of multiple actors. These games are based on differentiated, structured, goal-oriented interactions. From this perspective, large technical systems are the outcome of a multitude of atomic ecosystems, in which individual actors negotiate, bargain and make decisions as role players. Policy interaction and "gaming" takes place not in open fields, but within complex institutional configurations or rule systems, which were addressed in the chapter by Werle. He outlined and compared approaches that emphasised a diversity of social, economic, technical and political institutions in innovation processes and technology development. For instance, the varieties of capitalism (VoC) approach argues that firm-level innovation processes, as well as meso- and macro-level state intervention in support of research and development, unfold differently in liberal and coordinated market systems.

Last, but not least, technology development and innovation policy is also shaped by general public discourses, particularly in the mass media. Public discourse in the media can shape debates and decisions in policy arenas and may increase the range of participating actors in policy issues. This produces additional forms of intersystemic dependencies. For example, a debate on technological risk may transcend purely technocratic arenas, affecting not only economic choices in firms and political decisions in government, but also contribute to the mobilization of a variety of social movements.

This topic was dealt with in the chapter by Waldherr, where it was convincingly argued that the relationship between high technologies and media attention is complex, depending on the specific type of technology and the "policy type" to which public policy-making is related. Media interest can be low if a given technology is of low relevance to daily life, largely unrelated to social problems, and also when "distributive policies" (e.g., public subsidies) are applied to promote the technology. Media attention is generally higher in the context of regulative policies with more polarised actor constellations, where high risk is involved and many people may be affected. If a technical issue combines "drama," "conflict" and "personalization," it is prone to develop into media hype with potentially powerful effects not only on specific policy subsystems, but on economic and political processes in general.

13.3 National Systems of Innovation

The second part of this volume contained three detailed studies of national innovation systems (NIS) that focused on the role of governmental and private actors in innovation governance. Figures 13.1 and 13.2 depict public and private R&D expenditures for the Swiss, German and U.S. innovation system. All three NIS are characterised by high levels of R&D spending but vary in the sources of R&D funding. While all three countries can boast high private sector spending,

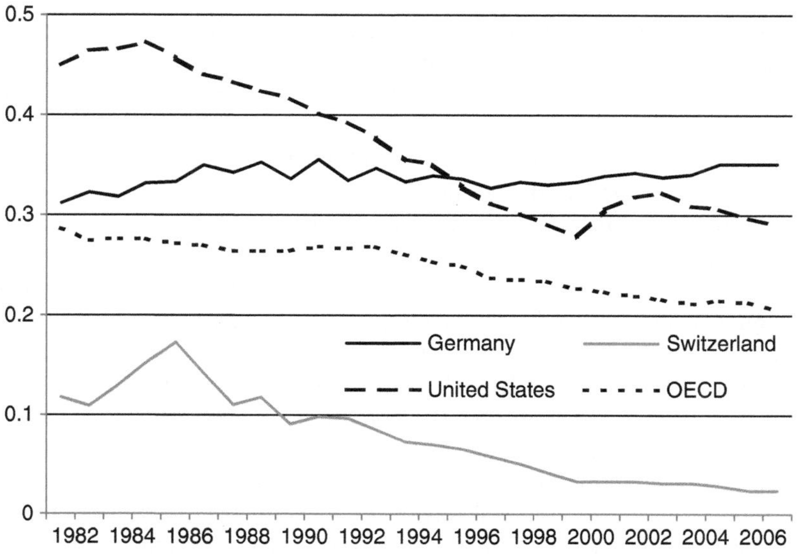

Fig. 13.1 Government R&D expenditures (in % of GDP)

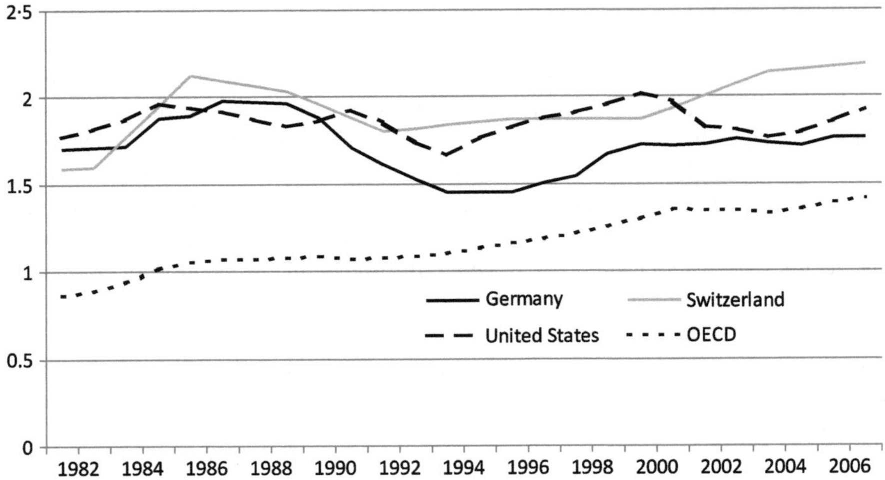

Fig. 13.2 Business R&D expenditures (in % of GDP)

government funding of R&D is much lower in Switzerland. This is somewhat compensated by the expenditures of Swiss businesses that are above the German and U.S. levels.

These findings are in line with the arguments presented by Hotz-Hart in his analysis of the Swiss national innovation system. Switzerland regularly ranks high in international comparisons of innovation and economic performance. Hotz-Hart pointed out the different strengths and weaknesses of this system and presented arguments for the high innovative performance. He argued that the Swiss NIS is characterised by a limited number of state interventions and by restriction to a few basic innovation principles and small scale measures. Furthermore, the Swiss government focuses on an economic strategy in which the promotion of innovation is supplementary to achieving high growth. Hotz-Hart identified this approach as a stark contrast to most European countries.

The restrictive use of government intervention in the Swiss innovation system is accompanied by liberalised and flexible markets and high competition even between public organizations. The Swiss government focuses on providing high-level infrastructure in R&D, science, secondary education and vocational training, which Hotz-Hart described as an implicit innovation policy, in contrast to the state interventionist and corporatist innovation policy in Germany.

Orlowski analysed the role of newly established councils and alliances in the coordination of the German High-Tech Strategy. In 2006, the German Federal Government established two advisory boards, the Council for Innovation and Growth and the Industry-Science Research Alliance, to improve coordination between companies, science organizations and politics. Orlowski contributed an analysis of the inter-organisational network surrounding the two advisory bodies, to determine which actors and industries occupy central positions in the national coordination network. Using several social network analytical techniques, he found that business interests play an important role in coordinating innovation policy. The coordination network consists predominantly of multinational companies, as well as small and medium-sized enterprises from various industries. Financial corporations are particularly active interlockers, together with corporations from the machinery and vehicle construction industry and the German power-supply industry. These industries provide the backbone of the German economy and are therefore disproportionately represented in coordination activities. However, most of these companies pertain to medium-tech industries while only one business association represents the interest of a truly high-tech sector (BITKOM, speaking for the information technology and telecommunications industry). Consumer associations, ecological groups and groups or institutions concerned with technological impact assessment are not participating in the coordination network. Orlowski furthermore pointed out that many actors are connected by membership in other advisory bodies or other linkages. The creation of the two advisory bodies added more redundancy than necessary for coordination.

In the U.S., the role of government in innovation policy has changed over time. Despite recent attempts to reverse the trend, government spending on R&D has declined substantially since the Cold War. The high innovation performance of the US economy is the outcome of a multifaceted system of government support for private sector initiative, a strong culture of entrepreneurship and risk-taking, and the historical alignment between national security interests and innovation in critical industries such as information and communications technology. In contrast to Germany and other industrialised peer countries, the U.S. until recently did not have a comprehensive national innovation strategy. The absence of national priorities and the reduced government R&D spending have probably contributed to the weaker innovation performance of the US during the past decade. It is too early to assess the effects of the "Strategy for American Innovation" adopted by the Obama Administration, although the increased public spending for R&D, the setting of national priorities, and the support for material and immaterial infrastructures should boost innovation. Yet, the complexities of coordinating between multiple policy-arenas remain challenging. Overall, the current U.S. innovation system is better suited to industries with modular innovation processes than sectors that require large-scale commitment and sustained investment.

13.4 Sectoral Innovation Systems

Tools derived from complexity theory were also employed in the analyses of sectoral innovation systems that made up part three of the book. Several of the chapters were related to the functional distinction between policy coordination and administrative coordination put forward by Peters (2006) and Braun (2008). Policy coordination designates efforts during the policy formulation process, while administrative coordination refers to the implementation phase, where concrete administrative procedures and policy instruments are applied and cooperative relations to external actors, who are supporting the various programmes, are activated.

Weyer and Schneider analysed the power struggles in the German and European space policy domain and observed a policy change in 2007 when the EU and the national governments returned to traditional state-driven governance. The failure to assure industry participation in the large satellite project Galileo was compensated for through the financial support of the European Union. In this context, major players in the German space policy domain opted for a strategy to re-nationalise space policy and to increase the budget for space research by putting specific emphasis on big technological projects with almost no known commercial impact. Policy discourse centred on prestige and international competition arguments, instead of economic feasibility and benefits. The EU Commission's white paper envisioned a readjustment in the European division of labour, limiting the European Space Agency to a service provider for technological know-how. The authors showed that innovation policy in the space domain consist of overlapping games played at the national, European and global level. In Germany, the power game is driven by conflicts between the research and economics ministry, both embedded in their respective advocacy coalitions. The national power game is over-determined by a European game in which core EU institutions strive for the extension of their competences and resources. This effort is fueled by global power games of strategic positioning with regard to emerging technologies and lead-markets.

Fink, discussing biotech policy, assessed the interaction of political factors, embryo research laws and the innovative ability of national economies. He demonstrated that embryo research turned into a commercially promising enterprise during the late 1990s, attracting the pharmaceutical industry and government regulators alike. Leftist governments and industry mobilised to liberalise existing regulatory systems. However, in several countries, the Catholic Church and Christian democratic parties opposed scientific and economic interests and, as a result, thwarted the liberalization of stem cell research. Fink employed a pluralist argument that strictness in embryo research laws is determined by the balance of power between secular and religious interests. Looking at the effects of embryo research laws on the innovativeness of national biotech industries, he concluded that the strictness of embryo research laws does not have a clear-cut effect on national innovative ability, but hinges on stable structural differences. His findings also cast doubt on the ability to steer scientific developments and sectors, and to stimulate innovations with policy instruments. Strict bioethics laws rarely increase the innovative ability of the national biotech industry. However, countries that have passed strict regulations were able to raise their biotechnology innovation quota, while some of the permissive regulators were able to achieve their objectives as well.

Fuchs and Wassermann argued that the emergence and development of the photovoltaic industry in Germany was based on the establishment of a protected niche market. This market, in turn, depended on the creation and success of advocacy coalitions supporting the photovoltaic industry. Photovoltaics are considered to be a technological innovation that might help transform the energy sector. Strong opposing forces with vested interests in the maintenance of the traditional energy mix therefore attempted to thwart its adoption and the use of environmentally friendly products and processes in general. A first step towards the creation of an advocacy coalition was made with the creation of specialised research departments and institutes, and with increased direct project funding. The main recipients of funds were public research institutes and two industrial actors, AEG-Telefunken and Siemens Solar. At the beginning, the photovoltaics advocacy coalition included local politicians, the Green party, researchers, environmental societies, and business associations of the infant photovoltaics industry. Despite the rather limited financial resources, the coalition managed to intensify its lobbying and achieved passage of the first feed-in law in the early 1990s. Broadening to include multinational companies (Siemens and AEG) increased the effectiveness of political pressure against the strong opposition of German utilities. During the formative stage, the photovoltaics advocacy coalition aimed at supporting the diffusion of the technology in order to reach a critical mass, at which point it was expected that the market would reach a sustainable state. Reaching the critical mass contributed to further consolidation of the advocacy coalition and solidified its success. After the 1998 elections, the Green party, together with the Social Democrats, took power in the government and the ministry of the environment. This placed the photovoltaics advocacy coalition in the center of political power.

Ronit tracked the development of the wind energy industry from the point of view of international efforts combating climate change. He found that national governments remain key players in environmental and innovation policies, but intergovernmental organizations have gained importance in coordinating the policies of states and in mitigating conflicts. Initially, wind energy was a small subdomain in national energy policy making. At that time, national and regional business associations, as well as environmental groups, had already been established, but were still exclusively linked to domestic politics and national systems of innovation. In contrast, agenda-setting and policy formulation had shifted to the international level. A large and increasing number of civil society organizations now take care of climate policy and wind energy.

Lang and Mertes focused on the policy and administrative coordination of a temporary policy network that transcended domain boundaries and was formed to implement the electronic health card (eHC) in Germany. They identified structural barriers to coordination and inconsistencies in goals and task settings that resulted from power asymmetries. The analysis revealed that different policy domains can hardly be accommodated without causing frictions among participating actors. These frictions result from different institutional logics inherent in different policy domains. The implementation of the eHC initially encouraged high hopes and aspirations for policy makers and public administration. However, the actual implementation was dominated by ruptures inherent in German health politics, namely cost transfer and cost reduction. This resulted in a bifurcation of actors: representatives of physicians, hospitals, and of statutory health insurance created advocacy coalitions that dominated information flows, while information technology associations and research institutes were relegated to the periphery of the network. The authors pointed out that traditional sectoral interest positions played a dominant role in the structuration of the implementation network, which also explains the slow and cumbersome process of carrying out the project. In the end, the health care providers have turned out to be the stalling element in the whole process.

Table 13.1 presents a summary of actor constellations and governance problems in the five sectoral systems of innovation. Innovation policy is often cutting across other policy domains. In the case of the implementation of the eHC, both affected policy domains – health care and technology policy – are well established. In wind energy and photovoltaics, domain innovation policy intersects with the newer environmental policy domain. In embryo research, biotech interests compete with much older ethics and religion policy actors. An exception is the space industry, which falls entirely in the innovation policy domain.

Policy coordination includes policy formulation and power games at various levels. In most of the examined innovation sectors, policy coordination takes place between actors belonging to different policy domains that provide input and influence one another. This is the case in the ehealth domain, the wind turbine industry and the photovaltaics domain, in which actors from the various involved domains struggle for dominance in the policy making game. In the case of the wind turbine industry, environmental issues gained ground through successful agenda setting at the international level, while the photovoltaic industry struggled to build an advocacy coalition at the domestic level. Until the association of several multinational corporations with the photovoltaic industry and the 1998 change in government, where the Green party gained control of the department for environmental protection, the incumbent power companies successfully blocked most attempts to promote the renewables. The implementation of the eHC in Germany provides an example of how actors from different policy domains remain separated even when faced with joint coordination tasks.

The policy coordination process in most cases involves numerous actors that form advocacy coalitions and play power games at different institutional and functional levels. Once the power games surrounding the policy formulation settle, administrative coordination takes place in a (sometimes only slightly) less complex setting. In the eHC domain, actors from both the health and the innovation policy domains carried out the administrative coordination in their own domain. Attempts to further integrate coordination by installing or redesigning advisory committees have failed so far. In both wind energy and photovoltaics, political interventions at the national and international level pushed economic activities. However, the photovoltaics industry continues to be much more dependent upon subsidies and asymmetric regulatory measures than the wind turbine industry. Accordingly, interdependence between environmental and innovation policy domains remains higher in the photovoltaic domain. In the biotech sector, market forces provide the coordination mechanism once the biotech regulations come into effect. The space industry is still very much dependent upon public funding of R&D, providing an example of a more state dependent industry.

In sum, complex actor constellations are the result of the multitude of actors participating in innovation policy and their multifaceted relationships and

Table 13.1 Actor constellations and governance problems

	eHealth card	Wind turbine industry	Photovoltaics	Embryo research	Space industry
Types of policy domains	Both established (technology and health)	Established (innovation) and new (environment)	Established (innovation) and new (environment)	Established (ethics) and new (biotech)	Established
Policy coordination					
Relations between policy domains	ICT domain and health policy provide inputs for one another	Energy and environmental policy provide inputs for one another	Energy and environmental policy provide inputs for one another	Dominance of ethics policy domain (some biotech actors involved)	Innovation and economic policy domain dominant
Number of actors	Large	Large	Large	Small	Small
Relations between actors	Severe cleavages (cost reduction vs. data security)	Creation of advocacy coalitions (conflict between incumbent and renewable power firms)	Creation of advocacy coalitions (conflict between incumbent and renewable power firms)	Power struggle between secular and catholic groups	Inter-ministerial conflicts of competence
Type of interdependence	Reciprocal	Reciprocal	Reciprocal	Pooled/ independent	Pooled/ independent
Administrative coordination					
Relations between policy domains	ICT and health policy domains provide inputs for one another	Energy and economic policy domain dominant	Energy and economic policy domain dominant	Innovation and economic policy domain dominant	Innovation and economic policy domain dominant
Number of actors	Large	Medium	Medium	Small	Small
Relations between actors	Separation of ICT and medical interests. Core: medical interests	Interventions of public authorities at the nation level and of international organizations	Government interventions in market structure (subsidies)	Basically market forces, independent decisions of companies	National ministries and industry
Type of interdependence	Reciprocal	Pooled/ independent	Reciprocal	Pooled/ independent	Pooled/ independent

interdependencies. In most cases, actors from different policy domains have to coordinate their activities in order to set coherent policy objectives and to ensure effective and efficient implementation. Policy integration (or lack thereof) emerges from power games in and between policy domains. Time and again, subsystem or domain cleavages transcend to the new policy domain. This lack of policy integration often leads to underdeveloped functional and role specification. Consequently, the complexity of the coordination task is increased rather than reduced.

13.5 Recommendations for High-Tech Policy

The analyses of national and sectoral systems of innovation have revealed two main challenges related to the complexity of high-tech policy. One set of problems is related to the technological and economic characteristics of high-tech industries. The other is related to the coordination of policy formulation and implementation. Often, this requires the management of diverse actor constellations. Traditional approaches to innovation policy are not well adapted to these conditions. It is not straightforward to devise policies that can overcome the challenges raised by the complexity of high-tech industries and their governance, and no single set of recommendations applies to all situations. Nonetheless, it is possible to identify the direction in which policy makers and practitioners in the field of high-tech innovation policy should move and we would like to conclude with seven core recommendations:

1. *Congruence*. High-tech policy requires an appropriate match between the technological and economic conditions of an industry and the governance mechanisms employed to shape its further development. In cases of modular technologies, as is characteristic for many Internet applications and services, measures that reduce transaction cost and support diverse market experimentation are important. This can be achieved through policies that facilitate interoperability and openness of the physical and logical platforms upon which further innovation rests. In market segments where significant infrastructure investment with long payback periods and substantial risk is required, such as the deployment of advanced broadband networks, the rollout of high-tech applications such as the eHC, or renewable energy, high-tech policy will require further-reaching interventions. This includes high-level coordination, support for the creation of niche markets, and possibly the commitment of public funds.
2. *Flexible coordination*. High-tech policy requires coordination across multiple domains. Often, no historical precedent exists for such inter-domain collaborations. Such coordination may emerge through repeated interaction of players in the affected domains, but this is not necessarily always the case, and deliberate action may be required to facilitate it. Even where such attempts are made (e.g., German High Tech Strategy, Strategy for American Innovation) successful outcomes cannot be guaranteed. Rather, complex systems open

windows of opportunity during which major changes can be effected (Brock and Colander 2000). Awareness of these contingencies may increase the chances of better coordination at high levels of the policy-making system.

3. *Product diversity*: The dominant global specialization approach to sustaining national competitive advantage may generate a temptation to concentrate on a few high-tech sectors. However, this strategy is risky from a macroeconomic perspective (e.g., the case of Ireland) and with respect to a country's innovation potential. Excessive specialization may lead to monocultures and reduce the economic potential of innovative recombination. Complexity research in growth and development economics has shown that diversity and product ubiquity, rather than sectoral specialization, are a key to competitiveness and successful economic development.
4. *Multiplicity of efforts*. Given the complexity of technology and markets and the challenges of coordination, multiplicity of efforts may actually enhance the chances of success. Rather than constituting a waste of resources, parallel and competing efforts may be a more effective way of exploring the adjacent possibilities of technological and policy opportunities. Considerable historical evidence and theoretical reasoning in the theory of complex adaptive systems points to the necessity of combining experimentation with forms of coordination. Thus, where possible, actor diversity and institutional diversity should be deliberately utilised to foster innovation.
5. *Support for adoption*. Many barriers to successful high-tech policy are related to the adoption of advanced technologies by individual and organizational users. Firms often face barriers that prevent them from adopting high-tech process, products, and business method innovations. In part, such adoption requires adaptation of education at all stages, from elementary school to the continuing education of the adult population. Measures such as tax policy and other complementary efforts may also support adoption.
6. *Importance of mass media discourses*. Innovation and processes of advanced technology development are not only shaped by the decisions of firms and governmental organizations, but also in public discourses. Particularly in technologies with potentially high impacts on the daily life of the general population, policy-makers need to be attentive to potentially adverse media effects. This implies not only a strong emphasis on professional communication strategies in business and politics, but also on the co-ordination of communication strategies among the various actors and intense cooperation with media actors. Flexible coordination thus has to be extended to the media domain, which can be highly influential in shaping a country's "technology acceptance".
7. *Policy as experimentation*. Complexity requires a fundamental rethinking of the ways public policy interacts with the systems it seeks to govern. At the level of technological frontiers, complexity often will result in unanticipated consequences, both positive (see the Internet) and negative (see obstacles faced by ehealth in Germany). Policy, to a certain degree, thus becomes experimentation rather than control of a system in the traditional. This is probably the most difficult challenge of complexity. It requires the commitment of resources

to projects with an uncertain outcome. To a certain degree, this is achieved through government funding for R&D, but more funds may have to be committed to highly risky endeavours. Another implication is that it will be important to continuously monitor outcomes and to adapt goals and the course of action if necessary. This all challenges basic notions of accountable government and will also require institutional innovation in the ways high-tech policy is carried out.

8. *Support for technology adoption.* National and international policy-makers would also be well advised to facilitate the adoption of advanced technologies in business and residential settings. High-tech often has its most decisive effect if embedded in other products and services. Policy-makers can adopt measures to assist existing businesses in this process. Moreover, they can shape educational measures that help users realise the potential benefits of new technologies and their applications more clearly. An increasing number of studies on the adoption of advanced communication technologies have revealed the lack of user skills as an impediment to diffusion.
9. *International coordination.* High-tech industries are part of global value networks. Not only is the production of high-tech products and components often moved abroad, research and development efforts have also migrated to foreign locations, particularly in Asia. As a considerable part of the innovation process is related to the practical knowledge generated in the production process, this may have unexpected effects on domestic economic performance. Moreover, the conditions for international trade, technology transfer and technology stripping are a far cry from the vision of free markets. Many forms of mercantilist interventions bias the game (Ezell and Atkinson 2010) and will need to be addressed by the international community. High-tech policy is not only a national, but a global coordination game.

Many of the most pressing global problems, including environmental challenges, the transition to sustainable economic growth, the more effective use of energy, and the provision of cybersecurity, may be alleviated by reliance on advanced technology. The contributions in this volume offer frameworks that allow a reassessment of existing and emerging policies toward high-tech industries. Much work remains to be done to put these notions into practical policy designs, but we hope to have provided a roadmap for future action.

References

Braun, D. (2008). Organising the political coordination of knowledge and innovation policies. *Science and Public Policy, 35*(4), 227–239.

Brock, W A. & Colander, D. (2000). Complexity and policy. In D. Colander (Ed.), *The complexity vision and the teaching of economics* (pp. 73–96). Cheltenham: Edward Elgar.

Bunge, M. (1996). *Finding philosophy in social science.* New Haven: Yale University Press.

Bunge, M. (2000). Systemism: The alternative to individualism and holism. *Journal of Socio-Economics, 29*(2), 147–157.

Ezell, S., & Atkinson, R. D. (2010). *The good, the bad, and the ugly of innovation policy.* Washington: The Information Technology & Innovation Foundation.

Frenken, K. (2006). *Innovation, evolution, and complexity theory.* Cheltenham/Northampton: Edward Elgar.

Kauffman, S. A. (1993). *The origins of order: Self-organization and selection in evolution.* New York: Oxford University Press.

Mitchell, M. (2009). *Complexity: A guided tour.* Oxford: Oxford University Press.

Peters, B. G. (2006). Concepts and theories of horizontal policy management. In B. G. Peters & J. Pierre (Eds.), *Handbook of public policy* (pp. 115–138). London/Thousand Oaks: Sage Publications.

Saavedra, R., Earley, P. C., & Van Dyne, L. (1993). Complex interdependence in task-performing groups. *Journal of Applied Psychology, 78*(1), 61–72.

Thompson, J. D. (1967). *Organizations in action.* New York: McGraw-Hill.

Index

J.M. Bauer et al. (eds.), *Innovation Policy and Governance in High-Tech Industries*,
DOI 10.1007/978-3-642-12563-8, © Springer-Verlag Berlin Heidelberg 2012

MIX
Papier aus verantwortungsvollen Quellen
Paper from responsible sources
FSC® C105338
FSC
www.fsc.org

If you have any concerns about our products,
you can contact us on
ProductSafety@springernature.com

In case Publisher is established outside the EU,
the EU authorized representative is:
Springer Nature Customer Service Center GmbH
Europaplatz 3, 69115 Heidelberg, Germany

Printed by Libri Plureos GmbH
in Hamburg, Germany